Groundwater Reactive Transport Models

Edited by

Fan Zhang

Institute of Tibetan Plateau Research
Chinese Academy of Sciences
China

Gour-Tsyh (George) Yeh

Department of Civil and Environmental Engineering
University of Central Florida
Presently at National Central University
Taiwan

Jack C. Parker

Department of Civil and Environmental Engineering
University of Tennessee
USA

Copy Editor: Xiaonan Shi

Institute of Tibetan Plateau Research
Chinese Academy of Sciences
China

CONTENTS

CHAPTERS

FOREWORD

Reliable assessment and prediction of contaminant fate and transport in the subsurface media requires close collaboration of specialists with expertise in different disciplines such as geology, hydraulics, chemistry, microbiology, environmental science, and mathematics. As the number of disciplines has increased and as each has become more complex and quantitative, the problem of integrating the knowledge and concepts of various specialists into a coherent overall interpretation has become progressively more difficult. To an increasing degree reactive transport simulation has emerged as an answer to this problem, and the reactive transport model has become a vehicle for integrating the inputs of specialists from a variety of backgrounds. Advances in computer hardware and software have led to remarkable increases in the number, variety, and complexity of the processes which can be represented in simulation. Thus the ideas and formulations of different specialists can be combined in a model to generate an overall result, usually a distribution of mass concentration in space and time, which can be compared with observed data to gain a better understanding of the physical, chemical and biological processes that affect contaminant transport and remediation.

The growing power of reactive transport simulation to incorporate complex processes and their interactions, and to address such issues as uncertainty, has of course come at the expense of increasingly complex models. More sophisticated solution approaches have been required, the need for ancillary mathematical tools to process and evaluate the results has increased, and in general the potential for numerical difficulty and model misuse has grown. To become an effective and well-informed user of a reactive transport model, one needs to be familiar with the basic theories of flow, transport and geochemical processes, understand the principal ideas behind a particular numerical solution technique and its implications for field applications, and know the limitations and constraints of different numerical models.

I am pleased to see that this book will allow the readers to fulfill some of the aforementioned needs. The book is a collection of self-contained papers by the developers of several state-of-the-art and widely used reactive transport models. Each paper describes the development history, mathematical background, numerical techniques, and application case studies of a particular model code. From the descriptions in these papers, the readers will gain useful information and insights to help them make decisions on the suitability of each model code for a particular application. The book will be an invaluable reference text for researchers and students interested in applying multicomponent and multiphase reactive transport modeling to a multitude of geological, hydrological and environmental problems, such as waste disposal, groundwater remediation, groundwater quality management, and carbon sequestration.

Chunmiao Zheng
University of Alabama and Peking University
Tuscaloosa, Alabama and Beijing, China

PREFACE

Groundwater has always played an important role in human history. Groundwater contamination has been a subject of intensive investigations since the mid-1980s. Contaminants in water environments undergo transformations and changes in concentration resulting from physical, chemical, and/or biological processes. A capability to understand and model these processes is at the core of water-quality management. It has long been appreciated that numerical models can enhance fundamental understanding of coupled physical and biogeochemical processes in geologic media as well as enable quantification of performance and risk assessment for engineering applications. Consideration of equilibrium and kinetic chemistry, thermal transport, and hydrologic transport and interactions between fluid flow, heat flow, and reactive transport is necessary to represent the complexity of real systems.

Reactive transport models have become essential tools to support investigations of the fate of chemicals in both pristine and contaminated hydrogeologic systems, to facilitate the testing of hypotheses and conceptual models, and to quantify the analysis of complex and non-linear interactions of migrating chemicals in laboratory experiments and field investigations. Groundwater reactive transport models are useful to assess and quantify contaminant precipitation, sorption and migration in subsurface media and to evaluate natural attenuation and alternative remediation strategies. Reactive transport phenomena in porous media have been studied in such diverse fields as chemistry, physics, engineering and geology. Simulations of reactive transport have been widely used in petroleum engineering, groundwater hydrology, environmental engineering and chemical engineering.

Many groundwater reactive transport models are available today which consider various complexities and exhibit various strengths and weaknesses. Selecting the most appropriate model to achieve the most accurate, efficient and economical solution possible is a challenging task. It is extremely important to know the models well to make the best choices. Comprehensive documentation of models and extensive education and training are critical. For this reason, numerical modeling of subsurface flow and reactive transport has attracted considerable attention and a variety of numerical techniques and computational tools have been proposed and developed, especially as computer power and processing capabilities have significantly increased in the past decades. Significant mathematical and computational challenges are posed by problems involving realistic representation of convection, diffusion, dispersion and chemical reactions. The development of mechanistically-based reactive chemical transport models has accelerated in the last two decades.

The groundwater reactive transport models introduced in this book have mostly been developed over decades, represent the state-of-the-art in subsurface flow and reactive transport modeling tools, and been widely used for numerical modeling of subsurface flow and reactive transport processes. Coupled modeling of subsurface multiphase fluid and heat flow, solute transport, and chemical reactions can be applied to many geologic systems and environmental problems, including geothermal systems, diagenetic and weathering processes, geological disposal of nuclear waste, CO_2 geological sequestration and storage, acid mine drainage remediation, contaminant transport, groundwater monitoring and remediation, and even energy related problems like in-situ production of oil shale, etc. Numerous examples have been introduced in this book to facilitate the transition from theory to field applications, which may use both United States Customary units and SI (Système international d'unités) units.

ABOUT THE BOOK

Chapter 1 presents the development of the latest version of **HYDROGEOCHEM** – a multi-dimensional numerical model of coupled fluid flow, thermal transport, hydrologic transport, and biogeochemical kinetic/equilibrium reactions in saturated/unsaturated media. It considers all reaction types (aqueous complexation, adsorption-desorption, precipitation-dissolution, ion-exchange, hydrolysis, and abiotic and biotic-mediated redox) as both fast/equilibrium and/or slow/kinetic processes, using a consistent definition of fast/equilibrium reaction rates. Input to the program includes the finite element numerical representation of the system, the properties of the media, reaction network, and initial and boundary conditions. Output includes the spatial distributions of pressure and total heads, velocity fields, moisture contents, temperature, and biogeochemical concentrations at user specified times and locations (finite element nodes).

Chapter 2 describes a number of efficient and locally conservative methods for subsurface flow and reactive transport that have been or are currently being implemented in the **IPARS** (Integrated Parallel and Accurate Reservoir Simulator). For flow problems, discontinuous Galerkin (DG) methods and mortar mixed finite element methods are considered. For transport problems, discontinuous Galerkin methods and Godunov-mixed methods are employed. For efficient treatment of reactive transport simulations, a number of state-of-the-art dynamic mesh adaptation strategies and implementations are presented. Operator splitting approaches and iterative coupling techniques are also discussed.

Chapter 3 focuses on **TOUGHREACT,** a program for chemically reactive non-isothermal flows of multiphase fluids in porous and fractured media. The program was written in Fortran 77 and developed by introducing reactive chemistry into the multiphase fluid and heat flow simulator TOUGH2. A variety of subsurface thermo-physical-chemical-biological processes are considered under a wide range of conditions of pressure, temperature, water saturation, ionic strength, and pH and Eh. Reactions among aqueous species and interactions between mineral assemblages and fluids can occur under local equilibrium or *via* kinetically controlled rates. The gas phase can be chemically active. Precipitation and dissolution reactions can change formation porosity and permeability. Intra-aqueous kinetics, biodegradation and surface complexation have recently been incorporated.

Chapter 4 introduces **RT3D,** a Fortran-based software for simulating three-dimensional, multi-species, reactive transport of chemical compounds (solutes) in groundwater. RT3D is a MODFLOW-based solute transport code derived from MT3DMS, but with greatly expanded capabilities, including simulation of inorganic reactions, geochemistry reactions, NAPL dissolution, mobile/immobile dual porosity, colloid transport, virus transport, heat transport, and risk analysis. With some degree of effort, RT3D can be linked to other codes to include time-varying porosity, interaction with the unsaturated zone, or full geochemistry. Commercial third-party graphical user interface software is typically used to define RT3D simulation model configurations and to visualize contours or isosurfaces of results. Results consist of whole-grid data sets at points in time and location-specific time series data sets.

Chapter 5 considers **ECKEChem**, a reactive transport module for the STOMP suite of multifluid subsurface flow and transport simulators. The ECKEChem module was designed to provide integrated reactive transport capabilities across the suite of STOMP simulator operational modes. The STOMP-ECKEChem solution approach to modeling reactive transport in multifluid geologic media is founded on an engineering perspective: 1) geochemistry can be expressed, input and solved as a system of coupled nonlinear equilibrium, conservation and kinetic equations, 2) the number of kinetic equation forms used in geochemical practice are limited, 3) sequential non-iterative coupling between the flow and reactive transport is sufficient, 4) reactive transport can be modeled by operator splitting with local geochemistry and global transport.

Chapter 6 discusses **PFLOTRAN**, a next-generation reactive flow and transport code for modeling subsurface processes, which has been designed from the ground up to run efficiently on machines ranging from leadership-class supercomputers to laptops. Based on an object-oriented design, the code is easily extensible to incorporate additional processes. It can interface seamlessly with Fortran 9X, C and C++ codes. Features of the code include a modular input file, implementation of high-performance I/O using parallel HDF5, ability to perform multiple realization simulations with multiple processors per realization in a seamless manner, and multiple modes for multiphase flow and multicomponent geochemical transport. Chemical reactions currently implemented in the code include homogeneous aqueous complexing reactions and heterogeneous mineral precipitation/dissolution, ion exchange, surface complexation and a multirate kinetic sorption model.

Chapter 7 depicts **CORE^{2D} V4**, a COde for modeling partly or fully saturated water flow, heat transport and multicomponent REactive solute transport under both local chemical equilibrium and kinetic conditions. It can handle abiotic reactions including acid-base, aqueous complexation, redox, mineral dissolution/precipitation, gas dissolution/exsolution, ion exchange and sorption reactions (linear K_d, Freundlich and Langmuir isotherms, and surface complexation using constant capacitance, diffuse layer and triple layer models) and microbial processes. Hydraulic parameters may change in time due to mineral precipitation/dissolution reactions. A sequential iterative approach is used to solve for the numerical solution of coupled reactive transport equations.

Chapter 8 explains **MIN3P,** a general purpose multicomponent reactive transport code for variably saturated media. The basic version of the code includes Richard's equation for the solution of variably-saturated flow, and solves mass balance equations for advective-diffusive solute transport and diffusive gas transport. Biogeochemical reactions are described by a partial equilibrium approach, using equilibrium-based law-of-mass-action relationships for fast reactions, and a generalized kinetic framework for reactions that are relatively slow in comparison to the transport time scale. The model formulation is based on the global implicit method (GIM) with direct substitution of the biogeochemical relationships into the transport equations. The capabilities of the basic code and three follow-up developments are demonstrated by several application examples.

Chapter 9 provides a general overview of **NUFT** code, which is a highly flexible computer software package for modeling multiphase, multi-component heat and mass flow and reactive transport in unsaturated and saturated porous media. An integrated finite difference method is used for numerical discretization. Several mathematical models are implemented in order to address various flow and reactive transport processes in porous media. The governing equations for each sub-model are solved by implicit time-integration. In particular a globally implicit approach is employed to solve transport and reaction equations simultaneously. The code is designed based on object-oriented principles, and equipped with efficient solvers and massively parallel computation capability.

ACKNOWLEDGEMENTS

This book would not exist without the contribution of all the chapter authors. We are grateful to these contributors. Furthermore, we would like to express our gratitude to Dr. Xiaonan Shi for editorial assistance.

Contribution of Dr. Fan Zhang to this book is supported partially by the Institute of Tibetan Plateau Research, Chinese Academy of Sciences and partially by the Environmental Sciences Division, Oak Ridge National Laboratory. Contribution of Dr. Gour-Tsyh (George) Yeh to this book is supported by Taiwan Typhoon and Flood Research Institute (TTFRI), National Applied Research Laboratory (NARL), Taiwan.

Fan Zhang

Institute of Tibetan Plateau Research
China

Gour-Tsyh (George) Yeh

National Central University
Taiwan

Jack C. Parker

University of Tennessee

List of Contributors

Amos, R. T.

Department of Earth and Environmental Sciences
University of Waterloo, Canada

Cheng, H. P.

ERDC
US Army Corps of Engineers
Vicksburg, MS 39180, USA

Cheng , J.-R. C.

ERDC
US Army Corps of Engineers
Vicksburg, MS 39180, USA

Clement, T. P.

Department of Civil Engineering
Auburn University, USA

Dai, Z.

Earth and Environmental Sciences
Division, Los Alamos National Laboratory, NM, USA

Fang, Y.

Hydrology Group
Pacific Northwest National Laboratory, USA

Gwo, J. P.

US Nuclear Regulatory Commission
Washington, D. C., USA

Gérard, F.

INRA-IRD-SupAgro
UMR 1222 Eco&Sols, Montpellier, France

Hammond, G. E.

Pacific Northwest National Laboratory, USA

Hao, Y.

Lawrence Livermore National Laboratory, USA

Lichtner, P. C.

Los Alamos National Laboratory, USA

Li, M. H.

Institute of Hydrological and Ocean Sciences
National Central University, Taiwan

Li, Y.

Ayres Associates
8875 Hidden River Parkway, Suite 200
Tampa, FL 33637, USA

Lu, C. (Chuan)
Energy and Geoscience Institute
University of Utah, USA

Lu, C. (Chuanhe)
Civil Engineering School
University of A Coruña, Spain

Mayer, K. U.
Department of Earth and Ocean Sciences
University of British Columbia, Canada

Mills, R. T.
Oak Ridge National Laboratory, USA

Montenegro, L.
Civil Engineering School
University of A Coruña, Spain

Molins, S.
Lawrence Berkeley National Laboratory
Earth Sciences Division, USA

Moreira, S.
Civil Engineering School
University of A Coruña, Spain

Nitao, J. J.
Lawrence Livermore National Laboratory, USA

Johnson, C. D.
Environmental Sustainability Division
Pacific Northwest National Laboratory, USA

Pruess, K.
Earth Sciences Division
Lawrence Berkeley National Laboratory, USA

Salvage, K. M.
Department of Geological Sciences
Binghamton University, USA

Samper, J.
Civil Engineering School
University of A Coruña, Spain

Siegel, M. D.
Sandia National Laboratories
Albuquerque, NM 87185, USA

Sonnenthal, E.
Earth Sciences Division
Lawrence Berkeley National Laboratory, USA

Spycher, N.
Earth Sciences Division
Lawrence Berkeley National Laboratory, USA

Sun, J. T.
Engineering & Applied Science, Inc.
13087 Telecom Parkway North,
Tampa, FL 33637, USA

Sun, S.
Earth Sciences and Engineering, Applied Mathematics and Computational Science
King Abdullah University of Science and Technology
Kingdom of Saudi Arabia.

Sun, Y.
Lawrence Livermore National Laboratory, USA

Thomas, S. G.
Institute for Computational Engineering and Sciences
University of Texas at Austin, USA

Tripathi, V. S.
EEESI
6801 Whittier Avenue
McLean, VA 22101, USA

Wheeler, M. F.
Institute for Computational Engineering and Sciences,
Department of Aerospace Engineering & Engineering Mechanics
Department of Petroleum and Geosystems Engineering,
The University of Texas at Austin, USA

White, M. D.
Hydrology Group,
Pacific Northwest National Laboratory, USA

Xu, T.
Earth Sciences Division
Lawrence Berkeley National Laboratory, USA

Yang, C.
Bureau of Economic Geology
University of Texas, University Station, USA

Yeh, G. T.
Department of Civil, Environmental and Construction Engineering
University of Central Florida, USA

Zhang, F.
Institute of Tibetan Plateau Research
Chinese Academy of Sciences, China

Zhang, G.
Shell International E&P Inc.
Houston, TX 77079, USA

Zheng, L.
Earth Sciences Division
Lawrence Berkeley National Laboratory, USA

CHAPTER 1

HYDROGEOCHEM: A Coupled Model of Variably Saturated Flow, Thermal Transport, and Reactive Biogeochemical Transport

G. T. Yeh[1*], V. S. Tripathi[2], J. P. Gwo[3], H. P. Cheng[4], J.-R. C. Cheng[4], K. M. Salvage[5], M. H. Li[6], Y. Fang[7], Y. Li[8], J. T. Sun[9], F. Zhang[10] and M. D. Siegel[11]

[1]Institute of Applied Geology, National Central University, Taiwan; [2]EEESI, 6801 Whittier Avenue, McLean, VA, USA; [3]US Nuclear Regulatory Commission, Washington, D. C., USA; [4]ERDC, US Army Corps of Engineers, Vicksburg, MS, USA; [5]Binghamton University, Department of Geological Sciences, Binghamton NY, USA; [6]Institute of Hydrological and Ocean Sciences, National Central University, Jhongli, Taiwan; [7]Pacific Northwest National Laboratory, P.O. Box 999, MS K9-36, Richland, WA, USA; [8]Environmental Consulting and Technology, Inc., 1408 N. Westshore Blvd., Suite 115, Tampa, FL, USA; [9]Engineering & Applied Science, Inc., 8909 Regents Park Drive, Tampa, FL, USA; [10]Institute of Tibetan Plateau Research, Chinese Academy of Sciences, Beijing, China and [11]Sandia National Laboratories, Albuquerque, NM, USA

Abstract: This chapter presents the development of the latest version of HYDROGEOCHEM a multi-dimensional numerical model of coupled fluid flow, thermal transport, hydrologic transport, and biogeochemical kinetic/equilibrium reactions in saturated/unsaturated media. It iteratively solves the Richards equation for fluid flow, the thermal transport equation for temperature fields, and reactive biogeochemical transport equations for concentration distributions. For the latter, the advective-dispersive-reactive transport equations are solved for mobile components and kinetic variables. The biogeochemical reaction equations along with the component- and kinetic-variable equations are solved for concentration distributions of all species. This version of HYDROGEOCHEM is designed for generic applications to reactive transport problems under non-isothermal conditions in subsurface media. It considers all types of reactions (aqueous complexation, adsorption-desorption, precipitation-dissolution, ion-exchange, hydrolysis, and abiotic and biotic-mediated redox) as both fast/equilibrium and/or slow/kinetic processes, using a consistent definition of fast/equilibrium reaction rates. Input to the program includes the finite element numerical representation of the system, the properties of the media, reaction network, and initial and boundary conditions. Output includes the spatial distributions of pressure and total heads, velocity fields, moisture contents, temperature, and biogeochemical concentrations at user specified times and locations (finite element nodes). Six examples are employed to demonstrate the design capabilities of HYDROGEOCHEM, to illustrate the calculations of fast/equilibrium reaction rates, and to highlight the non-intuitive notion that the rates of slow/kinetic reactions are not necessarily smaller than those of fast/equilibrium reactions.

Keywords: Reactive transport models, biogeochemical models, numerical biogeochemical transport model, variably saturated flow, thermal transport.

INTRODUCTION

Groundwater has always played an important role in human history. Groundwater contamination has been a subject of intensive investigations since the mid-1980s. Contaminants in the environment undergo transformations and changes in concentration resulting from physical, chemical, and/or biological processes. A capability to understand and model these processes is at the core of water-quality management [1]. Mathematical and numerical tools for reliable prediction of contaminant migration and transformation are necessary to support the task. Consideration of equilibrium and kinetic chemistry, thermal transport, and hydrologic transport and interactions between fluid flow, heat flow, and reactive transport is necessary to represent the complexity of real systems. For example, temperature and chemical concentrations will

***Address correspondence to G. T. Yeh:** Formerly at Dept. of Civil, Environmental, and Construction Engineering, University of Central Florida, USA; Tel: 886-3-427-9389; Email: gyeh@mail.ucf.edu or gyeh@ncu.edu.tw

Fan Zhang, Gour-Tsyh (George) Yeh, Jack C. Parker and Xiaonan Shi (Eds)

induce convective flow. Solid phase reactions including precipitation and dissolution can potentially plug pores or open fractures reducing matrix diffusion or promoting rapid flow through fractures.

The development of mechanistically-based reactive chemical transport models has accelerated in the last two decades [2-21]. These models have varied scopes. Many models simulate coupled transport and equilibrium geochemistry [22-30]. Some models couple transport with kinetic geochemistry for certain geochemical processes like precipitation-dissolution [31-34]. General reactive transport models capable of handling a complete suite of geochemical reaction processes (aqueous complexation, adsorption, precipitation-dissolution, acid-base hydrolysis, and reduction-oxidation phenomena) and allowing any individual reaction for any of these geochemical processes to be handled as either equilibrium or kinetic as appropriate for the system being considered have emerged since the late 1990's [17, 18, 35-36]. A number of models combine the ability to simulate transport with microbial growth and biodegradation reactions but do not encompass inorganic geochemistry [37-41]. There have been efforts to couple at least some purely geochemical reactions with microbially mediated processes and transport. Marzel *et al.* [42] include microbial reactions and equilibrium aqueous geochemical speciation. Rittman and Van Briesen [43] coupled equilibrium aqueous complexation and kinetic adsorption with microbiological degradation. Hunter *et al.* [44] coupled one-dimensional transport with microbial and geochemical reactions for a suite of reactions involving major redox and acid-base pair species. Smith and Jaffee [45] coupled one-dimensional transport with equilibrium speciation and kinetic biogeochemical reactions affecting trace metal transport. Coupled specific microbiological and generic geochemical processes have also been developed [46, 47]. Chilakapathi [7] provided significant flexibility in handling any type of kinetic biogeochemical reaction coupled with transport to represent equilibrium processes as well as fast reversible kinetic reactions, but requires problem specific modifications to the code for new simulations to be run. Recently, these types of reaction-based paradigms that are capable of simulating any number of reactions incorporating both geochemical and biological processes have gained popularity and have been implemented in a number of codes [19, 20, 48-50].

Development History

This chapter describes the development of the latest versions (4.5 and 5.5) of HYDROGEOCHEM [51, 52], their design capabilities and limitations, and several example problems typifying some applications. These versions have evolved from the previous versions in terms of their capability to simulate coupled processes of variably saturated fluid flow, thermal transport, and reactive biogeochemical transport with generic reaction networks. The original version of the reactive hydrogeochemical transport model developed by the senior authors at Oak Ridge National Laboratory has evolved into a comprehensive model of coupled fluid flow, thermal, and reactive chemical processes and it has incubated many reactive transport models throughout the world. The evolution reflects many contributions made by a dedicated cadre of colleagues. Dr. Hwai-Ping (Pearce) Cheng carried out the incorporation of multiple ion-exchange sites and multiple adsorbent components, integrated thermal transport with reactive geochemical transport, devised algorithms to deal with difficult problems of precipitate-dominant transport, and carried out extensive debugging. Dr. J.-R. C. Cheng contributed to and performed extensive debugging of the earlier versions. Dr. Karen Salvage initiated incorporation of microbiological reactions, coupled geochemical and microbiological processes, and was heavily involved in the implementation of mixed equilibrium and kinetic reactions. Dr. Ming-Hsu Li coded the initial version of the new paradigm of reaction networks, performed extensive debugging, implemented the version of coupled flow and reactive transport processes, and verified and applied it to many problems. Dr. Yilin Fang incorporated the removal of water due to precipitation-dissolution reactions and most importantly implemented the final version of the new paradigm of reaction-based biogeochemical processes and expanded it to multiphase reactive transport. Ms. Yuan Li incorporated a thermal module and made some important modifications in the two-dimensional versions. Mr. Jiangtao Sun contributed primarily to the development of the three-dimensional version and incorporated several numerical options to solve transport equations. Dr. Fan Zhang improved and implemented several numerical options to solve transport equations and three coupling strategies between hydrological transport and reactive biogeochemistry. She also implemented the algorithms in a watershed model. The development history of various versions of HYDROGEOCHEM is summarized in the following.

(1) HYDROGEOCHEM [3]: HYDROGEOCHEM was among the first comprehensive simulators of HYDROlogic Transport and GEOCHEMICAL reactions in saturated-unsaturated media. It iteratively solves the two-dimensional transport and geochemical equilibrium equations. A stand-alone version of geochemical speciation was developed for coupling with transport [53]. A version using the activity rather than the concentration of component species as the master variables and incorporating the Pitzer's activity coefficient model was subsequently developed [54-58].

(2) LEHGC 1.0 [59]: The computer program LEHGC is a hybrid Lagrangian-Eulerian finite element model of Hydro Geo Chemical (LEHGC) transport through Saturated-Unsaturated Media. It was a descendent of HYDROGEOCHEM, a strictly Eulerian finite element reactive transport code. The hybrid Lagrangian-Eulerian scheme improves on the Eulerian scheme by allowing larger time steps to be used in advection-dominant transport calculations, and also is more computationally efficient.

(3) LEHGC 1.1 [60]: LEHGC version 1.1 was a modification of Version 1.0. The modification included: (1) devising a tracking algorithm with the computational effort proportional to N where N is the number of computational grid nodes rather than N^2 as in the LEHGC Version 1.0, (2) adding multiple adsorbing sites and multiple ion-exchange sites, (3) using four preconditioned conjugate gradient methods for the solution of matrix equations, and (4) providing the capability of colloid transport.

(4) HYDROGEOCHEM 2.0 [61]: HYDROGEOCHEM 2.0 expanded the scope of LEHGC1.1 with mixed equilibrium and kinetic geochemical reaction models [62-64] and provided options in conventional finite element and hybrid Lagrangian-Eulerian finite element methods.

(5) HYDROGEOCHEM 2.1 [65]: HYDROGEOCHEM 2.1 was a minor modification of HYDROGEOCHEM 2.0. Version 2.1 incorporated a more flexible approach for simulating kinetically controlled reactions [66]. For Version 2.1, the contribution to the residual and Jacobian due to non-basic (non-canonical) kinetic reactions was done by chemical reactions rather than by chemical species as was done in Version 2.0. This allows for more efficient handling of non-basic reactions.

(6) HYDROBIOGEOCHEM [47, 67]: To enable modeling of microbial-mediated reaction processes, HYDROBIOGEOCHEM was developed by replacing the geochemical module, KEMOD [62], in HYDROGEOCHEM 2.1 with a newly developed mixed MicroBIOlogical and Chemical Kinetic and Equilibrium Reaction MODel – BIOKEMOD [46, 68]. Later, it was enhanced to include one-, two-, and three-dimensions – HBGC123D (http://hbgc. esd.-ornl.gov/). It was also greatly expanded to deal with reactive transport under multiphase flow and non-isothermal conditions [69], where a new paradigm of modeling reactions [70] was implemented.

(7) LEHGC 2.0 [17, 71]: LEHGC 2.0 was developed as a simulator for coupled density-dependent fluid flow and reactive geochemical transport. The coupling was achieved by combining the density-dependent fluid flow and solute transport model, 2DFEMFAT [72], with the reactive chemical transport model, HYDROGEOCHEM 2.1. It iteratively solves the two-dimensional fluid flow and reactive chemical transport equations. The Richards equation was solved for fluid flow and the advection-dispersion-reactive transport equations were solved for all chemical components and for kinetically controlled aqueous complexed species. This version was initiated at Penn State University and was successfully completed by Dr. Ming-Hsu Li working with Dr. Malcolm D. Siegel at Sandia National Laboratory.

(8) HYDROGEOCHEM 3.0 (Unpublished): HYDROGEOCHEM 3.0 was a very minor update of LEGHC 2.0. The update involved mainly the cleanup of the code. In keeping up with the tradition and spirit of HYDROGEOCHEM, the cleaned version was renamed HYDROGEOCHEM 3.0.

(9) HYDROGEOCHEM 3.1 (Unpublished): HYDROGEOCHEM 3.1 was an upgrade from HYDROGEOCHEM 3.0. The upgrade involved the removal/release of water in precipitation/dissolution reactions. This allows simulation of the genesis of geologic formations due to dewatering processes.

(10) HYDROGEOCHEM 3.2 (Unpublished): HYDROGEOCHEM 3.2 was a significantly modified version of HYDROGEOCHEM 3.1. The major modification was the formulation of reactive chemistry, in which a new paradigm of reaction-based approaches to biogeochemical processes [70] was implemented.

(11) HYDROGEOCHEM 4.0 [19, 73] and HYDROGEOCHEM 5.0 [20, 74] are the two- and three-dimensional versions, respectively. They incorporated heat transport that allows one to investigate the effect of temperature fields on reactive geochemical and biochemical transport. They were developed based on HYDROGEOCHEM 3.2.

(12) HYDROGEOCHEM 4.5 [51] and HYDROGEOCHEM 5.5 [52] are the most recent versions of the codes, where the rates of fast/equilibrium reactions as well as slow/kinetic reactions are calculated after all species concentrations are simulated.

(13) WASH123D [75-78]: The diagonalization approach to modeling reactive biogeochemistry has also been applied to modeling water quality transport in watersheds of integrated river network, overland regime, and subsurface media.

Features

HYDROGEOCHEM has the following main features that make it flexible and versatile for modeling a wide range of real-world problems.

(1) Irregularly-shaped three-dimensional domains can be faithfully represented. HYDROGEOCHEM contains three types of elements: hexahedral, triangular prism, and tetrahedral elements, which provide flexibility and convenience to discretize real-world problems. Fig. **1** demonstrates a combination of these three types of element. The only requirement for using more than one type of element to discretize the domain of interest is the consistency of geometry and base (shape) functions associated with the interface between connected elements.

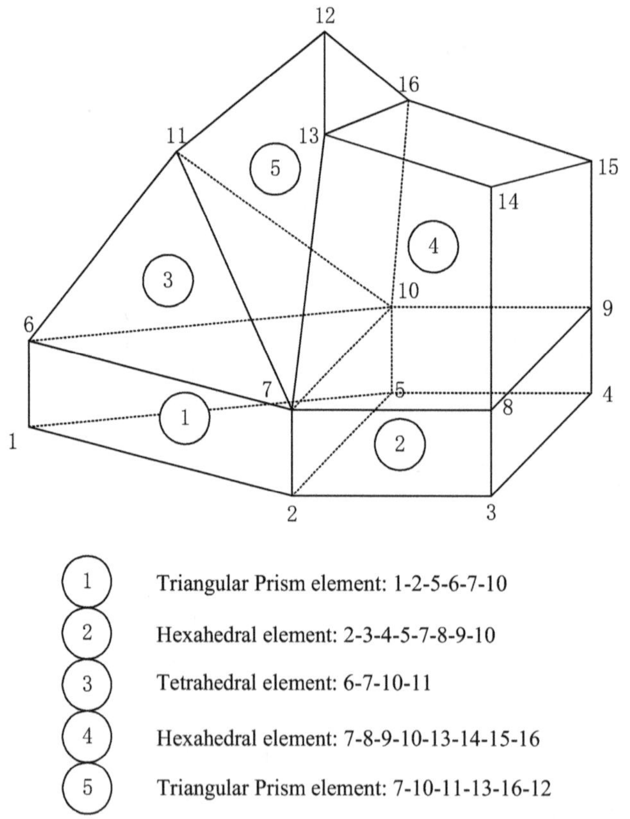

Figure 1: A combination of three types of elements.

(2) Heterogeneity and anisotropy of the subsurface media can be taken into account. HYDROGEOCHEM is designed to be capable of dealing with subsurface systems that contain multiple materials. For each material, one may provide values for the four or nine nonzero components of the saturated conductivity or permeability tensor in the Richards equation to account for anisotropy in two- and three-dimensional problems, respectively. In addition, a number of parameters for each material, including the bulk density, the tortuosity, the longitudinal and transverse dispersivities, the specific heat, the apparent thermal conductivity, and the saturated moisture content, are provided by users, for the related computations.

(3) Non-isothermal subsurface systems can be modeled. Temperature plays an important role in determining biogeochemical reactions and this is represented by the thermal transport module of HYDROGEOCHEM. Users can investigate the features of different transport scenarios by creating different heat transfer conditions.

(4) Both steady-state and transient simulations can be conducted. The model can simulate steady-state and/or transient problems under appropriate boundary and/or initial conditions. In both steady-state and transient simulations, weak coupling and strong coupling of flow with thermal and reactive transport are available.

(5) Initial conditions can be either prescribed or obtained by simulating a steady-state system. The initial conditions of a transient simulation cannot be arbitrary. An inappropriate initial condition may introduce either non-convergent or non-realistic solutions. It should be specified as close to the real situation as possible. Unfortunately, it may not be straightforward to prescribe natural initial conditions. One alternative to set up the initial conditions is to use the solution of a steady-state simulation with steady boundary conditions.

(6) Both spatially and temporally distributed element and point sources/sinks can be considered. The element sources/sinks simulate volumetric sources/sinks in the case of three-dimensional problems or areal sources/sinks in the case of two-dimensional problems. The point sources/sinks (*e.g.*, wells) simulate well injection/withdraw occurring in real-world problems. A specific composition of several point sources/sinks can be considered to represent a line or a volume source/sink and the source/sink intensity can be a function of time.

(7) Spatially- and/or temporally-dependent boundary conditions can be considered. Four types of boundary conditions are employed for simulations of subsurface flow and thermal and reactive chemical transport problems. These boundary conditions are Dirichlet, Cauchy, Neumann, and variable boundary conditions. In addition, a surface water boundary condition is employed to handle river-subsurface flow interaction. All boundary conditions can be prescribed as functions of time and space on the boundary.

(8) Appropriate variable boundary conditions are determined automatically. A variable boundary is one on which the boundary condition is not predetermined and needs to be set up so that consistent computational results can be obtained. For flow simulations, on a variable boundary either Dirichlet or flux-type conditions can prevail. For example, the land surface is a flux-type boundary during rainfall when complete infiltration of throughfall water occurs as well as during dry periods when evapotranspiration is simulated. Alternately, the boundary is a Dirichlet type when rainfall exceeds infiltration capacity and water accumulates on the ground surface as well as during very dry periods when a minimum pressure-type boundary is used to describe the allowed minimum pressure associated with the type of soil. For transport simulations, on a variable boundary the flow can either be directed into or out of the region. For the former case, heat or mass flux is prescribed. For the latter, heat or mass is carried out of the region by advection.

(9) The model is capable of simulating many types of chemical reactions, including: (a) aqueous complexation, (b) adsorption/ desorption, (c) ion-exchange, (d) precipi-tation /dissolution, (e) redox, (f) acid-base reactions, and (g) microbial-mediated reactions. The formulation of the production rate of any species and its associated parameters is the central challenge in biogeochemical modeling. *Ad hoc* and reaction-based formulations are two general means of formulating rates. In an *ad hoc* formulation, the production rate is obtained with empirical functions and does not consider the contribution of individual reactions. In other words, it is a lumped rate; it does not segregate reaction rates. In HYDROGEO-CHEM, a reaction-based

formulation is used. Each reaction can be treated as a fast/equilibrium or slow/kinetic reaction. Users also have the option to formulate rate equations for each reaction.

(10) Interaction among fluid flow and thermal and reactive transport is included. The model includes the effect of temperature and chemical concentrations on fluid flow. It also incorporates the effect of precipitation-dissolution on the change of pore sizes, hydraulic conductivity, and diffusion-dispersion.

(11) Multiple-adsorbing sites and/or multiple-ion exchange sites options are available for sorption reactions. To be capable of dealing with more complicated real-world problems, the model provides the options of multiple-adsorbing sites and multiple-ion exchange sites for chemical reactions. For adsorpsion, the surface complexation, the constant capacitance, or the triple-layer model can be used. For ion-exchange sites, cation- and/or anion-exchanged sites can be assigned as desired.

(12) Three numerical options are provided to solve the thermal and hydrologic transport equations. The first one is the conventional FEM (Finite Element Method). The second one is the hybrid Lagrangian-Eulerian FEM. The third one is hybrid Lagrangian-Eulerian FEM for interior and upstream boundary nodes plus FEM in advective form for downstream boundary nodes. The hybrid Lagrangian-Eulerian FEM has been broadly used in the past several years to solve transient transport equations, especially for advection-dominant transport problems. It attracts attention mainly because it is able to greatly reduce most numerical errors caused by the advection term [79]. In addition, if linear-group base functions (*i.e.*, linear, bi-linear, tri-linear base functions) are utilized, then the numerical results of the Lagrangian step are always non-negative. This is very important for the reactive chemical transport simulation because negative concentrations are neither allowed nor realistic to be used to compute chemical equilibrium and/or kinetics. This is why in addition to the conventional finite element method two optional hybrid Lagrangian-Eulerian FEMs are provided.

(13) Three schemes are available for numerically coupling hydrologic transport and reaction biogeochemistry: (a) fully implicit iteration approach, (b) operator splitting approach, and (c) predictor-corrector approach.

(14) The "in-element" particle tracking technique is used to accurately and efficiently perform particle tracking. The process of particle tracking is the principal issue of the Lagrangian step in the transient simulation of reactive chemical transport and thermal transport. An accurate particle tracking scheme may demand a significant amount of computational effort. The "in-element" particle tracking technique [80] is designed to consider both accurate and efficient particle tracking for real-world problems.

(15) Off-diagonal dispersion coefficient tensor components are included. The off-diagonal terms of the dispersion coefficient tensor are included to deal with cases when the coordinate system does not coincide with the principal directions of the dispersion coefficient tensor.

(16) The time step size can be reset when the boundary conditions and/or sources/sinks change abruptly. It is usually necessary to have a small time step size when a drastic change of either boundary conditions or sources/sinks occurs. Such a change, in general, produces a sharp front in the distribution of the state variables of interest. If the geometric discretization is not fine enough or the time step size is not small enough, then a convergent solution might be elusive under such a change. As the simulation continues, the sharp front is smoothed out and greater time steps can be used. In this model, the time step size can be originally set small and then increased gradually to a desired extent. It can also be reset to the original small value as many times as needed. The computer code will do the reset according to the prescribed values of time.

(17) Many options are available to both compose and solve matrix equations. As can be imagined, a reactive chemical transport system has the strong characteristics of nonlinearity from a mathematical point of view. Since there is no unique way to guarantee achieving convergent solutions for nonlinear systems from a numerical point of view, the best way to achieve the convergent solutions is to have as many approaches available as possible. In HYDRO-GEOCHEM, the following options are provided to attain convergent solutions: (a) three options (under, exact, and over-relaxation) are available to estimate the matrix in the

nonlinear loops (*i.e.*, the rainfall-evaporation loop in the subsurface flow and the solute transport-biogeochemical reactions coupling loop in the reactive chemical transport); (b) two options (consistent and lumping) are available to treat the mass matrix; and (c) six options (block iteration method, successive point iterations, and four preconditioned conjugate gradient methods) are available to solve the linearized matrix equations.

Limitations

Despite a long history of development, there are many limitations in HYDROGEOCHEM. These include, but are not limited to (1) applications are only for single-fluid phase flows, (2) multiple-scale media (*e.g.*, the simultaneous presence of micro-, meso-, and macro-scale heterogeneity) cannot be effectively dealt with, (3) geomechanical processes are not explicitly modeled, (4) high performance parallel computing has not been implemented, and (5) graphical interfaces of pre-and post-processors have not been integrated with the simulator. Further modifications of the model to relax these limitations in its design capabilities are needed to improve the model flexibility for a wider range of applications.

MATHEMATICAL FOUNDATION

Governing equations of HYDROGEOCH-EM 4.5 and 5.5 include four types: a flow equation, a thermal transport equation, a set of reactive transport equations, and a set of reactive biogeochemical equations.

Flow Equations

A modified Richards equation describes density-dependent fluid flow in variably-saturated media. It can be derived based on continuity of fluid, continuity of solid, motion of fluid (Darcy's law), consolidation of the media, and compressibility of water [81] as:

$$\frac{\rho}{\rho_o} F \frac{\partial h}{\partial t} = \nabla \cdot \left[\mathbf{K} \cdot \left(\nabla h + \frac{\rho}{\rho_o} \nabla z \right) \right] + \frac{\rho^*}{\rho_o} q \,;\; F = \alpha' \frac{\theta}{n_e} + \beta' \theta + n_e \frac{dS}{dh};\;\; \mathbf{K} = \frac{(\rho/\rho_o)}{(\mu/\mu_o)} \mathbf{K}_{so} k_r \quad \textbf{(1)}$$

where ρ is the fluid density (M/L^3), ρ_o is the referenced fluid density at zero chemical concentration and at the reference temperature (M/L^3), F is the generalized storage coefficient (1/L), h is the pressure head (L), t is the time (T), $\mathbf{K}$ is the hydraulic conductivity tensor (L/T), z is the potential head (L), ρ^* is the fluid density of either injections ($=\rho^{inj}$) or withdrawals ($=\rho$) (M/L^3), q is the source or sink representing the artificial injection or withdrawal of fluid [(L^3/L^3)/T], α' is the modified compressibility of the media (1/L), θ is the effective moisture content (L^3/L^3), n_e is the effective porosity (L^3/L^3), β' is the modified compressibility of the liquid (1/L), S is the degree of saturation of water, μ_o is the fluid dynamic viscosity at zero chemical concentration and at the reference temperature (M/L/T), μ is the fluid dynamic viscosity (M/L/T), $\mathbf{K}_{so}$ is the referenced saturated hydraulic conductivity tensor (L/T), and k_r is the relative permeability or conductivity (dimensionless).

To complete the mathematical formulation of the variably saturated density dependent flow problem, Eq. (1) is supplemented with initial conditions and five types of boundary conditions: Dirichlet, Cauchy, Neumann, Variable, and Surface-water boundary conditions [19, 20].

On a ***Dirichlet boundary***, the pressure head is prescribed as a function of time as:

$$h = h_d(\mathbf{x}, t) \text{ on } B_d(\mathbf{x}) = 0 \quad \textbf{(2)}$$

where $h_d(\mathbf{x}, t)$, a function of space $\mathbf{x}$ and time t, is the prescribed pressure head on the Dirichlet boundary specified by $B_d(\mathbf{x}) = 0$ [L].

On a ***Cauchy boundary***, the volumetric flux is prescribed as function of time as:

$$-\mathbf{n}\cdot\mathbf{K}\cdot\left(\frac{\rho_o}{\rho}\nabla h+\nabla z\right)=q_c(\mathbf{x},t)\text{ on }B_c(\mathbf{x})=0 \qquad \textbf{(3)}$$

where **n** is an outward unit vector normal to the boundary surface; $q_c(\mathbf{x},t)$ is the prescribed volumetric flux [$L^3/L^2/T$] on the Cauchy boundary specified by $B_c(\mathbf{x})=0$.

On a ***Neumann boundary***, the flux due to a pressure gradient is prescribed as function of time as:

$$-\mathbf{n}\cdot\mathbf{K}\cdot\left(\frac{\rho_o}{\rho}\nabla h\right)=q_n(\mathbf{x},t)\text{ on }B_n(\mathbf{x})=0 \qquad \textbf{(4)}$$

where $q_n(\mathbf{x},t)$ is the prescribed volumetric flux [$L^3/L^2/T$] due to the pressure gradient on the Neumann boundary specified by $B_n(\mathbf{x})=0$.

On a ***Variable boundary***, either the flux or Dirichlet boundary conditions can be prescribed depending on the infiltration capacity of the media and the rainfall intensity during precipitation periods or on the evaporative capacity of the media and evaporation potential of the atmosphere during non-precipitation periods. These physical configurations can be described mathematically as follows. During *precipitation* periods, the boundary conditions on the Variable boundary are stated as:

$$h=h_p(\mathbf{x},t)\;\; \mathit{iff}\;\; -\mathbf{n}\cdot\mathbf{K}\cdot\left(\frac{\rho_o}{\rho}\nabla h+\nabla z\right)\geq q_p(\mathbf{x},\mathrm{t})\;\text{ on }B_v(\mathbf{x})=0 \qquad \textbf{(5)}$$

or

$$-\mathbf{n}\cdot\mathbf{K}\cdot\left(\frac{\rho_o}{\rho}\nabla h+\nabla z\right)=q_p(\mathbf{x},\mathrm{t})\;\; \mathit{iff}\;\; h\leq h_p(x,t)\text{ on }B_v(\mathbf{x})=0 \qquad \textbf{(6)}$$

where $h_p(\mathbf{x},t)$ is the ponding depth [L], $q_p(\mathbf{x},t)$ is the flux due to precipitation (negative values) [$L^3/L^2/T$], and $B_v(\mathbf{x})=0$ is the Variable boundary. Eq. (5) states that if infiltration capacity is less than rainfall intensity, a ponding depth on the surface is present. Eq. (6) states that if infiltration capacity is greater than rainfall intensity, ponding on the surface is not present and a flux boundary condition with rainfall intensity is applied to the surface. During *non-precipitation* periods, the boundary conditions are mathematically stated as:

$$h=h_p(x,t)\;\; \mathit{iff}\;\; -\mathbf{n}\cdot\mathbf{K}\cdot\left(\frac{\rho_o}{\rho}\nabla h+\nabla z\right)\geq 0\;\text{ on }B_v(\mathbf{x})=0 \qquad \textbf{(7)}$$

or

$$h=h_m(x,t)\;\; \mathit{iff}\; -\mathbf{n}\cdot\mathbf{K}\cdot\left(\frac{\rho_o}{\rho}\nabla h+\nabla z\right)\leq q_e(\mathbf{x},\mathrm{t})\;\text{ on }B_v(\mathbf{x})=0 \qquad \textbf{(8)}$$

or

$$-\mathbf{n}\cdot\mathbf{K}\cdot\left(\frac{\rho_o}{\rho}\nabla h+\nabla z\right)=q_e(\mathbf{x},\mathrm{t})\;\; \mathit{iff}\;\; h\geq h_m(x,t)\text{ on }B_v(\mathbf{x})=0 \qquad \textbf{(9)}$$

where $h_m(\mathbf{x},t)$ is the minimum pressure head (which normally is pressure corresponding to the wilting point) on the surface [L] and $q_e(\mathbf{x},t)$ is the potential evaporation allowed by the atmosphere [$L^3/L^2/T$]. Eq. (7) states that a seepage condition still prevails after precipitation stops. Eq. (8) states that when the evaporative capacity (evaporative capacity is defined as the evaporation rate when the pressure at the surface is at minimum) of the media is smaller than the potential evaporation of the atmosphere, a minimum pressure is maintained on the surface. Eq. (9) states if the evaporative capacity of the media is greater than the potential evaporation, the actual evaporation rate is equal to the potential evaporation.

On a ***Surface-water boundary*** (*e.g.*, rivers, large lakes, *etc.*), the flux is both a function of the surface water depth and the pressure head of the subsurface media. This boundary condition is employed when there is a thin layer of medium separating the surface water and the subsurface media and this thin layer is not included as part of the subsurface media. This boundary condition is mathematically posed as:

$$-\mathbf{n}\cdot\mathbf{K}\cdot\left(\frac{\rho_o}{\rho}\nabla h+\nabla z\right)=\frac{K_L}{b_L}\left(h_s-h\right)\text{ on }B_s(\mathbf{x})=0 \qquad \textbf{(10)}$$

where K_L is the hydraulic conductivity of the thin layer [L/T], b_L is the thickness of the thin layer [L], h_s is the depth of the surface water [L], and $B_s(\mathbf{x})=0$ is the interface between the surface water and the subsurface media.

Thermal Transport Equations

The governing equations of thermal transport in a subsurface system can be derived based on the principles of conservation of energy and the law of thermal flux. Yeh and Luxmoore [82] derived the governing equations describing the moisture and temperature fields in an unsaturated aquifer system. Assuming that the thermal flux due to moisture gradient is small compared to that due to temperature, one can write the governing equation in conservative form for thermal transport as:

$$\frac{\partial\left(\rho C_w+\rho_b C_m\right)T}{\partial t}+\alpha'\frac{\partial h}{\partial t}\left(\rho C_w+\rho_b C_m\right)T+\nabla\cdot\left(\rho C_w\mathbf{V}\nabla T\right)=\nabla\cdot\left(\mathbf{D}_T\cdot\nabla T\right)+S_H \qquad \textbf{(11)}$$

where T is the temperature [°K]; ρ_b is the bulk density of the medium [M/L^3]; C_w and C_m are the specific heats of the groundwater and the dry medium, respectively [(L^2/T^2)/ °K]; $\mathbf{V}$ is the Darcy's velocity of groundwater (L/T), $\mathbf{D}_T$ is the thermal dispersion coefficient tensor [(ML/T^3)/°K]; and S_H is the source/sink term of thermal energy that may be due to artificial injection and withdrawal as well as chemical reactions [(M/L)/T^3].

To enable the simulation of thermal transport over a wide range of problems, in addition to initial conditions, four types of boundary conditions (Dirichlet, Cauchy, Neumann, and Variable (inflow-outflow) are implemented in HYDROGEOCHEM.

On a ***Dirichlet boundary***, the temperature is prescribed as function of time as:

$$T=T_d(\mathbf{x},t)\text{ on }B_d(\mathbf{x})=0 \qquad \textbf{(12)}$$

where $T_d(\mathbf{x},t)$, a function of space $\mathbf{x}$ and time t, is the prescribed temperature on the Dirichlet boundary specified by $B_d(\mathbf{x})=0$ [°K]. It should be noted that the Dirichlet boundary for thermal transport does not necessarily coincide with that for fluid flow.

On a ***Cauchy boundary***, the thermal flux is prescribed as function of time as:

$$\mathbf{n}\cdot\left(\rho C_w \mathbf{V}T - \mathbf{D}_T \nabla T\right) = H_c(\mathbf{x},t) \text{ on } B_c(\mathbf{x}) = 0 \tag{13}$$

where $H_c(\mathbf{x},t)$ is the prescribed heat flux on the Cauchy boundary specified by $B_c(\mathbf{x}) = 0$ [E/L^2/T], where E denotes the unit of energy. Usually this type of boundary condition is applied to open upstream boundaries.

On a ***Neumann boundary***, the thermal flux due to a temperature gradient is prescribed as function of time as:

$$-\mathbf{n}\cdot\mathbf{D}_T \cdot \nabla T = H_n(\mathbf{x},t) \text{ on } B_n(\mathbf{x}) = 0 \tag{14}$$

where $H_n(\mathbf{x},t)$ is the prescribed heat flux due to the temperature gradient on the Neumann boundary specified by $B_n(\mathbf{x}) = 0$. This type of boundary condition is normally applied to the boundary where the conductive heat flux is known.

On a ***Variable boundary***, the flow direction can change with time during simulations. Two cases are considered depending on flow direction. When the flow is directed into the region of interest, the heat flux is normally coming into the region *via* fluid flow and the boundary condition is mathematically described as:

$$\mathbf{n}\cdot\left(\rho C_w \mathbf{V}T - \mathbf{D}_T \nabla T\right) = \mathbf{n}\cdot\rho C_w \mathbf{V}T_v(\mathbf{x},t) \text{ on } B_v(\mathbf{x}) = 0 \ \text{ if } \mathbf{n}\cdot\mathbf{V} \le 0 \tag{15}$$

where $T_v(\mathbf{x},t)$ is the temperature of the incoming fluid through the variable boundary $B_v(\mathbf{x}) = 0$ [^{o}K]. When the flow is directed out of the region, it is assumed that the heat is transported to the outside world *via* advection. The boundary condition is mathematically stated as:

$$-\mathbf{n}\cdot\mathbf{D}_T \nabla T = 0 \text{ on } B_v(\mathbf{x}) = 0 \ \text{ if } \mathbf{n}\cdot\mathbf{V} > 0 \tag{16}$$

Species Transport Equations

The transport/mass balance equation for any species subject to a reaction network can be derived based on the conservation law of material mass stating that the rate of mass change is due to advective-dispersive transport, geochemical reactions, and source/sinks as:

$$\frac{\partial(\theta C_i)}{\partial t} + \theta\alpha'\frac{\partial h}{\partial t}C_i = \iota_i L\left(C_i\right) + \theta r_i,\ i \in \left\{M\right\} \tag{17}$$

in which

$$L\left(C_i\right) = -\nabla\cdot(\mathbf{V}C_i) + \nabla\cdot\left[\theta\mathbf{D}\cdot\nabla C_i\right] + M_{C_i}{}^{as} \tag{18}$$

where C_i is the concentration of the i-th species in units of chemical mass per water volume [M/L^3]; ι_i is an indicator, $\iota_i = 1$ if C_i is a mobile species, $\iota = 0$ if C_i is an immobile species; r_i is the production rate of the i-th species due to biogeochemical reactions in chemical mass per water volume per unit time [M/L^3/T]; $\left\{M\right\} = \left\{1,2,..,M\right\}$ in which M is the number of biogeochemical species; **D** is the dispersion coefficient tensor [L^2/T]; and $M_{C_i}{}^{as}$ is the source/sink (other than sources/sinks due to chemical reactions) of the i-th species in chemical mass per unit volume of media [M/L^3/T].

As in thermal transport, to enable the simulation of reactive transport over a wide range of problems, in addition to initial conditions, four types of boundary conditions (Dirichlet, Cauchy, Neumann, and Variable inflow-outflow) are implemented in HYDROGEOCHEM.

On a ***Dirichlet boundary***, the species concentration is prescribed as function of time as:

$$C_i = C_{id}(\mathbf{x},t) \text{ on } B_d(\mathbf{x}) = 0 \qquad \textbf{(19)}$$

where $C_{id}(\mathbf{x},t)$, a function of space **x** and time *t*, is the prescribed concentration of the *i*-th species on the Dirichlet boundary specified by $B_d(\mathbf{x}) = 0$ [M/L^3]. It should be noted that the Dirichlet boundary for reactive transport does not necessarily coincide with that for fluid flow or for thermal transport.

On a ***Cauchy boundary***, the mass flux is prescribed as function of time as:

$$\mathbf{n}\cdot\left(\mathbf{V}C_i - \mathbf{D}\nabla C_i\right) = M_{ic}(\mathbf{x},t) \text{ on } B_c(\mathbf{x}) = 0 \qquad \textbf{(20)}$$

where $M_{ic}(\mathbf{x},t)$ is the prescribed mass flux of the *i*-th species on the Cauchy boundary specified by $B_c(\mathbf{x}) = 0$ [M/L^2/T]. Usually this type of boundary condition is applied to open upstream boundaries.

On a ***Neumann boundary***, the mass flux due to a concentration gradient is prescribed as function of time as:

$$-\mathbf{n}\cdot\mathbf{D}\cdot\nabla C_i = M_{in}(\mathbf{x},t) \text{ on } B_n(\mathbf{x}) = 0 \qquad \textbf{(21)}$$

where $M_{in}(\mathbf{x},t)$ is the prescribed mass flux of the *i*-th species due to the concentration gradient on the Neumann boundary specified by $B_n(\mathbf{x}) = 0$ [M/L^2/T]. This type of boundary condition is normally applied to boundaries where the dispersive mass flux is known.

On a ***Variable boundary***, the transport direction can change with time during simulations. Two cases are considered depending on flow direction. When the flow is directed into the region of interest, the mass flux is normally coming into the region *via* fluid flow and the boundary condition is mathematically described as:

$$\mathbf{n}\cdot\left(\mathbf{V}C_i - \mathbf{D}\nabla C_i\right) = \mathbf{n}\cdot\mathbf{V}C_{iv}(\mathbf{x},t) \text{ on } B_v(\mathbf{x}) = 0 \text{ if } \mathbf{n}\cdot\mathbf{V} \le 0 \qquad \textbf{(22)}$$

where $C_{iv}(\mathbf{x},t)$ is the concentration of the *i*-th species in the incoming fluid through the variable boundary $B_v(\mathbf{x}) = 0$ [M/L^3]. When the flow is directed out of the region, it is assumed the mass is carried to the outside world *via* advection. The boundary condition is mathematically stated as:

$$-\mathbf{n}\cdot\mathbf{D}\nabla C_i = 0 \text{ on } B_v(\mathbf{x}) = 0 \text{ if } \mathbf{n}\cdot\mathbf{V} > 0 \qquad \textbf{(23)}$$

Biogeochemical Equations

The model uses a reaction-based formulation, in which a reaction network is conceptualized as:

$$\sum_{i\in\{M\}} \mu_{ik}\hat{C}_i \leftrightarrow \sum_{i\in\{M\}} \nu_{ik}\hat{C}_i, \; k\in\{N\} \qquad \textbf{(24)}$$

where $\hat{C}_i$ is the chemical formula of the *i*-th species, μ_{ik} is the reaction stoichiometry of the *i*-th species in the *k*-th reaction associated with the reactants, ν_{ik} is the reaction stoichiometry of the *i*-th species in the *k*-th reaction associated with the products, and $\{N\} = \{1, 2, ..N\}$ in which N is the number of reactions. The production rate of any species is obtained based on the principle of reaction kinetics as:

$$r_i = \sum_{k\in\{N\}} \left(\nu_{ik} - \mu_{ik}\right)R_k, \; i\in\{M\} \qquad \textbf{(25)}$$

where R_k is the rate of the *k*-th reaction in moles of chemical per unit volume of water per unit time [M/L^3/T].

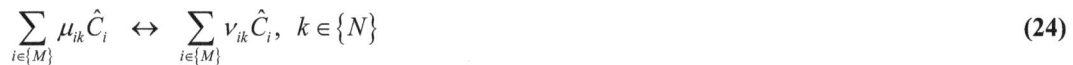

The substitution of Eq. (25) into Eq. (17) results in the following system of equations.

$$\frac{\partial(\theta C_i)}{\partial t}+\theta\alpha'\frac{\partial h}{\partial t}C_i=\iota_i L(C_i)+\theta\sum_{k\in\{N\}}(\nu_{ik}-\mu_{ik})R_k,\ i\in\{M\} \quad \textbf{(26)}$$

There are $(M+N)$ unknowns (M C_i's and N R_k's) but only M equations in Eq. (26). The system needs to be closed. Let us assume that there are N_E fast/equilibrium reactions (which must be all independent) and N_K slow/kinetic reactions, *i.e.*, $N=N_E+N_K$. For N_K kinetic reactions, an explicit rate equation for each reaction is usually formulated as:

$$R_k=f_k\left(C_i:i\in\{M\}\right),\ k\in\{N_K\} \quad \textbf{(27)}$$

where $f_k\left(C_i:i\in\{M\}\right)$ is an explicit function of M C_i's that represents the rate equation of the k-th reaction and $\{N_K\}\subset\{N\}$ is the set index of kinetic reactions having N_K members. Substitution of Eq. (27) into Eq. (26) yields

$$\frac{\partial(\theta C_i)}{\partial t}+\theta\alpha'\frac{\partial h}{\partial t}C_i=\iota_i L(C_i)+\theta\sum_{k\in\{N_K\}}(\nu_{ik}-\mu_{ik})f_k\left(C_i:i\in\{M\}\right)+\theta\sum_{k\in\{N_E\}}(\nu_{ik}-\mu_{ik})R_k,\ i\in\{M\} \quad \textbf{(28)}$$

where $\{N_E\}\subset\{N\}$ is the set index of equilibrium reactions having N_E members.

Eq. (28) still involves more unknowns (M C_i's and N_E R_k's) than the number of equations (M). To close the system, for every equilibrium reaction, thermodynamic approaches are used where a consistent algebraic equation, for example a mass action equation, can be used to describe the corresponding fast/equilibrium reaction as:

$$F_k\left(C_i:i\in\{M\}\right)=0,\ e.g.\ \beta_k^E-\prod_{i\in M}(C_i)^{\nu_{ik}}/\prod_{i\in M}(C_i)^{\mu_{ik}}=0,\ k\in\{N_E\} \quad \textbf{(29)}$$

where $F_k\left(C_i:i\in\{M\}\right)=0$ is a nonlinear algebraic equation and β_k^E is the modified equilibrium constant of the k-th equilibrium reaction [2]. Eqs. (28) and (29) constitute a system of M nonlinear transport equations and N_E nonlinear algebraic equations for M C_i's and N_E R_k's of unknowns. Simultaneous solutions of this system of M reactive transport equations and N_E nonlinear algebraic equations over the entire domain of interest would demand excessive computational time and computer storage [2] even with today's computer capacities.

Decomposition of Reactive Transport Equations

Decoupling of the N_E R_k's of unknowns from the M C_i's of unknowns can be done *via* the Gauss-Jordan reduction of reaction networks [49, 70, 83]. Performing matrix decomposition of the reaction matrix in Eq. (26) can reduce it into three subsets of equations as described below.

1) $M_{KI}\left(=N_K-N_{KD}\right)$ transport-equations for N_{KI} independent kinetic variables:

$$\frac{\partial\theta E_i}{\partial t}+\alpha'\theta\frac{\partial h}{\partial t}E_i=L\left(E_i^a\right)+\theta D_{ik}f_k\left(C_\ell:\ell\in\{M\}\right)+\theta\sum_{j\in\{N_{KD(k)}\}}D_{ij}f_i\left(C_\ell:\ell\in\{M\}\right),\ i\in\{M_{KI}\},$$
$$k\in\{N_{KI}\};\text{in which } E_i=\sum_{j\in\{M\}}a_{ij}C_j \quad \textbf{(30)}$$

where N_{KI} and N_{KD} are the number of independent kinetic reactions and linearly dependent kinetic reactions, respectively, and $N_{KI} = N_K - N_{KD}$; E_i is the *i*-th reaction extent [M/L^3], which shall be called a kinetic variable since it corresponds to an independent kinetic reaction on the right hand side of Eq. (30); E_i^a is the portion of E_i which contains the linear combination of only mobile aqueous species; D_{ik} is the non-zero entry of the *i*-th row and the *k*-th column of the decomposed reaction matrix; $\{M_{KI}\} \subset \{M\}$ having $M_{KI} = N_{KI}$ members is the set index of reaction extents for kinetic variables; $\{N_{KD(k)}\} \subset \{N_k\}$ having $N_{KD(k)}$ members is the set index of kinetic reactions that depend on the independent reaction *k*, in which $N_{KD(k)}$ is the number of linearly dependent kinetic reactions that depend on the independent reaction *k*; D_{ij} is an element of decomposed reaction matrix; $\{N_{KI}\} \subset \{N_K\}$ having N_{KI} members is the set index of independent kinetic reactions; and a_{ij} is the element of an $M \times M$ matrix resulting from the matrix decomposition of a unit matrix and represents the coefficient of the linear combination of C_j in E_i. It should be noted that $\{N\} = \{N_K\} \cup \{N_E\}$, $\{N_K\} = \{N_{KI}\} \cup \{N_{KD}\}$, and $\{N_{KD}\} = \bigcup_{k \in \{N_{KI}\}} \{N_{KD(k)}\}$ where $\{N_{KD}\} \subset \{N_K\}$ is the set index of linearly dependent kinetic reactions.

2) $M_C \left(= M - N_E - N_{KI}\right)$ transport-equations for N_C components:

$$\frac{\partial \theta E_i}{\partial t} + \alpha' \frac{\partial h}{\partial t} \theta E_i = L\left(E_i^a\right),\ i \in \{M_C\} \tag{31}$$

where N_C is the number of components; E_i, the *i*-th reaction extent, shall be called a component since there are no reaction terms on the right hand side of Eq. (31); and $\{M_C\} \subset \{M\}$ having $M_C = N_C$ members is the set index of reaction extents for components. The term 'components' is used to refer to the ions or molecules as the building blocks which other species contain. (*e.g.* for a system containing calcite, Ca^{2+} and CO_3^{2-} can be used as 'components' which combine to form $CaCO_3$).

3) $M_E \left(= N_E\right)$ rate-equations for N_E fast or equilibrium reactions:

$$\theta D_{ik} R_k = \frac{\partial \theta E_i}{\partial t} + \alpha' \frac{\partial h}{\partial t} \theta E_i - L\left(E_i^a\right) - \theta \sum_{j \in \{N_{KD(k)}\}} D_{ij} f_j\left(C_\ell : \ell \in \{M\}\right),\ k \in \{N_E\},\ i \in \{M_E\} \tag{32}$$

where E_i, the *i*-th reaction extent, shall be called an equilibrium variable since it corresponds to an equilibrium reaction and $\{M_E\} \subset \{M\}$ having $M_E = N_E$ members is the set index of reaction extents for equilibrium variables. It should be noted that $\{M\} = \{M_{KI}\} \cup \{M_C\} \cup \{M_E\}$.

The initial and boundary conditions for species transport can be transformed from those in terms of C_i's into those in terms of E_i's in a straightforward manner.

Before proceeding further, several comments are in order. Firstly, instead of solving *M* nonlinear reactive transport equations in Eq. (28) and N_E nonlinear algebraic equations in Eq. (29) simultaneously, one solves M_{KI} nonlinear reactive transport equations in Eq. (30) and M_C component transport equations in Eq. (31) one by one for $N_{KI} \left(= M_{KI}\right)$ kinetic variables and $N_C \left(= M_C\right)$ components. Having obtained these variables over the domain, one then solves Eq. (33) below, which consists of M_{KI} and M_C linear algebraic equations and $M_E\left(=N_E\right)$ nonlinear algebraic equations for M species concentrations, C_j's, in the discretized domain of interest point by point as:

$$F_k\left(C_\ell : \ell \in \{M\}\right) = 0,\ k \in \{N_E\};\ \sum_{\ell\in\{M\}} a_{i\ell}C_\ell = E_i,\ i \in \{M_{KI}\};\ \text{and} \sum_{\ell\in\{M\}} a_{i\ell}C_\ell = E_i,\ \ i \in \{M_C\} \qquad \textbf{(33)}$$

Secondly, the definition of the rate for an equilibrium reaction is very controversial. It has been argued that the rate of an equilibrium reaction "can be mathematically abstracted as infinity" for the convenience of decoupling equilibrium reactions from kinetic reactions [70]. It has also been argued that the rate of an equilibrium reaction is indefinite [31]. This controversy need not have arisen at all since, by definition, an equilibrium reaction should not be associated with a rate per se. We can associate a rate to an equilibrium reaction only if we treat it as a fast kinetic reaction; then the rate of an equilibrium reaction is clearly defined by Eq. (32) and the controversy over the definition of the rate for an equilibrium reaction is settled. Simply stated, the rate of the equilibrium reaction is that rate which is "necessary" to assure that the thermodynamic equations remain fulfilled, *i.e.*, that the solution $C(\mathbf{x},t)$ remains on the manifold defined by the thermodynamic equations. As a result, these rates are finite and definite, which makes the modeling of mixed fast/equilibrium and slow/kinetic reactions consistent.

Thirdly, although among those N_{KD} linearly dependent kinetic reactions that depend on only equilibrium reations are irrelevant to the simulation of all species concentrations [70], they are relevant in the calculation of the N_E rates of equilibrium/fast reactions.

Fourthly, the N_E equilibrium reactions must be independent to avoid over or under determining the system. If they are not independent, then they can be divided into $N_E = N_{EI} + N_{ED}$ where N_{EI} is the number of independent equilibrium reactions and N_{ED} is the number of linearly dependent equilibrium reactions. Under such circumstances, N_C would be $N_C = M - N_{KI} - N_{EI}$. Superficially, Eqs. (28) and (29) still constitute a system of M nonlinear transport equations and N_E non-linear algebraic equations for M C_i's and N_E R_k's of unknowns. However, after decomposition of the reaction matrix, the system of equations in Eqs. (29) through (31) [equivalently, the system of equations in Eq. (33)] is decoupled from the system of equations in Eq. (32). It is obviously seen that the former system constitutes $\left(M+N_{ED}\right)$ equations for M unknowns. The system is over-determined and singularities may occur. On the other hand, the latter system [Eq. (32)] becomes:

$$\theta D_{ik}R_k = \frac{\partial \theta E_i}{\partial t} + \alpha'\frac{\partial h}{\partial t}\theta E_i - L\left(E_i^a\right) - \theta \sum_{j\in\{N_{KD(k)}\}} D_{ij}f_i\left(C_\ell :\ \ell \in \{M\}\right),\ k \in \{N_{EI}\},\ i \in \{M_{EI}\} \qquad \textbf{(34)}$$

where $\{N_{EI}\} \subset \{N_E\}$ has N_{EI} members and $\{M_{EI}\} \subset \{M\}$ has $M_{EI} = N_{EI}$ members. This subsystem constitutes N_{EI} equations in N_E unknowns, N_E R_k's, which is under-determined yielding infinite sets of solutions. Thus, N_E equilibrium reactions must be independent and any redundant linearly dependent equilibrium reactions must be removed [70].

Coupling between Flow and Thermal and Reactive Transport

Dependence of Groundwater Density and Dynamic Viscosity on Species Concentration and Temperature. The density of groundwater is a function of biogeochemical-species concentration and temperature, which can be derived as follows [29].

$$\rho = \rho_w + \sum_{i=1}^{M^d} m_i C_i \left(1 - \frac{\rho_w}{\rho_i}\right) \qquad \textbf{(35)}$$

where ρ_w is the density of water at zero concentration, which is a function of temperature [M/L^3]; M^d is the number of dissolved mobile species; and C_i, m_i, and ρ_i are the molar concentration, molecular weight, and intrinsic density of the i-th dissolved species, respectively. The dynamic viscosity of groundwater is assumed to take the following form [29].

$$\mu = \mu_w + \sum_{i=1}^{M^d} a_i m_i C_i \qquad \textbf{(36)}$$

where μ_w is the dynamic viscosity of pure water [M/L/T], which is a function of temperature; and a_i is the i-th weighting parameter representing the dependence of dynamic viscosity on the concentration of the i-th species. It should be noted that Eq. (35) can be derived theoretically, whereas Eq. (36) is an empirical equation and can be replaced with any other empirical formula to match the situation being considered. The modification in the computer code is straightforward. Eq. (35) is used to describe groundwater density as a function of the concentration of dissolved species. For some cases, however, it may not be sufficient or the intrinsic densities of dissolved species or aqueous components might not be available. To overcome this, an empirical formula can be used to characterize the groundwater density as a function of temperature and concentrations of all species.

Effect of Precipitation/Dissolution on Porosity, Hydraulic Conductivity, and Water Capacity. The effective moisture content is theoretically described in terms of the pressure head (*via* the degree of saturation *S*) and the concentrations of biogeochemical species as [17, 71].

$$\theta = \frac{S\theta_{so}}{1+S\varphi_p}; \quad \varphi_p = \sum P_i V_i \qquad \textbf{(37)}$$

where θ is the effective moisture content [dm^3 of water/dm^3 of pore]; θ_{so} is the effective saturated moisture content when solid or surface biogeochemical species are not present [dm^3 of pores/dm^3 of medium]; φ_p is the volume of precipitated species per unit volume of water [dm^3 of precipitates/dm^3 of water]; P_i is the precipitated concentration of the i-th mineral [mole/dm^3 of water]; V_i is the molar volume of the i-th mineral [dm^3 of mineral/mole of mineral]; and i stands for the i-th precipitated mineral. The hydraulic conductivity and the water capacity, respectively, in Eq. (1) are modified to incorporate the effect of precipitation-dissolution as the first and second equations, respectively, in Eq. (38) below:

$$\mathbf{K} = \frac{\left(\dfrac{\rho}{\rho_o}\right)}{\left(\dfrac{\mu}{\mu_o}\right)} \mathbf{K}_{so} k_r \left(\frac{1}{1+S\varphi_p}\right)^n; \quad \psi = \frac{\theta_{so}}{1+S\varphi_p}\frac{dS}{dh} \qquad \textbf{(38)}$$

where n is the fractional exponent depending on particle size and packing structure and $\psi \equiv n_e \times dS/dh$ is the water capacity.

The Effect of Precipitation/Dissolution on Hydrodynamic Dispersion. The effect of precipitation/dissolution on hydrodynamic dispersion has been discussed elsewhere [17]. From Archie's law, the paper of Steefel and Lichtner [84], and the assumption that dispersion is affected by precipitation/dissolution in a similar way as diffusion, we can derive the hydrodynamic dispersion as [19, 20].

$$\theta\mathbf{D} = \left[\alpha_T |\mathbf{V}|\boldsymbol{\delta} + (\alpha_L - \alpha_T)\mathbf{V}\mathbf{V}/|\mathbf{V}| + \theta D_w \tau \boldsymbol{\delta}\right] \left[\theta^{m-1}(1-\varphi_p)^m\right] \qquad \textbf{(39)}$$

where a_L and a_T are the longitudinal and transverse diffusivity, respectively [L]; $\boldsymbol{\delta}$ is Kronecker delta tensor; $|\mathbf{V}|$ is the magnitude of the Darcy's velocity [L/T]; D_w is the molecular diffusion coefficient of pure water [L^2/T]; τ is tortuosity; m is the cementation exponent [85] with the reported values between 1.3 and 2.5; and θ is either the input moisture content or equal to the moisture content calculated in the coupled flow simulation. Eq. (39) provides the adjustment to both the dispersion and diffusion parts in performing the transport simulations. Even though $\mathbf{V}$ and θ are both computed with the effect of precipitation in the flow module, we still need to adjust the dispersion coefficient (computed based on $\mathbf{V}$ and θ) to reflect the effect of precipitation on the hydrodynamic dispersion. Since the precipitation might be changed in each nonlinear reactive transport loop accounting for precipitation and/or adsorption, this adjustment should be performed during each nonlinear transport loop. For the case without any effect of precipitation-dissolution on the dispersion coefficient, $m = 1.0$ and $V_i = 0$ should be set for all possible precipitated minerals.

Dependence of Chemical Reactions on Temperature. The Van't Hoff relationship is used to adjust ΔH and K using the following equation

$$\ell n\frac{K^e}{K_o^e} = \frac{\Delta H_o^e}{R}\left(\frac{1}{T_o} - \frac{1}{T}\right) \Rightarrow K^e = K_o^e \exp\left[\frac{\Delta H_o^e}{R}\left(\frac{T - T_o}{TT_o}\right)\right] \tag{40}$$

and

$$K^f = K_o^f \exp\left[\frac{\Delta H_o^f}{R}\left(\frac{T - T_o}{TT_o}\right)\right] \text{ and } K^b = K_o^b \exp\left[\frac{\Delta H_o^b}{R}\left(\frac{T - T_o}{TT_o}\right)\right] \tag{41}$$

where K^e and K_o^e are the temperature dependent and reference equilibrium constants for equilibrium reactions, respectively; ΔH_o^e is the reference enthalpy of the associated equilibrium reaction; R is the ideal gas constant; T_o is the reference temperature; K^f and K_o^f are the temperature dependent and reference forward rate constants for kinetic reactions, respectively; K^b and K_o^b are the temperature dependent and reference backward rate constants for kinetic reactions, respectively; and ΔH_o^f and ΔH_o^b are the referenced forward and the backward enthalpy of the kinetic reaction, respectively. In Eqs. (40) and (41), K_o^e or K_o^f and K_o^b, R, and T_o are all input by users; ΔH_o^e, ΔH_o^f, and ΔH_o^b are calculated from each species' enthalpy input by users. Any consistent system of units can be used to perform the computation.

NUMERICAL APPROXIMATION

HYDROGEOCHEM solves a system of equations describing fluid flow, thermal transport and reactive hydrologic transport in variably saturated conditions. In each time step, an iterative procedure is used to solve the equations for coupled processes as shown in Fig. **2**.

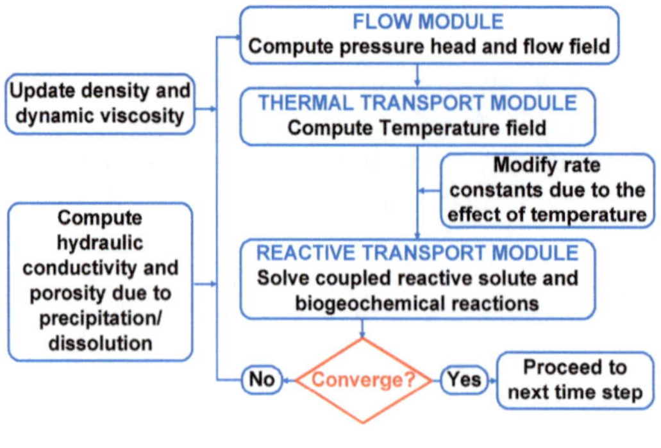

Figure 2: Schematic of solving coupled flow and thermal and reactive transport equations.

First, the pressure head and flow field are obtained in the flow module which solves Eq. (1) with initial and boundary conditions of Eqs. (2) through (10). Second, the temperature field is obtained in the thermal transport module which solves Eq. (11) with initial and boundary conditions of Eqs. (12) through (16). Third, the reaction rate constants and parameters are updated to account for the effect of temperature using Eqs. (40) and (41). Fourth, the concentrations of all species and rates of all reactions are obtained in the reactive transport module which solves Eqs. (30), (31), and (33) with initial and boundary conditions that are transformed from Eqs. (19) through (23). Fifth, convergence of the solutions is checked. If solutions are convergent, the program goes go to the next time step. If not, fluid density and viscosity are updated using Eqs. (35) and (36) and hydraulic parameters using Eqs. (37) and (38), and then goes to the next non-linear loop of coupled processes. Once all species concentrations are obtained, Eq. (32) is used to do posterior calculations of all rates of fast/equilibrium reactions.

In all three modules, Finite Difference (FD) methods are employed for temporal discretization of the governing partial differential equations. The Galerkin finite element method is employed for spatial discretization of the modified Richards equation governing the distribution of pressure fields. For scalar transport equations (including both thermal and reactive transport) either conventional finite element methods or the hybrid Lagrangian-Eulerian finite element methods are used for spatial discretization. Three options are available to solve reactive transport equations: (1) conventional FEM applied to the advective form of the equation, (2) hybrid Lagrangian-Eulerian FEM; and (3) hybrid Lagrangian-Eulerian FEM for interior and downstream boundary nodes plus conventional FEM in the advective form of the equation for upstream boundary nodes. The same three options are employed to approximate the thermal transport equation, but the conservative form of the equation is used in the conventional finite element option. Details of numerical procedures of these options and the implementation of boundary conditions can be found in Yeh *et al.* [19, 20].

Numerical discretization of reactive transport problems poses a great challenge because of the strong nonlinearity between transport and reactions. Ideally, one would like to use a numerical approach that is accurate, efficient, and robust. Depending on the specific problem at hand, different numerical approaches may be more or less suitable. For research applications, accuracy is the primary requirement, because one does not want to distort physical processes due to numerical errors. On the other hand, for large-scale problems, efficiency and robustness are primary concerns as long as accuracy remains within the bounds of uncertainty associated with model parameters. Thus, to provide accuracy for research applications and efficiency and robustness for practical applications, three coupling strategies [86, 87] were investigated to deal with reactive chemistry. They are: (1) a fully-implicit scheme, (2) a mixed predictor-corrector/operator-splitting method, and (3) an operator-splitting method. For each time-step, the advective-dispersive transport equations are solved first with or without reaction terms. Then the reactive biogeochemical system is solved node-by-node to yield concentrations of all species.

Solving Advective-Dispersive-Reactive Transport Equations

Fully-Implicit Sequential Iteration Approach. In the fully-implicit approach, the biogeochemical reaction subsystem of equations and the hydrologic transport subsystems of equations are solved sequentially and iteratively. The concentrations of all immobile components and kinetic variables are computed in the biogeochemical subsystem using all species concentrations computed from the prior iteration. The concentrations of mobile components and kinetic variables are determined in the hydrologic module using the species concentrations and the reaction terms evaluated from species concentrations of the current iteration. Using this approach, the systems of transport equations that are solved are:

$$\frac{\partial \theta E}{\partial t}+\alpha'\theta\frac{\partial h}{\partial t}E-L(E)= -L\left(\left(E^{s}\right)^{(r)}+\left(E^{p}\right)^{(r)}\right)+\theta R^{(r)} \tag{42}$$

in which

$$R^{(r)}=D_{ik}f_k\left(C_\ell^{(r)}:\ell\in\{M\}\right)+\sum_{j\in\{N_{KD(k)}\}}D_{ij}f_i\left(C_\ell^{(r)}:\ell\in\{M\}\right),\ i\in\{M_{KI}\}, \tag{43}$$

and $k\in\{N_{KI}\}$ if E is a kinetic variable; $R^{(r)}=0$ if E is a component where E denotes any of the M_{KI} kinetic variables or M_C components, E^s is the sorbed content of E, E^p is the precipitated content of E, and the superscript (r) denotes the previous iteration. The solution procedure is outlined as:

Step 1 Provide initial estimate for E, *i.e.*, $E^{(r)}$;

Step 2 With known $E^{(r)}$, solve Eq. (33) to yield the concentrations of all species, from which $\left(E^s\right)^{(r)}$, $\left(E^p\right)^{(r)}$, and $R^{(r)}$ are calculated;

Step 3 With $\left(E^s\right)^{(r)}, \left(E^p\right)^{(r)}$, and $R^{(r)}$ known, solve Eq. (42) to update E and obtain new estimate $E^{(r+1)}$

Step 4 Check convergence by comparing the newly obtained $E^{(r+1)}$ and $E^{(r)}$ in Step 1;

Step 5 If a convergent solution is obtained, update the coefficients in the governing equations and proceed to next time step computation; if the solution is not convergent, update $E^{(r)}$ with new $E^{(r+1)}$, then repeat Step 2 through 4.

Operator Splitting Approach. In this approach, the concentrations of mobile kinetic variables and components at the new time step are obtained with two systems of equations.

$$\frac{\partial \theta E^a}{\partial t} + \alpha' \theta \frac{\partial h}{\partial t} E^a - L\left(E^a\right) = 0 \tag{44}$$

and

$$\frac{E^{(n+1)} - \left[\left(E^a\right)^{(n+1/2)} + \left(E^s\right)^{(n)} + \left(E^p\right)^{(n)}\right]}{\Delta t} + \left[\frac{\partial \ell n(\theta)}{\partial t} + \alpha' \frac{\partial h}{\partial t}\right]\left[\left(E^s\right)^{(n+1)} + \left(E^p\right)^{(n+1)}\right] = R^{(n+1)} \tag{45}$$

where the superscript $(n+1)$ denotes the new time step, the superscript $(n+1/2)$ denotes the intermediate time step, the superscript (n) denotes the old time step, and $R^{(n+1)}$ is defined similarly to $R^{(r)}$ in Eq. (43). The solution procedure is outlined as:

Step 1 Solve Eq. (44) for the intermediate values $\left(E^a\right)^{(n+1/2)}$ and

Step 2 With known $\left(E^a\right)^{(n+1/2)}$, solve Eq. (45) along with the first equation in Eq. (33) to yield the concentrations of all species at new time step, from which $\left(E^a\right)^{(n+1)}, \left(E^s\right)^{(n+1)}$, and $\left(E^p\right)^{(n+1)}$, are calculated, which serve as the old time step values for the next time step.

In this approach, Steps 1 and 2 are repeated for every time step. No convergence check or iteration is needed.

Predictor-Corrector Approach. In this approach, the concentrations of mobile kinetic variables and components at the new time step are obtained with two systems of equations.

$$\frac{\partial \theta E^a}{\partial t} + \alpha' \theta \frac{\partial h}{\partial t} E^a - L\left(E^a\right) = R^{(n)} \tag{46}$$

and

$$\frac{E^{(n+1)} - \left[\left(E^a\right)^{(n+1/2)} + \left(E^s\right)^{(n)} + \left(E^p\right)^{(n)}\right]}{\Delta t} + \left[\frac{\partial \ell n(\theta)}{\partial t} + \alpha' \frac{\partial h}{\partial t}\right]\left[\left(E^s\right)^{(n+1)} + \left(E^p\right)^{(n+1)}\right] = R^{(n+1)} - R^{(n)} \tag{47}$$

The solution procedure is outlined as:

Step 1 Solve Eq. (46) for the intermediate value $\left(E^a\right)^{(n+1/2)}$ and

Step 2 With known $\left(E^a\right)^{(n+1/2)}$, solve Eq. (47) along with the first Eq. in Eq. (33) to yield the concentrations of all species at new time step, from which $\left(E^a\right)^{(n+1)}, \left(E^s\right)^{(n+1)}$, $\left(E^p\right)^{(n+1)}$, and $R^{(n+1)}$ are calculated, which serve as the old time step values at the next time step.

In this approach, Steps 1 and 2 are repeated for every time step. No convergence check or iteration is needed.

DEMONSTRATIVE EXAMPLES

Thirty-seven example problems (twenty-one for 2 Dimensions and sixteen for 3 Dimensions) were used to demonstrate the ability of HYDROGEOCHEM to simulate a variety of mixed subsurface flow, thermal transport, solute transport, and chemical reaction problems [19, 20]. These problems could serve as templates for applications to hypothetical and real-world problems. Here only six problems are chosen to illustrate the application of HYDROGEOCHEM to a variety of situations.

Example 1

This problem considers the release and migration of uranium from a simplified uranium mill tailings pile [19, 20]. A schematic two-dimensional vertical cross section of the hypothetical site is shown in Fig. **3**. The mill tailings pile is located adjacent to a surface that slopes down to a river.

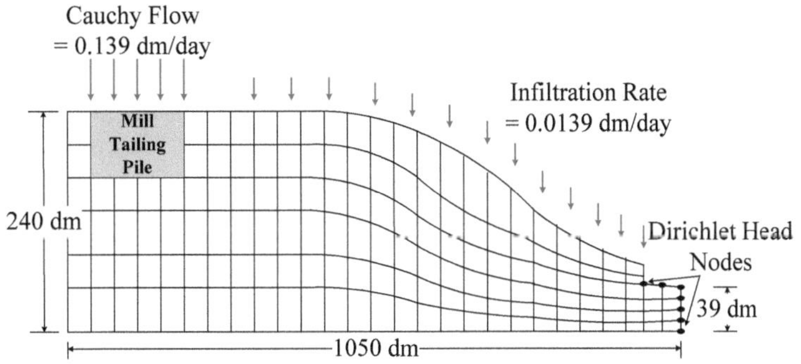

Figure 3: Problem definition with a schematic two-dimensional vertical cross section.

The vertical left edge, front edge, back edge and horizontal bottom of the region are impermeable no-flow boundaries. The sloping region on the top right is a variable flow boundary with either zero ponding depth or a net rainfall rate of 0.0139 dm/day. The horizontal region on the top of the mill tailings pile is a Cauchy flow boundary with an infiltration rate of 0.139 dm/day. The ten nodes on the vertical line on the right side and the four nodes on the river bottom are specified as Dirichlet known-head conditions. Total head at the vertical nodes is 39 dm. The total head on the left and right nodes of the river bottom are 45 dm and 40 dm, respectively. The region is discretized as shown in Fig. **3** (one element thick) with 158 elements and 192 ×

2 = 384 nodes. From the steady-state flow simulations, pressure head contours and velocity fields are depicted in Fig. **4**.

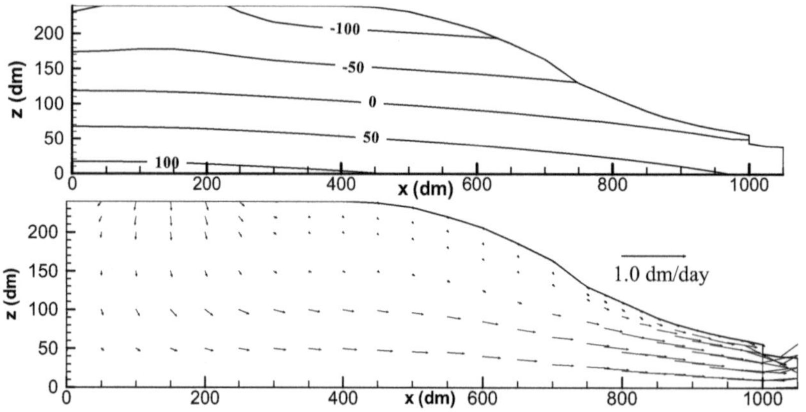

Figure 4: Pressure head contours (top) and velocity fields (bottom).

For reactive hydrogeochemical transport, the problem consists of seven components: Ca^{2+}, CO_3^{2-}, UO_2^{2+}, SO_4^{2-}, PO_4^{3-}, Fe^{2+}, and H^+ (pH). The pH is simulated on the basis of the total excess hydronium designated by TOTH (total analytical excess or deficit of protons). A total of 35 aqueous species and 14 minerals ($CaSO_4$, $CaCO_3$, $Ca(OH)_2$, $Ca_5(OH)(PO_4)_3$, $Ca(UO_2)_2(PO_4)_2$, $Ca_4H(PO_4)_3$, $CaHPO_4$, $FeCO_3$, $Fe(OH)_2$, $Fe_3(PO_4)_2$, $(UO_2)(OH)_2$, $(UO_2)(CO_3)$, $Fe(UO_2)_2(PO_4)_2$, and $H(UO_2)PO_4$) are defined for the problem and redox reactions were not considered. Table **1** lists chemical reactions involved in this example.

The composition of pore water in the tailings pile, pore water outside of the tailing pile and the recharge water are given in Table **2**. The mill tailings pile is represented by Dirichlet nodes in order to hold the total concentrations of the chemical components constant throughout the simulation. The sloping area to the right of the tailings pile, the river bank, the river and right-hand side of the domain below the river are variable boundary nodes. Only advective transport is considered in this example. A total of 300 days is simulated with a step size of 2 days.

Table 1: Reaction network for example 1.

Reaction	**No**	**Log K**
Aqueous Complexation Reactions		
$H_2O \leftrightarrow H^+ + OH^-$	(R1)	-14.00
$Ca^{2+} + CO_3^{2-} \leftrightarrow CaCO_3$ (aq)	(R2)	3.22
$Ca^{2+} + H^+ + CO_3^{2-} \leftrightarrow CaHCO_3^+$	(R3)	11.43
$Ca^{2+} + SO_4^{2-} \leftrightarrow CaSO_4$ (aq)	(R4)	2.31
$Ca^{2+} + 2H^+ + PO_4^{3-} \leftrightarrow CaH_2PO_4^+$	(R5)	20.96
$Ca^{2+} + PO_4^{3-} \leftrightarrow CaPO_4^-$	(R6)	6.46
$Ca^{2+} + H^+ + PO_4^{3-} \leftrightarrow CaHPO_{4\,(aq)}$	(R7)	15.08
$Ca^{2+} + H_2O \leftrightarrow H^+ + CaOH^+$	(R8)	-12.58
$Fe^{2+} + SO_4^{2-} \leftrightarrow FeSO_{4\,(aq)}$	(R9)	2.20
$Fe^{2+} + H_2O \leftrightarrow H^+ + FeOH^+$	(R10)	-9.5
$Fe^{2+} + 2H_2O \leftrightarrow 2H^+ + Fe(OH)_{2\,(aq)}$	(R11)	-20.57
$Fe^{2+} + 3H_2O \leftrightarrow 3H^+ + Fe(OH)_3^-$	(R12)	-31.0
$Fe^{2+} + 4H_2O \leftrightarrow 4H^+ + Fe(OH)_4^{2-}$	(R13)	-43.0
$UO_2^{2+} + H_2O \leftrightarrow H^+ + UO_2OH^+$	(R14)	-5.3

Table 1: cont….

$2UO_2^{2+} + 2H_2O \leftrightarrow 2H^+ + (UO2)_2(OH)_2^{2+}$	(R15)	-5.68
$3UO_2^{2+} + 4H_2O \leftrightarrow 4H^+ + (UO2)_3(OH)_4^{2+}$	(R16)	-11.88
$3UO_2^{2+} + 5H_2O \leftrightarrow 5H^+ + (UO2)_3(OH)_5^+$	(R17)	-15.82
$4UO_2^{2+} + 7H_2O \leftrightarrow 7H^+ + (UO2)_4(OH)_7^+$	(R18)	-21.90
$3UO_2^{2+} + 7H_2O \leftrightarrow 7H^+ + (UO_2)_3(OH)_7^-$	(R19)	-28.34
$UO_2^{2+} + CO_3^{2-} \leftrightarrow UO_2CO_{3(aq)}$	(R20)	9.65
$UO_2^{2+} + 2CO_3^{2-} \leftrightarrow UO_2(CO_3)_2^{2-}$	(R21)	17.08
$UO_2^{2+} + 3CO_3^{2-} \leftrightarrow UO_2(CO_3)_3^{4-}$	(R22)	21.7
$2UO_2^{2+}+CO_3^{2-}+3H_2O \leftrightarrow 3H^+ + (UO_2)_2CO_3(OH)_3^-$	(R23)	-1.18
$UO_2^{2+} + SO_4^{2-} \leftrightarrow UO_2SO_{4\,(aq)}$	(R24)	2.95
$UO_2^{2+} + 2SO_4^{2-} \leftrightarrow UO_2(SO_4)_2^{2-}$	(R25)	4.00
$2H^+ + UO_2^{2+} + PO_4^{3-} \leftrightarrow H_2UO_2PO_4^+$	(R26)	23.2
$3H^+ + UO_2^{2+} + PO_4^{3-} \leftrightarrow H_3UO_2PO_4^{2+}$	(R27)	22.9
$Ca^{2+}+4H^++UO_2^{2+}+ 2PO_4^{3-} \leftrightarrow CaH_4UO_2(PO_4)_2^{2+}$	(R28)	45.24
$Ca^{2+}+5H^++UO_2^{2+}+ 2PO_4^{3-} \leftrightarrow CaH_5UO_2(PO_4)_2^{3+}$	(R29)	46.0
$H^+ + CO_3^{2-} \leftrightarrow HCO_3^-$	(R30)	10.32
$2H^+ + CO_3^{2-} \leftrightarrow H_2CO_{3(aq)}$	(R31)	16.67
$H^+ + SO_4^{2-} \leftrightarrow HSO_4^-$	(R32)	1.99
$H+ + PO_4^{3-} \leftrightarrow HPO_4^{2-}$	(R33)	12.35
$2H^+ + PO_4^{3-} \leftrightarrow H_2PO_4^-$	(R34)	19.55
$3H^+ + PO_4^{3-} \leftrightarrow H_3PO4$	(R35)	21.74
Precipitation-Dissolution Reactions		
$Ca^{2+} + SO_4^{2-} \leftrightarrow CaSO_{4(s)}$	(R36)	4.62
$Ca^{2+} + CO_3^{2-} \leftrightarrow CaCO_{3(s)}$	(R37)	8.48
$5Ca^{2+} + 3PO_4^{3-} + H_2O \leftrightarrow H^+ + Ca_5(OH)(PO_4)_{3(s)}$	(R38)	40.47
$Fe^{2+} + CO_3^{2-} \leftrightarrow FeCO_{3(s)}$	(R39)	10.50
$Ca^{2+} + 2UO_2^{2+} + 2PO_4^{3-} \leftrightarrow Ca(UO_2)_2(PO_4)_{2(s)}$	(R40)	48.61
$4Ca^{2+} + H^+ + 3PO_4^{3-} \leftrightarrow Ca_4H(PO_4)_{3(s)}$	(R41)	48.20
$Ca^{2+} + H^+ + PO_4^{3-} \leftrightarrow CaH(PO_4)_{(s)}$	(R42)	19.3
$Ca^{2+} + 2H_2O \leftrightarrow 2H^+ + Ca(OH)_{2(s)}$	(R43)	-21.9
$3Fe^{2+} + 2PO_4^{3-} \leftrightarrow Fe_3(PO_4)_{2(s)}$	(R44)	33.30
$Fe^{2+} + 2H_2O \leftrightarrow 2H^+ + Fe(OH)_{2(s)}$	(R45)	-12.1
$UO_2^{2+} + 2H_2O \leftrightarrow 2H^+ + (UO_2)(OH)_{2(s)}$	(R46)	-5.40
$UO_2^{2+} + CO_3^{2-} \leftrightarrow (UO_2)(CO_3)_{(s)}$	(R47)	14.11
$Fe^{2+} + 2UO_2^{2+} + 2PO_4^{3-} \leftrightarrow Fe(UO_2)_2(PO_4)_{2\,(s)}$	(R48)	46.00
$H^+ + UO_2^{2+} + PO_4^{3-} \leftrightarrow H(UO_2)(PO_4)_{(s)}$	(R49)	25.00

Table 2: Compositions of recharge water and pore water in and outside the tailings.

Species	Inside the Tailings	Outside the Tailings	Recharge Water
Ca^{2+}	1.0×10^{-2}	1.0×10^{-2}	1.0×10^{-3}
CO_3^{2-}	1.0×10^{-2}	1.5×10^{-3}	1.5×10^{-3}
UO_2^{2+}	5.0×10^{-4}	1.0×10^{-7}	1.0×10^{-8}

Table 2: cont….

PO_4^{3-}	1.0 x 10^{-6}	1.0 x 10^{-6}	1.0 x 10^{-6}
SO_4^{2-}	2.0 x 10^{-1}	2.0 x 10^{-2}	1.0 x 0^{-4}
H^+	3.5 x 10^{-2}	1.0 x 10^{-7}	1.0 x 10^{-7}
Fe^{2+}	2.0 x 10^{-1}	1.0 x 10^{-3}	1.0 x 10^{-3}

Fig. **5** shows contour plots of computed pH at various times. Initially, acidity is confined to the tailing pile. Simulations at 100 and 300 days illustrate transport of acidity outside the tailings; however, due to reaction with $CaCO_3$ the pH is neutralized rapidly in the reaction zone.

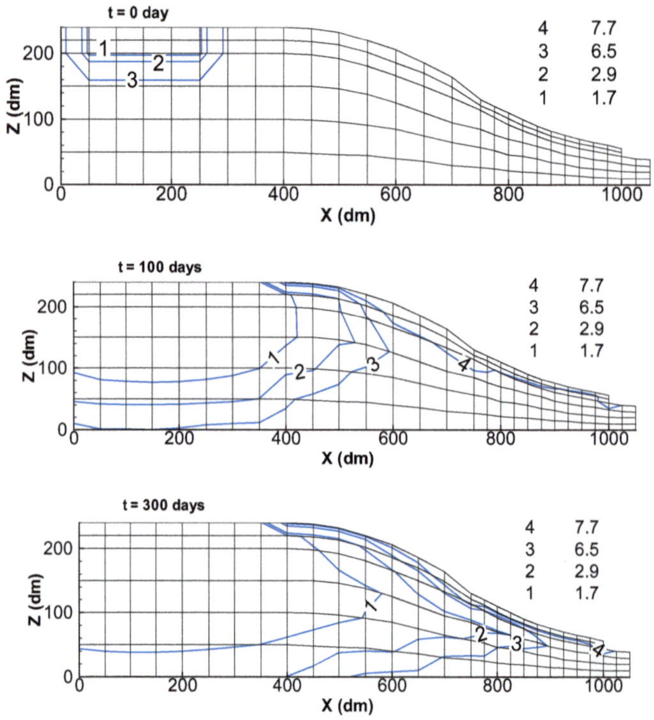

Figure 5: Distribution of pH at various times.

Figs. **6** and **7** depict contour plots of precipitated carbonate and sulfate, respectively. The two figures show the displacement of calcite by gypsum as the simulation proceeded. $CaCO_3$, which was initially present everywhere outside of the tailings pile (t = 0 day), disappears as acidity is transported out of the tailings pile. At t = 300 days $CaCO_3$ is present only in a small zone near the right edge of the cross section. Gypsum concentrations spread from the tailings pile due to transport of sulfate and calcium. However, below the hill slope, gypsum is dissolved by incoming rainwater with low sulfate concentration. The release of calcium by dissolution of the gypsum leads to some precipitation of calcite by reaction with the carbonate transported from the tailings pile.

The distribution of uranium is significantly different from that of sulfate and calcium (Fig. **8**). The concentration of uranium steadily increases with time in the region surrounding the tailings pile due to advective transport. Unlike calcium and sulfate that are present in high concentrations throughout the cross section, the difference between the uranium concentration within and outside the tailings is more than an order of magnitude.

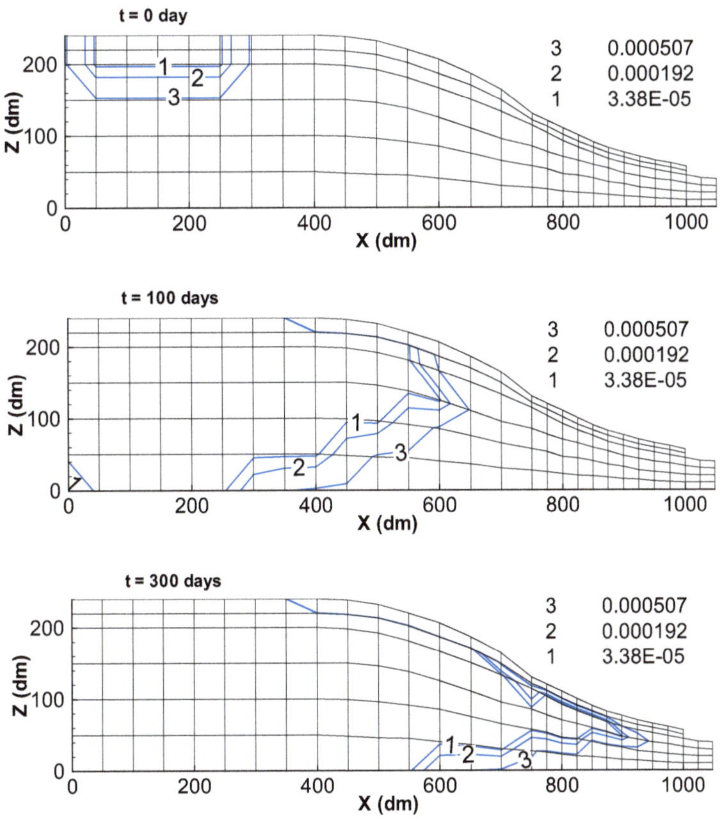

Figure 6: Distribution of precipitated carbonate (calcite) at various times.

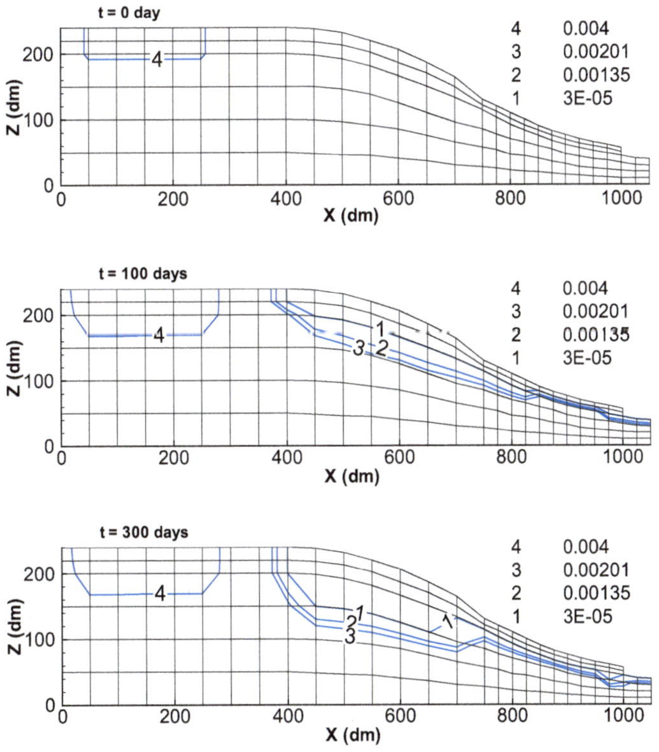

Figure 7: Distribution of precipitated sulfate (gypsum) at various times.

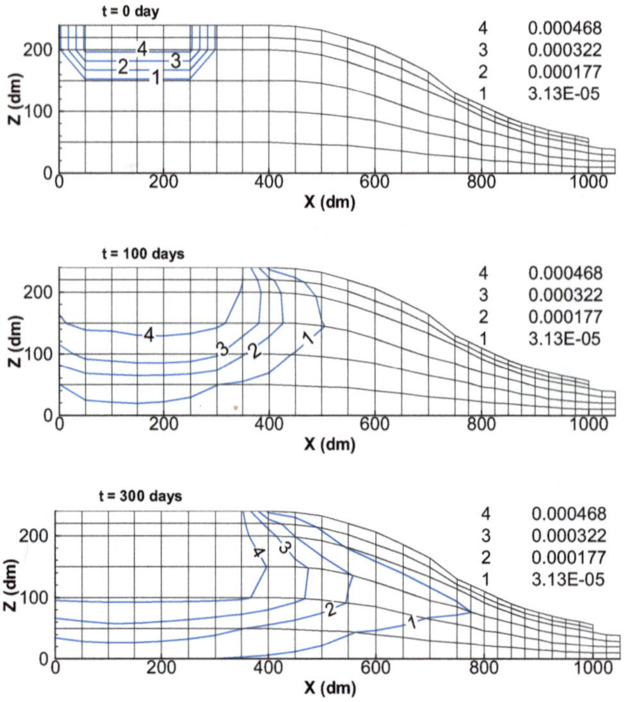

Figure 8: Distribution of dissolved uranium at various times.

Example 2

This problem simulates monitored natural attenuation of uranium. In addition to the aqueous complexation and precipitation processes of the previous problem, adsorption-desorption processes are also included. The objective is to conduct a parametric study of an adsorption-desorption hysteresis of uranium. Three cases are studied: fast reversible equilibrium sorption, kinetic-limited sorption with slow uptake, and rapid adsorption with slow desorption.

The domain of interest is a rectangular region having a size of 20 dm × 10 dm and containing soil with a porosity of 0.3, a longitudinal dispersivity of 0.1 dm, a lateral dispersivity of 0.05 dm, and a diffusion coefficient of 0.01 dm^2/day (Fig. **9**).

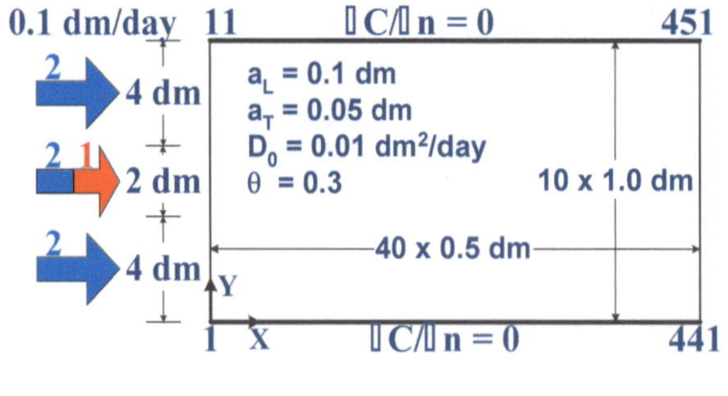

Figure 9: Problem definition.

The flow field is steady and uniform with a velocity of 0.1 dm/day. Initial total concentrations of pore water inside the system are listed in Table **3**. Only incoming solution 1 contains uranium and is assumed to enter the middle segment of the upstream boundary (*i.e.*, y = 4 dm to 6 dm at x = 0 dm) for the first 50 days as shown in Fig. **9**. Incoming solution 2 is assumed to enter both the rest of the upstream boundary (x = 0 dm) for 100 days and the segment involving uranium solution for the next 50 days. The total concentrations of incoming solutions 1 and 2 are also listed in Table **3**. The top (*i.e.*, y = 10 dm) and bottom (*i.e.*, y = 0 dm) boundaries are assumed to be impermeable. The downstream boundary (x = 20 dm) is assigned with zero gradient condition. A total simulation of 100 days is simulated using a time step size 0.25 days.

Table 3: Initial total and incoming total dissolved concentrations.

Unit: M	Initial Condition	Incoming Solution 1	Incoming Solution 2
Ca^{2+}	1.0E-2	1.0E-3	1.0E-3
CO_3^{2-}	1.5E-3	1.5E-3	1.5E-3
UO_2^{2+}	1.0E-8	1.0E-4	1.0E-8
PO_4^{3-}	1.0E-6	1.0E-7	1.0E-7
SO_4^{2-}	2.0E-2	4.0E-4	9.0e-4
H^+	1.0E-3	1.4E-3	2.8E-3
Fe^{2+}	1.0E-7	1.0E-7	1.0E-7
pH	7.58	7.39	5.52

In addition to 49 reactions as presented in Table **1**, two types of sorption sites, SOH and TOH, are used to describe equilibrium and kinetic sorption reactions, respectively. All sorption reactions are described as simple surface complexation. Three cases are simulated with different sorption reactions (Table **4**). Case 1 contains fast sorption only and has a SOH site concentration of 0.002 M throughout the system. Four equilibrium controlled sorption reactions are considered.

Case 2 is a mixture of fast and slow sorption containing a combination of equilibrium and kinetic reactions. It is assumed to have site concentration of 0.001 M for both SOH and TOH. Sorption reactions of SOH in Case 2 are considered to be the same as those of Case 1 [*i.e.*, Reactions (R50) through (R53)]. The sorption of uranium on TOH is kinetically controlled as indicated in Table **4**.

Table 4: Adsorption-desorption processes.

Reaction	No	Log K
Case 1 Fast Adsorption Only $TOT_{SOH} = 0.002$ M		
$SOH - H^+ \leftrightarrow SO^-$	(R50)	K_e=-10.3
$SOH + H^+ \leftrightarrow SOH_2^+$	(R51)	K_e = -5.4
$SOH - 2H^+ + UO_2^{2+} \leftrightarrow (UO_2)(OH)_2(SOH)$	(R52)	K_e = -7.1
$SOH - 8H^+ + 3UO_2^{2+} \leftrightarrow (UO_2)_3(OH)_8(SOH)$	(R53)	K_e=-31.0
Case 2 Mixed Fast and Slow Adsorption = Case 1 Plus the Following Reactions and $TOT_{SOH} = 0.001$, $TOT_{TOH} = 0.001$		
$TOH - H^+ \leftrightarrow TO^-$	(R54)	K_e=-10.3
$TOH + H^+ \leftrightarrow TOH_2^+$	(R55)	K_e = -5.4
$TOH - 2H^+ + UO_2^{2+} \leftrightarrow (UO_2)(OH)_2(TOH)$	(R56)	k_f– 10.0 k_b = 17.1
Case 3 Mixed Reversible and Irreversible Adsorption = Case 2 but different rate constants and $TOT_{SOH} = 0.0019$, $TOT_{TOH} = 0.0001$		
$TOH - 2H^+ + UO_2^{2+} \leftrightarrow (UO_2)(OH)_2(TOH)$	(R56)	k_f = 24.7 k_b = 17.1

Case 3 is designed to have a mixture of reversible and irreversible sorption. The site concentrations of SOH and TOH are 0.0019 M and 0.0001 M, respectively. Sorption reactions of SOH and TOH are considered the same as those of Case 2. The only difference is the kinetic sorption of uranium on TOH. In order to describe irreversible uranium sorption, the sorption and desorption rate constants of uranium are estimated with a fixed pH value of 7.35 in association with Reaction (R56).

Figs. **10** through **12** show the migration of dissolved uranium at different times for Cases 1 through 3. The migration of the dissolved uranium plume can be affected by non-chemical factors like advection and dispersion, such as in this problem. The intensity and size of uranium plumes of Case 3 (Fig. **12**) are much smaller than other two cases because the mobility of uranium is greatly reduced by irreversible sorption to TOH sites. Since less amount of uranium is sorbed in Case 2 (Fig. **11**) than Case 1 (Fig. **10**), the dissolved uranium of Case 2 moves faster than Case 1 and the uranium plume of Case 2 is more dispersed than in Case 1.

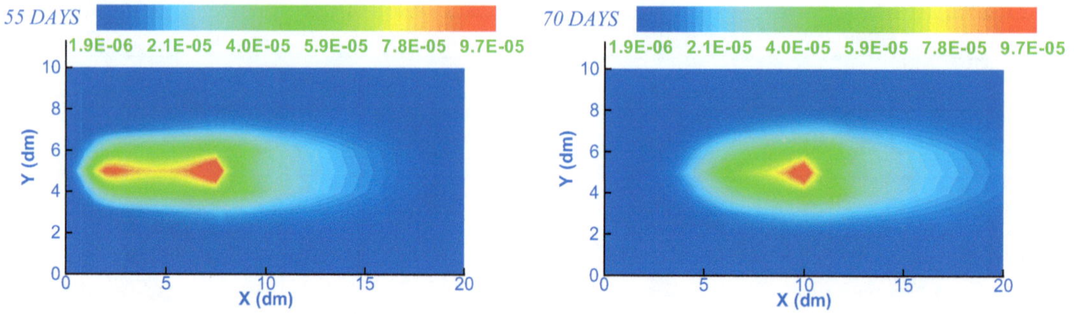

Figure 10: Migration of Uranium plume for case 1.

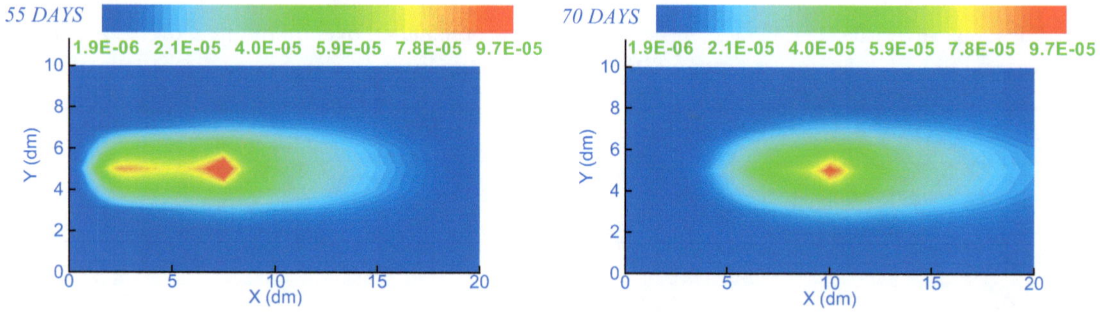

Figure 11: Migration of Uranium plume for case 2.

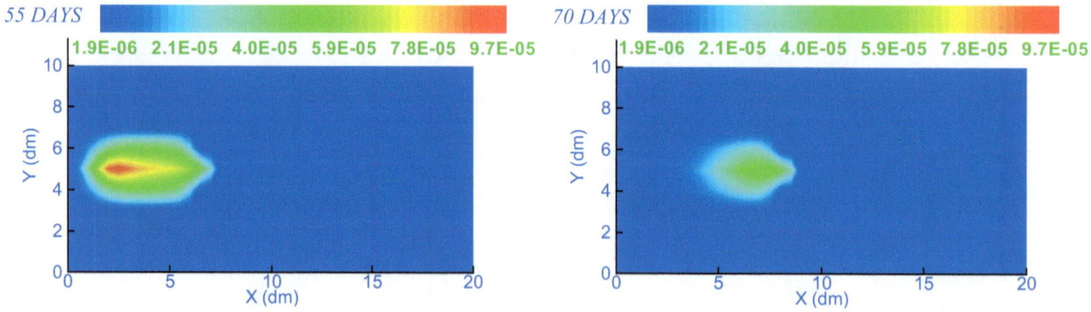

Figure 12: Migration of Uranium plume for case 3.

A similar calculation with the same proportions but on a larger scale (3,000 dm x 12,000 dm) was also carried out. The flow rate is 0.045 dm/yr and the simulation was run for 150 years. The initial uranium concentrations of pore water inside the source area was 10^{-4} M and was assumed to enter the middle 600 dm segment of the upstream boundary for the first 50 days. The incoming solution assumed to enter both the rest of the upstream boundary (x = 0 dm) for 100 days and the segment involving uranium solution for the next 50 days had a uranium concentration of 10^{-8} M. Fig. **13** shows the movement of the center-of-mass of dissolved uranium plumes for the equilibrium case and for 3 cases in which the rate of desorption is much slower than the rate of adsorption. Note that when 50 - 100% of the sites have slow desorption rates, the center of mass for the dissolved uranium actually begins to move upstream (in the –X direction) after about 50 years.

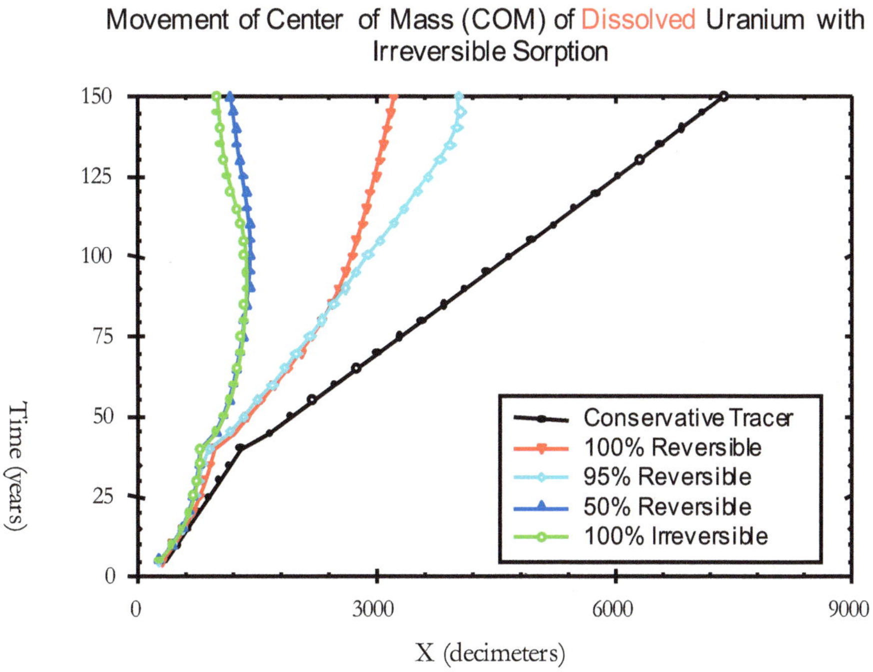

Figure 13: Movement of center of mass of dissolved uranium for different ratios of equilibrium (reversible) and nonequilibrium ("irreversible") sorption sites.

Example 3

This example is used to demonstrate the need for species switching in dealing with stiff reactive systems under transport and to illustrate the computation of fast/equilibrium rates [88]. It involves the transient simulation of advection-dispersion and equilibrium complexation and precipitation reactions. The domain is 10 dm × 5,000 dm and is discretized with 200 equally sized elements. It consists of three heterogeneous zones with moisture content of 0.3, a constant flow field of 155.5 dm/day, and a longitudinal dispersivity of 150 dm (Fig. **14**).

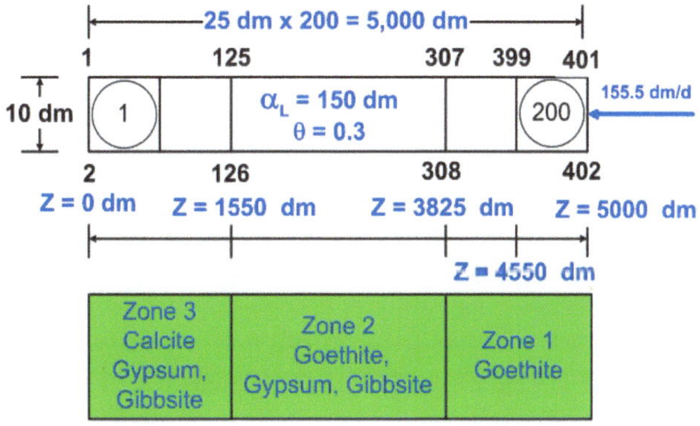

Figure 14: Problem domain and discretization.

Initially there are seven chemical components present in the system: Ca^{2+}, CO_3^{2-}, Al^{3+}, SO_4^{2-}, H^+, Fe^{3+}, and Na^+ (Table **5**). The incoming flow has different concentrations of these chemicals (Table **6**).

Table 5: Initial conditions for example **3**.

Total Con. (M)	Sub-Region			
	$0 \le x < 1{,}550$	$1{,}550 \le x < 3{,}825$	$3{,}825 \le x < 4{,}550$	$4{,}550 \le x < 5{,}000$
Ca^2	9.250E-1	6.350E-1	9.670E-3	7.845E-3
CO_3^{2-}	9.550E-1	6.360E-1	1.000E-4	1.000E-4
Al^{3+}	4.262E-5	4.262E-5	1.240E-2	4.000E-2
SO_4^{2-}	2.429E-2	2.429E-2	1.330E-1	2.000E-1
H^+	2.045E-2	2.045E-2	-1,266E-1	-3.452E-3
Fe^{3+}	1.235E-2	1.235E-5	3.707E-2	3.259E-2
Na^+	2.352E-2	2.352E-2	9.465E-2	1.400E-1

Table 6: Incoming species concentrations for variable boundary conditions.

Unit: M	$0 \le t \le 2.5\ d$	$2.6 \le t \le 120\ d$
Ca^{2+}	7.4854×10^{-3}	3.94×10^{-3}
CO_3^{2-}	1.0×10^{-4}	2.5×10^{-3}
Al^{3+}	4.0×10^{-2}	9.3077×10^{-9}
SO_4^{2-}	2.0×10^{-1}	4.4424×10^{-3}
H^+	1.7071×10^{-3}	2.7628×10^{-3}
Fe^{3+}	2.093×10^{-3}	3.0788×10^{-7}
Na^+	1.4×10^{-1}	2.3518×10^{-2}

The key to conducting reactive transport simulations is to transform the understanding of the system into a reaction network. For this problem, the conceptual model is transformed into a reaction network of 24 fast/equilibrium reactions (20 aqueous complexation and 4 precipitation-dissolution reactions) in Yeh *et al.* [71]. The dynamics of this reaction network is simulated for a total of 120 days with a time step size of 0.025 days.

At the start of the simulation, the preprocessor in HYDROGEOCHEM 4.5 [51] automatically chooses Ca^{2+}, CO_3^{2-}, Al^{3+}, SO_4^{2-}, H^+, Fe^{3+}, and Na^+ as 7 components. Without using species switching, the simulation stops at node 299 while checking the mass balance of the Al^{3+} component. With the species switching algorithm, the dominant aqueous complexed species $Al(OH)_4^-$ is selected to replace Al^{3+} as the new component and the simulation is successfully completed [71]. $Al(OH)_4^-$ is chosen because it has the largest concentration among aluminum complexes.

Simulation results of the dynamics of all species have been presented elsewhere [17, 19, 71] and will not be presented here. The newly obtained simulations of reaction rates and their implications were reported in Yeh *et al.* [88]. It was shown that the simulated rates of all 24 fast/equilibrium reactions were finite and definite. A consistent system for simulating concentrations of all species and rates of all reactions was obtained through the diagonalization approach. The fast/equilibrium reaction rate for calcite is included here as an example (Fig. **15**).

It is seen that the reaction rate at z = 3,700 dm is approximately −0.95 M/day. The corresponding calcite concentration quickly changes from approximately 0.6 M to almost zero. Calcite mineral concentrations will change with time at locations where the corresponding precipitation reaction rates are not zero. In other words, measured mineral concentrations versus time at a particular location can be used to characterize their corresponding reaction rates. This is so because all mineral species are considered immobile in a reactive transport system, and as a result, the reaction rates are calculated with

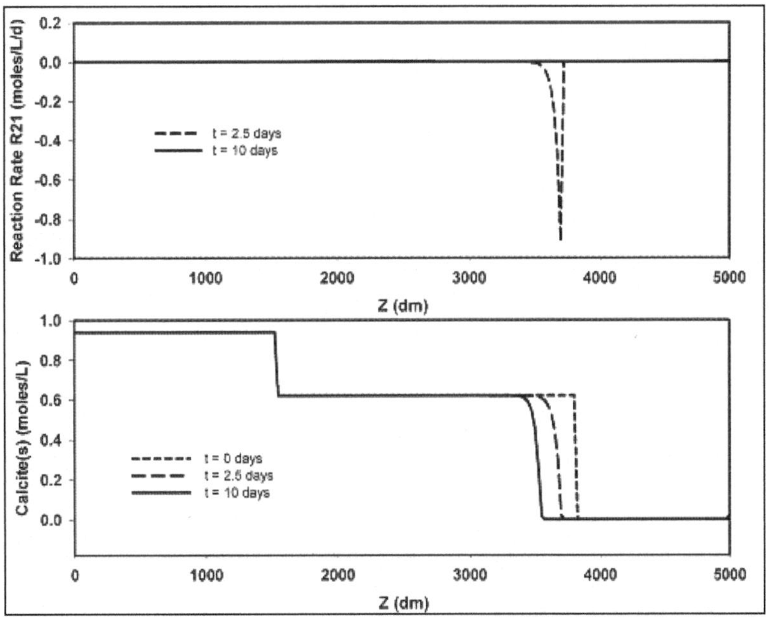

Figure 15: Rates of the calcite precipitation reaction (Top) and concentration profiles of calcite (bottom).

$$\theta R_i = \frac{\partial \theta P_i}{\partial t} \quad \textbf{(48)}$$

where P_i is the concentration of the *i*-th mineral species and R_i is the rate of the *i*-th precipitation reaction that produces the mineral [88].

On the other hand, for an aqueous species, the concentration-versus-time curve at a particular location cannot be used to characterize a reaction even if there is one and only one reaction that produces the species. The reaction rate for an aqueous species can be calculated with

$$\theta R_i = \frac{\partial \theta C_i}{\partial t} - L(C_i) \quad \textbf{(49)}$$

where C_i is the concentration of the *i*-th aqueous product species and R_i is the rate of the *i*-th aqueous complexation reaction that produces the species. It has been observed that unless one can segregate the transport rate $L(C_i)$ from the local rate $\partial C_i/\partial t$, the slope of the concentration-versus-time curve cannot be used to characterize the reaction rate.

Example 4

This example demonstrates the capability of the model to simulate coupled flow dynamics, temperature fields, and reactive chemical evolutions of comprehensive reaction network in three-dimensional space. The complete problem description and solutions are reported in Sun [74], Yeh *et al.* [20], and Yeh *et al.* [88].

Flow Field The domain is 200 dm × 200 dm × 100 dm. The boundary conditions are: no flux on the left (x = 0), on the back (y = 200 dm), and at the bottom (z = 0) sides of the region; pressure head varied from zero at z = 30 dm to 30 dm at the bottom (z = 0) on the right (x = 200 dm) and front (y = 0) sides; and variable boundary conditions are used elsewhere. Ponding depth is zero for all variable boundary surfaces. The incoming fluxes on the variable boundary are equal to 1 dm/day for the top surface (variable B.C.-I in Fig. **16**) and equal to zero for the front and right sides above the water surface (variable B.C.-II in Fig. **16**). Simulated steady state pressure and velocity was reported in Yeh *et al.* [20].

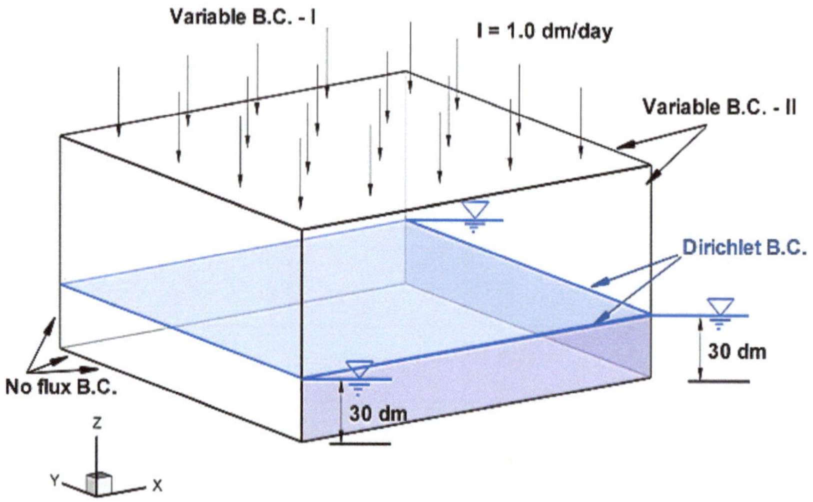

Figure 16: Problem definition: flow boundary conditions.

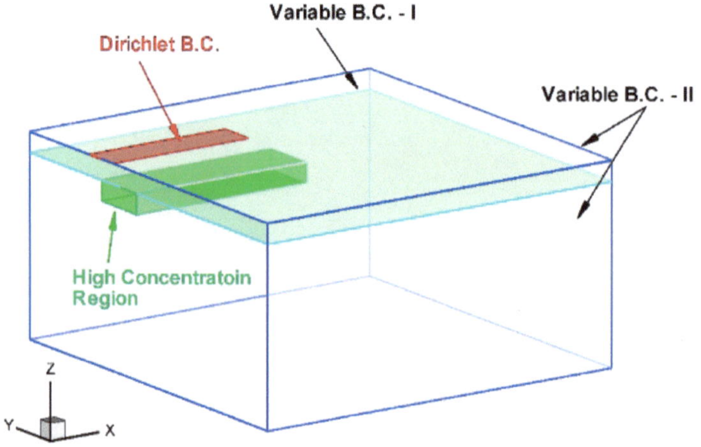

Figure 17: Boundary conditions for thermal and reactive transport and a high initial concentration region for reactive transport.

Thermal Transport Given the flow field, transient thermal transport is simulated followed by reactive chemical transport. For the simulation of the temperature field, the initial temperature is set at 298 ^{o}K. The types of boundary conditions are described in Fig. **17**.

Fig. **18** depicts the temperature distribution at various times.

Reactive Chemical Transport The network is complex involving both kinetic and equilibrium reactions (8 and 25, respectively) for every reaction type (aqueous complexation, adsorption-desorption, ion-exchange, and precipitation-dissolution). The mineral dissolution reactions and 4 of the 22 aqueous reactions are included in Table **7**. Details on the user-specified reaction for formation of surface adsorption sites can be found in Yeh *et al.* [88].

For numerical simulations, low initial concentrations are given throughout the entire domain, except for a small region- which has high initial concentrations (Fig. **17**).

The regions of boundary conditions for reactive transport are identical to those for thermal transport. The species concentrations with non-zero values on the boundaries as well as the initial conditions were given by Yeh *et al.* [88].

Table 7: Partial list of reaction network.

Reaction	**No**	**Parameter**
Mineral Dissolution and Surface Site Formation Reactions		
$Fe(OH)_3(s) \leftrightarrow Fe^{3+} - 3\ H^+$	(KR1)	$k_1^f = 0.05$
$Fe(OH)_3(s) \leftrightarrow FeOH$	(EqR1)	*
Aqueous Complexation Reactions		
$CoEDTA^{2-} \leftrightarrow Co^{2+} + EDTA^{4-}$	(KR3)	$Logk_3^f = 2.03$ $Log\ k_3^b = 20.00$
$Ca^{2+} + EDTA^{4-} \leftrightarrow CaEDTA^{2}$	(EqR2)	$Log\ K_4^e = 12.32$
$H^+ + EDTA^{4-} + Ca^{2+} \leftrightarrow CaHEDTA^-$	(EqR3)	$Log\ K_5^e = 15.93$
$Ca^{2+} \leftrightarrow H^+ + Ca(OH)^+$	(EqR4)	$LogK_6^e = -12.60$

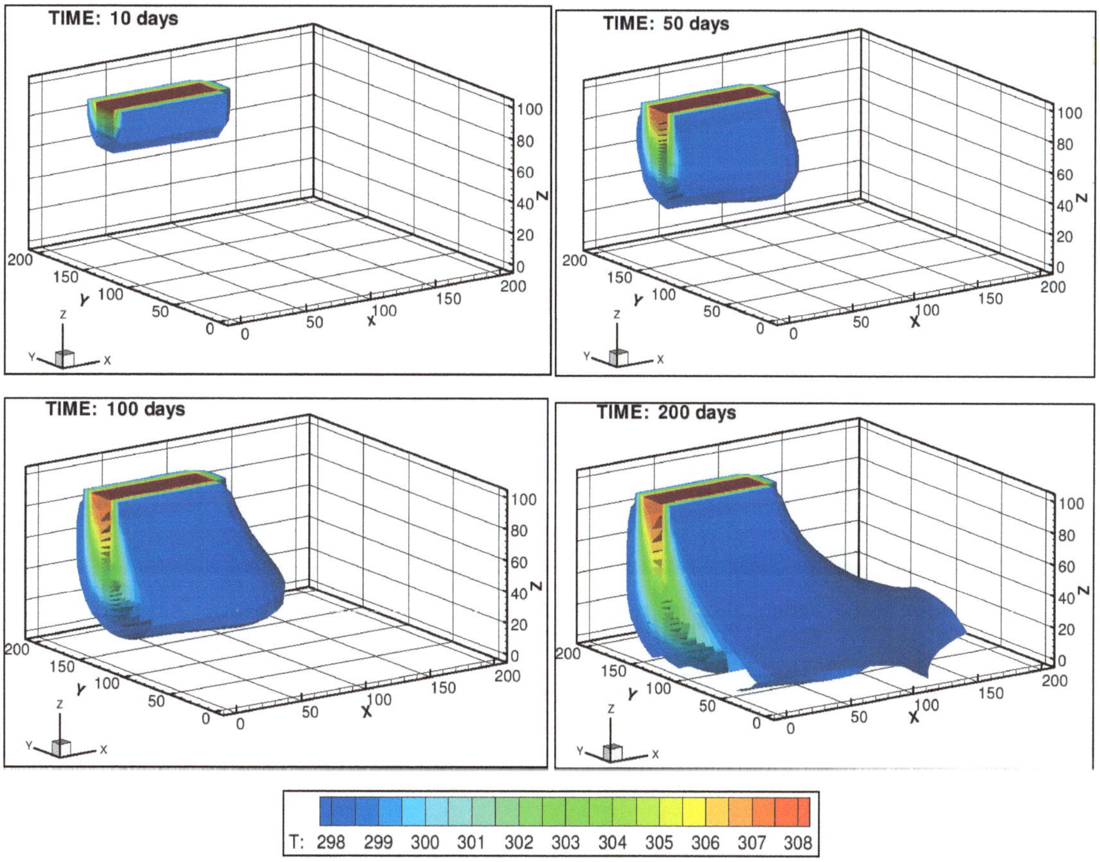

Figure 18: Three-dimensional temperature distributions at various times.

The simulated rates of all 25 equlibrium reactions were again finite and definite. Furthermore, it was demonstrated that the rates of slow/kinetic reactions are not necessarily smaller than those of fast/-equilibrium reactions [88]. As an example, the rates of the 6 reactions in Table **7** are examined.

The rates of equilibrium reactions and kinetic reactions at (x, y, z) = (30, 120, 30), which is located in the region of initially low concentration, are shown in Fig. **19**.

Intuitively one would suspect that the rate of a fast/equilibrium reaction is greater than that of a slow/kinetic reaction. It is clearly seen from Fig. **19** that this is not the case. For example, the kinetic rates for Reactions (K R1) and (K R3) are greater than the equilibrium rates for Reactions (Eq R1) and (Eq R4). Thus, rates of slow/kinetic reactions are not necessarily smaller than those of fast/equilibrium reactions.

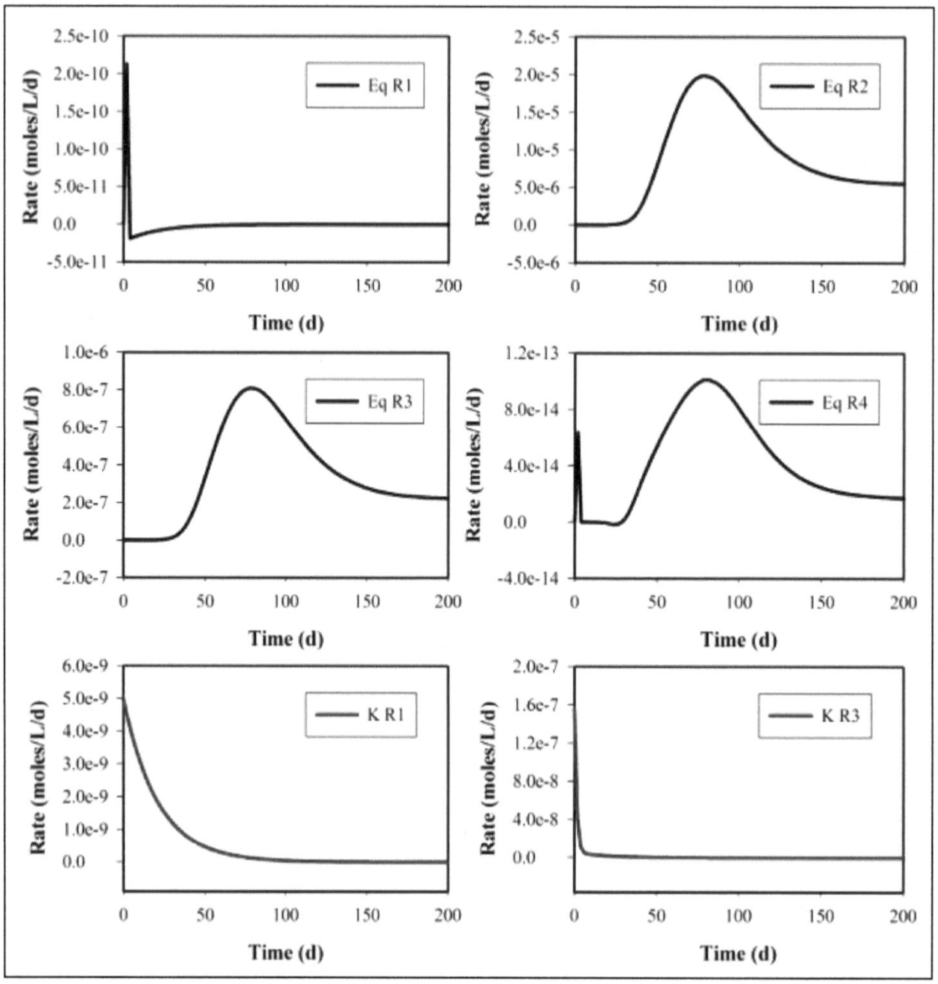

Figure 19: Rates of equilibrium and kinetic reactions.

Example 5

This example involves modeling of laboratory column experiments involving extremely high concentrations of uranium, technetium, aluminum, nitrate, and toxic metals. The experiment set up is described in Zhang *et al.* [89]. The experiment is modelled with a reaction network of 92 equilibrium and 5 kinetic reactions involving 138 chemical species [89]. The conceptual model involves 12 chemical components for the experiments: NO_3, Na, K, Al, Si, SO_4, Ca, Mg, Mn, U, Co, and Ni. An equilibrium reaction model that considers 72 aqueous complexation reactions is first used to perform speciation calculation for each data point so as to determine the concentrations of individual species given pH and the total aqueous concentrations of the 12 components. Based on the calculated aqueous species concentrations, the saturation index is calculated for each of the 26 minerals considered and a decision is made to include only five precipitationdissolution reactions. Finally soil buffering capacity is modeled with twelve equilibrium ion-exchange reactions in addition to eight equilibrium ionization reactions of a polyprotic acid H_4X and a polyprotic base $Y(OH)_2$. Simulation results indicate good agreement between experiments and theoretical predictions using the proposed reaction network. The caveat in this example is one's ability to come up with a reaction network. Once this is done, HYDROGEOCHEM can automatically set up transport equations for components and kinetic variables and simulations of all species concentrations and reaction rates follow.

Example 6

This example models bioremediation field experiments conducted at a Uranium Mill Tailings Remedial Action (UMTRA) site using acetate amendment to stimulate microbially mediated immobilization of uranium in the unconfined aquifer [90]. Field experiments at the Rifle site showed that the growth of acetate-oxidizing Fe(III)-reducers dominated by Geobacter *sp.* is accompanied by significant uranium removal from groundwater [91, 92]. An important feature of these field experiments is the eventual onset of sulfate reduction, which is characterized by a decrease in aqueous sulfate, near complete consumption of acetate, and less efficient U(VI) removal from groundwater [93-95]. Abiotic reactions are enabled to address key issues during biostimulation including the relative importance of uranium adsorption and desorption processes and the fate of Fe(II) and sulfide produced in the biologically mediated Terminal Electron Accepting Process (TEAP) reactions. The conceptual model is developed to understand the principal processes and properties controlling uranium biogeochemistry [90]. The model includes four TEAPs involving two pools of bioavailable Fe(III) minerals (phyllosilicate, oxide), aqueous U(VI), and aqueous sulfate, two distinct functional microbial populations (iron, sulfate reducers), as well as aqueous and surface complexation and mineral precipitation and dissolution. The conceptualization is transformed into a reaction network of 75 reactions (67 equilibrium reactions and 8 kinetic reactions) involving 95 species. Among the 67 equilibrium reactions, 23 account for aqueous complexation and 6 for surface complexation involving uranium. The 7 kinetic reactions include: 3 precipitation-dissolution reactions of calcite ($CaCO_{3(s)}$), siderite ($FeCO_{3(s)}$), and iron sulfide ($FeS_{(s)}$); 4 microbe-mediated reduction of phyllosilicate iron ($Fe(III)_{(ls)}$), goethite ($FeOOH_{(s)}$), uranium, and sulfate; and one abiotic redox of goethite and sulfide.

The proposed reaction network is calibrated with data from a 2002 experiment. Simulated and measured breakthrough at several locations down gradient of the injection gallery show consistently good agreement indicating adequate calibration [90]. The same reaction network is then applied to a second experiment that is carried out in the summer of 2003 at the same Rifle field plot and the 2007 field biostimulation experiment, which is performed in a new, unperturbed site to successfully test the robustness of the reaction parameterization [90]. The validation also shows consistent agreement between simulations and experiments. This supports the notion that in a reaction-based model the rate formulation is descriptive of a specific biogeochemical reaction and therefore may be applicable to a widerange of environmental conditions beyond the specific condition originally modeled [70, 96].

SUMMARY AND CONCLUSION

HYDROGEOCHEM, originally developed at Oak Ridge National Laboratory for simulating coupled HYDROlogical transport and GEOCHEMical equilibrium in two-dimensional subsurface media, has evolved into a comprehensive model for simulating coupled fluid flow, thermal transport, and reactive biogeochemical transport in both two- and three-dimensional spaces while it has incubated many other complete models of similar types. The model includes four basic modules: flow, thermal transport, hydrologic transport, and biogeochemistry. The flow module solves Richards equation with Galerkin finite element methods to yield the spatial-temporal distribution of pressure and total heads, Darcy velocity, and moisture content. The thermal transport module solves the energy conservation equation to render the temperature field as a function of time as well as thermal flux and energy balance throughout the entire domain. The hydrologic transport module solves transport equations of components and kinetic variables to create the spatial-temporal distributions of components and kinetic variables. It also solves equilibrium rate equations with finite element methods to yield the spatial-temporal distribution of reaction rates of fast/equilibrium reactions. The biogeochemical module solves a system of mass balance equations (component equations and kinetic variable equations) and mass action equations. The effects of thermal transport on flow and reactions are included. The effects of reactive transport on fluid flow *via* the change of porosity, hydraulic conductivity, and diffusion-dispersion are also considered.

State-of-the-art conceptualizations on reactive transport have been incorporated into the code. For example, a concept of defining the reaction rates of fast reactions has been recently proposed, which makes the modeling of mixed fast/equilibrium and slow/kinetic reactions consistent. The rate of an equilibrium reaction is then defined as the rate which is 'necessary' to assure that the corresponding thermodynamic equation remains fulfilled. With this definition, the rates of equilibrium reactions are shown to be finite and definite.

The challenge in applying the reactive transport model HYDROGEOCHEM to real-world problems is the transformation of our understandings and/or hypotheses of the system into reaction networks. This is by no means easy. However, if we do not or can not, our understandings are incomplete or false. Six example problems were used to demonstrate the flexibility and range of capabilities of HYDROGEOCHEM. The examples included all reaction types (aqueous complexation, adsorption-desorption, precipitation-dissolution, ion-exchange, hydrolysis, and both abiotic and biotic-mediated redoxes) as both fast/equilibrium and/or slow/kinetic processes. The dynamic interplay of fluid flow, hydrologic transport, biogeochemical reactions, and thermal transport were demonstrated. Example 3 exemplified the need of a species switching scheme to solve stiff geochemical equilibrium problems. Examples 3 and 4 both demonstrated that rates of all fast/equilibrium reactions are finite and definite, while Example 4 also showed that rates of slow/kinetic reactions are not necessarily smaller than those of fast/equilibrium reactions.

Example 5 demonstrated the ability of the model to generate good agreement between experiments and theoretical predictions using the appropriate reaction network. It clearly showed the importance of one's ability to generate an adequate reaction network for a given system. In Example 6, a reaction network was proposed and calibrated with one set of experimental results and then applied to two other different experiments. Despite different experimental locations and field conditions, a single set of reactions generally captured complex biogeochemical dynamics. This example demonstrated the ability of mechanistic-based reactive transport models to carry the simulations over to different environmental settings.

The modeling capabilities of HYDROGEOCHEM are far beyond those illustrated in the six examples. HYDROGEOCHEM is not limited to simulating subsurface groundwater problems but can model any system involving single phase fluid in porous media. For example, it was used to simulate oil shale pyrolysis and retorting gas production when the parameters of soil and water properties were substituted by those of oil shale rock matrix and releasing gas [97]. HYDROGEOCHEM can also be used to simulate Mobile-Immobile dual continuum problems by treating the diffusion between mobile and immobile zones in terms of reactions and considering the flow in the mobile part of the total porosity [98].

AFTER THOUGHTS

Since HYDROGEOCHEM's inception, our aim has been to provide the capability of simulating coupled processes in a way that will encompass a host of scenarios including "what-if" simulations, laboratory column studies, small scale field experiments and ultimately practical real world problems. It cannot be overemphasized that the accuracy of HYDROGEOCHEM simulations can not surpass the uncertainties in hydrologic parameters, thermodynamic, and chemical and biochemical kinetic data. While it is beyond the scope of this book chapter to present an in-depth discussion of various uncertainties, users should strive to get as much accurate input data as possible and to use appropriate experience-based caution while interpreting results.

We recommend exploring error propagation or interval analysis in future of versions of HYDROGEOCHEM. However, we recognize that such an endeavor may affect the stability and/or convergence of this multi-way-coupled modeling approach, in addition to potentially increasing computational demand by orders of magnitude.

ACKNOWLEDGEMENTS

The preparation of this book chapter is supported by National Science Council, in part, under Contract No. NSC 99-2116-M-008-020 with National Central University and, in part, under Contract No. NSC 99-1903-02-05-03 with Taiwan Typhoon Flood Research Institute (TTFRI), National Applied Research Laboratory, Taiwan.

REFERENCES

[1] D. A. Chin, *Water-Resources Engineering.* Prentice Hall, 2000.

[2] G. T. Yeh, and V. S. Tripathi, "A Critical Evaluation of Recent Developments in Hydrogeochemical Transport Models of Reactive Multichemical Components", *Water Resources Research*, vol. 25, pp. 93-108, 1989.

[3] G. T. Yeh and V. S. Tripathi, HYDROGEOCHEM: *A Coupled Model of HYDROlogical Transport and GEOCHEMical Equilibrium of Multi component Systems.* ORNL-6371. Oak Ridge National Laboratory, Oak Ridge. 1990.

[4] P. Nienhuis, C.A.T. Appelo, and A. Willemsen, *Program PHREEQM: Modified from PHREEQE for use in mixing cell flow tube.* Amsterdam: Free University, 1991.

[5] P. Engesgaard and K. L. Kipp, "A geochemical transport model for redox-controlled movement of mineral fronts in groundwater flow systems: A case of nitrate removal by oxidation of pyrite", *Water Resources Reearch.*, vol. 20, pp. 2829-2843, 1992.

[6] J. Šimunnek and D. L. Squares, "Two-dimensional model for variably saturated porous media with major ion chemistry", *Water Resources Research.*, vol. 30, pp. 1115-1133, 1994.

[7] A. Chilakapati, *RAFT: A Simulator for ReActive Flow and Transport Groundwater Contaminants*, PNL Report 10636. Richland: Pacific Northwest Laboratory, 1995.

[8] C. I. Steefel and S. B. Yabusaki, *OS3D/GIMRT, Software for Modeling Multi-Component-Multidimensional Reactive Transport, User's Manual and Programmer's Guide*, PNL-11166. Richland: Pacific Northwest Laboratory, 1996.

[9] P. C. Lichtner and M. S. Seith, *User's Manual for MULTIFLO: Part II MULTIFLO 1.0 and GEM 1.0, Multicomponent-multiphase reactive transport model*, CNWRA96-019. San Antonio: Center for Nuclear Waste Regulatory Analyses, 1996.

[10] C. I. Steefel and P. Van Cappellen, "Special Issue: Reactive Transport Modeling of Natural Systems", *Journal of Hydrology*, vol. 209, pp. 1-388, 1998.

[11] C. Tebes-Stevens, A.J. Valocchi, J.M. VanBriesen, and B. E. Rittmann, "Multicomponent transport with coupled geochemical and microbiological reactions: model description and example simulations", *Journal of Hydrology*, vol. 209, pp. 8-26, 1998.

[12] J. J. Nitao, *Reference Manual for the USNT Flow and Transport Code Version 2.0*, UCRL-MA-130651. Livermore: Lawrence Livermore National Laboratory, 1998.

[13] D. L. Parkhurst and C. J. Appelo, *User's guide to PHREEQC (version 2)-A computer program for speciation, batch-reaction, one-dimensional transport, and inverse geochemical calculations*, WRIR99-4259. U.S. Geological Survey, 1999.

[14] U. Mann, "New design formulation of chemical reactors with multiple reactions: I. basic concepts", *Chemical Engineering Science*, vol. 55, pp. 991-1008, 2000.

[15] D. H. Bacon, M. D. White, and B. P. McGrail, *Subsurface Transport Over Reactive Multiphases (STORM): A General, Coupled, Nonisothermal Multiphase Flow, Reactive Transport, and Porous Medium Alternation Simulator*, Version 2, User's Guide, PNNL-13108. Richland: Pacific Northwest national Laboratory, 2000.

[16] C. I. Steefel, *GIMRT, version 1.2: Software for modeling multicomponent, multidimensional reactive transport. User's Guide*, UCRL-MA-143182. Livermore, California: Lawrence Livermore National Laboratory, 2001.

[17] G. T. Yeh, M. D. Siegel, and M. H. Li, "Numerical modeling of coupled fluid flows and reactive transport including fast and slow chemical reactions", *Journal of Contaminant Hydrology*, vol. 47, pp. 379-390, 2001.

[18] T. Xu, E. Sonnenthal, N. Spycher, and Pruess, *TOUGHREACT User's Guide: A Simulation Program for Non-Isothermal Multiphase Reactive Geochemical Transport in Variably Saturated Geologic Media.* Berkeley: Lawrence Berkeley National Laboratory, 2003.

[19] G. T. Yeh, J. T. Sun, P. M. Jardine, *et al.*, *HYDROGEOCHEM 4.0: A Coupled Model of Fluid Flow, Thermal Transport, and HYDROGEOCHEMical Transport through Saturated-Unsaturated Media: Version 4.0*, ORNL/TM-2004/103. Oak Ridge: Oak Ridge National Laboratory, 2004a.

[20] G. T. Yeh, J. T. Sun, P. M. Jardine, *et al.*, *HYDROGEOCHEM 5.0: A Three-Dimensional Model of Coupled Fluid Flow, Thermal Transport, and HYDROGEOCHEMical Transport through Variably Saturated Conditions - Version 5.0*, ORNL/TM-2004/107. Oak Ridge: Oak Ridge National Laboratory, 2004b.

[21] C. Yang, J. Samper, and L. Montenegro, "CORE2D V4: A Code for water flow, heat and solute transport and geochemical reactions: Simulations of chemical interactions of clays and concrete", in *International Workshop - Modeling Reactive Transport in Porous Media*, 2008.

[22] C. W. Miller and L. V. Benson, "Simulation of Solute Transport in a Chemically Reactive Heterogeneous System: Model Development and Application", *Water Resources Research*, vol. 19, pp. 381-391, 1983.

[23] G. A., Cederberg, R. L. Street, and J. O. Leckie, "A Groundwater Mass Transport and Equilibrium Chemistry Model for Multicomponent Systems", *Water Resources Research*, vol. 21, pp. 1095-1104, 1985.

[24] J. C. Hostetler and R. L. Erickson, *FASTCHEM Package 5*, Report EA-5870-CCM. Electric Power Research Institute, 1989.

[25] T. N. Narasimhan, A. F. White, and T. Tokunaga, "Groundwater contamination from an inactive uranium mill tailings pile: 2. Application of a dynamic mixing model", *Water Resources Research*, vol. 22, pp. 1820-1834, 1986.

[26] C. W. Liu and T. Narasimhan, "Redox-controlled Multiple Species Reactive Transport, 1. Model Development", *Water Resources Research*, vol.25, pp. 869-882, 1989.

[27] J. Griffioen, "Multicomponent Cation Exchange Including Alkalinization-Acidification Following Flow Through a Sandy Sediment", *Water Resources Research*, vol. 29, pp. 3005-3019, 1993.

[28] G. T. Yeh and V. S. Tripathi, "A model for simulating transport of reactive multispecies components: Model development and demonstration", *Water Resources Research*, vol. 27, pp. 3075-3094, 1991.

[29] H. P. Cheng, PhD Dissertation: *Development and Application of a Three-Dimensional Finite Element Model of Subsurface Flow, Heat Transfer, and Reactive Chemical Transport*. The Pennsylvania State University, 1995.

[30] D. L. Parkhurst, *User's Guide to PHREEQC, A Computer Model for Speciation, Reaction-Path, Advective-Dispersive Transport and Inverse Geochemical Calculations*, WRIR95-4227. U.S. Geological Survey, 1995.

[31] P. C. Lichtner, "Continuum Formulation of Multicomponent-Multiphase Reactive Transport", in *Reactive Transport in Porous Media, Reviews in Mineralogy*, vol. 34, P. C. Lichtner, C. I. Steefel, and E. H. Oelkers, Ed. Washington, D.C.: Mineralogical Society of America, 1996, pp. 1-79.

[32] D. Suarez and J. Šimunnek, "Solute Transport Modeling Under Variably Saturated Water Flow Conditions", in *Reactive Transport in Porous Media, Reviews in Mineralogy*, vol. 34, P. C. Lichtner, C. I. Steefel, and E. H. Oelkers, Ed. Washington, D.C.: Mineralogical Society of America, 1996, pp. 229-268.

[33] T. L. Theis, D. J. Kirkner, and A. A. Jennings, *Multi-Solute Subsurface Transport Modeling for Energy Solid Wastes*, C00-10253-3. University of Notre Dame, 1982.

[34] H. J. Lensing, M. Voyt, and B. Herrling, "Modeling of Biologically Mediated Redox Processes in the Subsurface", *Journal of Hydrology*, vol. 159, pp. 125-143, 1994.

[35] G. T. Yeh, K. Salvage, and W. H. Choi, "Reactive Multispecies-Multicomponent Chemical Transport Controlled by both Equilibrium and Kinetic Reactions", in *XI-th International Conference on Numerical Methods in Water Resources*, pp. 585-592, 1996.

[36] C. I. Steefel, *CRUNCH. A Computer Program for Multicomponent Reactive Transport in Porous Media.* http://www-esd/lbl.gov/ESD-staff/steefel/WebCrunch.htm

[37] M. A., Widdowson, F. J. Molz, and L. D. Benefield, "A Numerical Transport Model for Oxygen- and Nitrate-Based Respiration Linked to Substrate and Nutrient Availability in Porous Media", *Water Resources Research*, vol. 24, pp. 1553-1565, 1988.

[38] J. S. Kindred and M. A. Celia, "Contaminant Transport and Biodegradation 2. Conceptual Model and Test Simulations", *Water Resources Research*, vol. 25, pp. 1149-1159, 1989.

[39] S. W. Taylor and P. R. Jaffee, "Substrate and Biomass Transport in a Porous Medium", *Water Resources Research*, vol. 26, pp. 2181-2194, 1990.

[40] B. D. Wood, C. N. Dawson, J. E. Szecsody, and G. P. Streile, "Modeling Contaminant Transport and Biodegradation in a Layered Porous Media System", *Water Resources Research*, vol. 30, pp. 1833-1845, 1994.

[41] B. D. Dykaar and P. K. Kitanidis, "Macrotransport of a Biologically Reacting Solute through Porous Media", *Water Resources Research*, vol. 32, pp. 307-320, 1996.

[42] P. Marzel, A. Seco, J. Ferrer, and C. Gambaldón, "Modeling Multiple Reactive Solute Transport with Adsorption Under Equilibrium and Nonequilibrium Conditions", *Advances in Water Resources*, vol. 17, pp. 363-374, 1994.

[43] B. E. Rittmann, and J. M. VanBriesen, "Microbiological Processes in Reactive Modeling", in *Reactive Transport in Porous Media, Reviews in Mineralogy*, vol. 34, P. C. Lichtner, C. I. Steefel, and E. H. Oelkers, Ed. Washington, D.C.: Mineralogical Society of America, 1996, pp. 311-334.

[44] K. S. Hunter, Y. Wang, and P. Van Cappellen, "Kinetic modeling of microbially-driven redox chemistry of subsurface environments: coupling transport, microbial metabolism and geochemistry", *Journal of Hydrology*, vol. 209, pp. 53-80, 1998.

[45] S. L. Smith and P. R. Jaffee, “Modeling the Transport and Reaction of Trace Metals in Water Saturated Soils and Sediments”, *Water Resources Research*, vol. 34, pp. 3135-3147, 1998.

[46] K. M. Salvage and G. T. Yeh, “Development and application of a numerical model of kinetic and equilibrium microbiological and geochemical reactions (BIOKEMOD)”, *Journal of Hydrology*, vol. 209, pp. 27-52, 1998.

[47] G. T. Yeh, K. M. Salvage, J. P. Gwo, J. M. Zachara, and J. E. Szecsody, *HYDROBIOGEOCHEM: A Coupled Model of Hydrologic Transport and Mixed Biogeochemical Kinetic/ Equilibrium Reactions in Saturated-Unsaturated Media*, ORNL/TM-13668. Oak Ridge: Oak Ridge National Laboratory, 1998.

[48] S. Kräutle and P. Knabner, “A new numerical reduction scheme for fully coupled multicomponent transport-reaction problems in porous media”, *Water Resources Research*, vol. 41, W09414. 2005.

[49] S. Kräutle and P. Knabner, “A reduction scheme for coupled multicomponent transport-reaction problems in porous media: Generalization to problems with heterogeneous equilibrium reactions”, *Water Resources Res*earch, vol. 43, W03429, 2007.

[50] F. Zhang, G. T. Yeh, J. C. Parker, *et al.*, “A reaction-based paradigm to model three-dimensional reactive chemical transport in groundwater”, *Journal of Contaminant Hydrology*, vol. 93, pp. 10-32, 2007.

[51] G. T. Yeh, and Y. L. Fang, *HYDROGEOCHEM 4.5: A Coupled Model of Fluid Flow, Thermal Transport, and HYDROGEOCHEMical Transport through Saturated Unsaturated Media Version 4.5*. University of Central Florida, 2008a.

[52] G. T. Yeh and Y. L. Fang, *HYDROGEOCHEM 5.5: A Three Dimensional Model of Coupled Fluid Flow, Thermal Transport, and HYDROGEOCHEMical Transport through Variably Saturated Conditions Version 5.5*. University of Central Florida, 2008b.

[53] G. T. Yeh, *EQMOD: A Chemical Equilibrium Model of Complexation, Adsorption, Ion-Exchange, Precipitation/ Dissolution, Redox, and Acid-Base Reaction*. The Pennsylvania State University, 1990.

[54] K. S. Pitzer, “Thermodynamics of Electrolytes. I. Theoretical basis and governing equations”, *Journal of Physical Chemistry*, vol. 77, pp. 268-277, 1973.

[55] K. S. Pitzer And G. Mayorga, “Thermodynamics of Electrolytes. II. Activity and osmotic coefficients for strong electrolytes with one or both ions univalent”, *Journal of Physical Chemistry*, vol. 77, pp. 2300-2308, 1973.

[56] K. S. Pitzer and J. Kim, “Thermodynamics of Electrolytes. IV. Activity and Osmostic coefficients for mixed electrolytes”, *Journal of the American Chemical, Society*, vol. 96, pp. 5701-5707, 1974.

[57] K. S. Pitzer, “Thermodynamics of Electrolytes. V. Effects of high-order electrostatic terms”, *Journal of Solution Chemistry*, vol. 4, pp. 249-265, 1975.

[58] W. Yan, M.S. Thesis: *Numerical Modeling of Geochemical Equilibrium with Pitzer's Activity Coefficient Model*. The Pennsylvania State University, 1992.

[59] G. T. Yeh, S. L. Carpenter, P. L. Hopkins, and M. D. Siegel, *Users' Manual of LEGHC: A Lagrangian-Eulerian Finite Element Model of HydroGeoChemical Transport through Saturated-Unsaturated Media - Version 1-0*, SAND93-7081. Albuquerque, Sandia National Laboratory, 1995a.

[60] G. T. Yeh, S. L. Carpenter, P. L. Hopkins, and M. D. Siegel, *Users' Manual of LEGHC: A Lagrangian-Eulerian Finite Element Model of HydroGeoChemical Transport through Saturated-Unsaturated Media - Version 1-1*, SAND95-1121. Albuquerque, Sandia National Laboratory, 1995b.

[61] G. T. Yeh and K. M. Salvage, *HYDROGEOCHEM 2.0: A Coupled Model of HYDROlogic Transport and Mixed GEOCHEMical Kinetic/Equilibrium Reactions in Saturated Unsaturated Media*. The Penn State University, 1997a.

[62] G. A. Iskra, M.S. Thesis: *The Verification and Validation of a General Mixed Chemical Kinetic and Equilibrium Model (KEMOD)*. the Pennsylvania State University, 1994.

[63] G. T. Yeh, G. A. Iskra, J. E. Szecsody, J. M. Zachara, and G. P. Streile, *KEMOD: A Mixed Chemical Kinetic and Equilibrium Model of Aqueous and Solid Phase Reactions*, PNL-10380. Richland: Pacific Northwest Laboratory, 1995.

[64] G. T. Yeh and K. Salvage, *KEMOD 2.0: A Mixed Chemical Kinetic and Equilibrium Model of Complexation, Adsorption-Desorption, Ion-Exchange, Precipitation-Dissolution, Redox, and Acid-Base Reactions*. The Pennsylvania State University, 1996a.

[65] G. T. Yeh and K. M. Salvage, *HYDROGEOCHEM 2.1: A Coupled Model of HYDROlogic Transport and Mixed GEOCHEMical Kinetic/ Equilibrium Reactions in Saturated Unsaturated Media*. The Penn State University, University Park, 1997b.

[66] G. T. Yeh and Karen Salvage, *KEMOD 2.1: A Mixed Chemical Kinetic and Equilibrium Model of Complexation, Adsorption-Desorption, Ion-Exchange, Precipitation-Dissolution, Redox, and Acid-Base Reactions*. The Pennsylvania State University, 1996b.

[67] K. M. Salvage, PhD Dissertation: *Reactive Contaminant Transport in Variably Saturated Porous Media: Biogeochemical Model Development, Verification, and Application.* The Pennsylvania State University, 1998.

[68] G. T. Yeh and K. M. Salvage, *BIOKEMOD 1-0: A Mixed MicroBIOlogical and Chemical Kinetic and Equilibrium Reaction MODel.* The Pennsylvania State University, 1998.

[69] Y. L. Fang, PhD Dissertation: *Reactive Chemical Transport under Multiphase System.* The Pennsylvania State University, 2003.

[70] Y. L. Fang, G. T. Yeh, and W. D. Burgos, "A New Paradigm to Model Reaction-Based Biogeochemical Processes", *Water Resources Res*earch, vol. 39, pp. 1083-1108, 2003.

[71] G. T. Yeh, M. H. Li and M. D. Siegel, *Users' Manual for LEHGC: A Lagrangian-Eulerian Finite Element Model of Coupled Fluid Flows and HydroGeoChemical Transport through Variably saturated Media – Version 2.0.* The Penn State University, 1999.

[72] G. T. Yeh, J. R. Cheng, and H. P. Cheng, *2DFEMFAT: User's Manual of a 2-Dimensional Finite Element Model of Flow and Transport through Saturated-Unsaturated Media* in Course Notes on Modeling of Flow and Contaminants in the Subsoil. Delft: Engineering and Mine Surveying, 1993.

[73] Y. Li, M.S. Thesis: *A Coupled Model of Fluid Flow, Thermal Transport, and Reactive Chemical Transport through Variably saturated Media.* University of Central Florida, 2003.

[74] J. Sun, M.S. Thesis: *A Three-Dimensional Model of Fluid Flow, Thermal Transport, and Hydrogeochemical Transport through Variably Saturated Conditions.* University of Central Florida, 2004.

[75] F. Zhang, PhD Dissertation: *A New Paradigm of Modeling Watershed Water Quality.* University of Central Florida, 2005.

[76] G. T. Yeh and F. Zhang, *A General Paradigm of Modeling One-dimensional River/Stream Watershed Water Quality.* University of Central Florida, 2005a.

[77] G. T. Yeh and F. Zhang, *A General Paradigm of Modeling Two-dimensional Overland Watershed Water Quality.* University of Central Florida, 2005b.

[78] G. T. Yeh and F. Zhang, *A General Paradigm of Modeling Three-dimensional Subsurface Water Quality.* University of Central Florida, 2005c.

[79] G. T. Yeh, "A Lagrangian-Eulerian method with zoomable hidden fine-mesh approach to solving advection-dispersion equations", *Water Resources Research*, vol. 26, pp.1133-1144, 1990.

[80] H. P. Cheng, J. R. Cheng, G. T. Yeh, "A particle tracking technique for the Lagrangian-Eulerian finite element method in multi-dimensions". *International Journal for Numerical Methods in Engineering*, vol. 39, pp. 1115-1136, 1996.

[81] G. T. Yeh, J. R. Chang, J. P. Gwo, H. C. Lin, W. Martin, and D. Richards, *3DSALT: A Three-dimensional Salt Intrusion Model in Saturated-Unsaturated Media*, HL-94-1. Vicksburg: U. S. Army Corps of Engineers, 1994.

[82] G. T. Yeh and R. J. Luxmoore, *MATTUM: A Multi dimensional Model for Simulating Moisture and Thermal Transport in Unsaturated Porous Media*, ORNL-5888. Oak Ridge: Oak Ridge National Laboratory, 1983.

[83] A. Chilakapati, T. Ginn and J. Szecsody, An Analysis of Complex Reaction Networks in Groundwater Modeling, *Water Resources Research*, vol. 34, pp. 1767-1780, 1998..

[84] C. I. Steefel, and P. C. Lichtner, "Diffusion and Reaction in Rock Matrix Bordering a Hyperalkaline Fluid-filled Fracture", *Geochimica Cosmochimica Acta*, vol. 58, pp. 3595-3612, 1994.

[85] F. A. L. Dullien, *Porous Media.* Academic Press, 1979.

[86] G. T. Yeh, G. B. Huang, F. Zhang, H. P. Cheng, and H. C. Lin, *WASH123D: A Numerical Model of Flow, Thermal Transport, and Salinity, Sediment, and Water Quality Transport in WAterSHed Systems of 1-D Stream-River Network, 2-D Overland Regime, and 3-D Subsurface Media.* University of Central Florida, 2006.

[87] G. E. Hammond, A. J. Valocchi, and P. C. Lichtner, "Application of Jacobian-free Newton–Krylov with physics-based preconditioning to biogeochemical transport", *Advances in Water Resources*, vol. 28, pp. 359-376, 2005.

[88] G. T. Yeh, Y. L. Fang, F. Zhang, J. T. Sun, Y. Li, M. H. Li, and M. D. Siegel, "Numerical Modeling of Coupled Fluid Flow and Thermal and Reactive Biogeochemical Transport in Porous and Fractured Media", *Computational Geosciences*, vol. 14, pp. 149-170, 10.1007/s10596-009-9140-3, 2009.

[89] F. Zhang, W. Luo, J. C. Parker, *et al.*, "Modeling of uranium and technetium transport in acidic contaminated groundwater with pH adjustment", *Personal Communication*, 2009.

[90] Y. L. Fang, S. B. Yabusaki, S, J. Morrison, J. P. Amonette, and P. E. Long, "Multicomponent Reactive Transport Modeling of Uranium Bioremediation Field Experiments", *Geochimica et Cosmochimica Acta*, revision in progress, 2009.

[91] K. T. Finneran, R. T. Anderson, K. P. Nevin, and D. R. Lovley, "Potential for Bioremediation of uranium-contaminated aquifers with microbial U(VI) reduction", *Soil & Sediment Contamination*, vol. 11, pp. 339-357, 2002.

[92] D. E. Holmes, K. T. Finneran, R. A. O'Neil, and D. R. Lovley, "Enrichment of members of the family Geobacteraceae associated with stimulation of dissimilatory metal reduction in uranium-contaminated aquifer sediments", *Applied and Environmental Microbiology*, vol. 68, pp. 2300-2306, 2002.

[93] R. T. Anderson, H. A. Vrionis, I. Ortiz-Bernad, *et al.*, A. Peacock, D. C. White, M. Lowe, and D. R. Lovley, "Stimulating the in situ activity of Geobacter species to remove uranium from the groundwater of a uranium-contaminated aquifer", *Applied and Environmental Microbiology*, vol. 69, pp. 5884-5891, 2003.

[94] D. R. Lovley, E. E. Roden, E. J. P. Phillips, and J. C. Woodward, "Enzymatic Iron and Uranium Reduction by Sulfate-Reducing Bacteria", *Marine Geology*, vol. 113, pp. 41-53, 1993.

[95] I. Ortiz-Bernad, R. T. Anderson, H. A. Vrionis, and D. R. Lovley, "Resistance of solid-phase U(VI) to microbial reduction during in situ bioremediation of uranium-contaminated groundwater", *Applied and Environmental Microbiology*, vol. 70, pp.7558-7560, 2004.

[96] G. T. Yeh, W. D. Burgos, and J. M. Zachara, "Modeling and measuring biogeochemical reactions: system consistency, data needs, and rate formulations", *Advances in Environmental Research*, vol. 5, pp.219-237, 2001.

[97] F. Zhang and J. C. Parker, "An Efficient Modeling Approach to Simulate Heat Transfer between Fracture and Matrix Regions for Oil Shale Retorting", *Transport in Porous Media*, revision in progress, 2009.

[98] F. Zhang, J. C. Parker, P. M. Jardine, *et al.*, "Scale Effects on Apparent Reaction Kinetics at the Oak Ridge FRC", in *Environmental Remediation Sciences Program Fall Meeting*, 2006.

CHAPTER 2

Modeling of Flow and Reactive Transport in IPARS

M. F. Wheeler[1], S. Sun[2] and S. G. Thomas[3*]

[1]Institute for Computational Engineering and Sciences, Department of Aerospace Engineering & Engineering Mechanics, and Department of Petroleum and Geosystems Engineering, The University of Texas at Austin, Austin, TX, USA; [2]Earth Sciences and Engineering, Applied Mathematics and Computational Science, King Abdullah University of Science and Technology, Thuwal, Kingdom of Saudi Arabia and [3] Institute for Computational Engineering and Sciences, The University of Texas at Austin, Austin, TX, USA

Abstract: In this work, we describe a number of efficient and locally conservative methods for subsurface flow and reactive transport that have been or are currently being implemented in the IPARS (Integrated Parallel and Accurate Reservoir Simulator). For flow problems, we consider discontinuous Galerkin (DG) methods and mortar mixed finite element methods. For transport problems, we employ discontinuous Galerkin methods and Godunov-mixed methods. For efficient treatment of reactive transport simulations, we present a number of state-of-the-art dynamic mesh adaptation strategies and implementations. Operator splitting approaches and iterative coupling techniques are also discussed. Finally, numerical examples are provided to illustrate the capability of IPARS to treat general biogeochemistry as well as the effectivity of mesh adaptations with DG for transport.

Keywords: Multiphase Darcy flow, reactive transport, multiblock, mixed finite element methods, discontinuous Galerkin finite element methods, operator splitting.

1. INTRODUCTION

1.1. Numerical Modeling of Reactive Transport

Reactive transport phenomena in porous media have been studied in such diverse fields as chemistry, physics, engineering and geology. Simulations of reactive transport have widely used in petroleum engineering, groundwater hydrology, environmental engineering and chemical engineering. Significant mathematical and computational challenges are imposed by realistic simulations for simultaneous convection, diffusion, dispersion and chemical reaction [1-12]. For example, transport simulations often demand a high accuracy of numerical solutions. Many realistic transport problems in porous media involve a very long period of time. Small errors in each time step may cause substantial accumulated inaccuracy, sometimes leading to physically meaningless results. This also necessitates numerical schemes being locally conservative, as transport phenomena are based on the principle of mass conservation. In addition, effective adaptivities are often required to efficiently simulate reactive transport, because the transport systems often exhibit rich time-dependent local behaviors, such as concentration plumes, sharp fronts, shocks and layers. In this chapter, we describe simulations of reactive transport coupled with flow using efficient and locally conservative schemes based on Mixed Finite Element Methods (mixed FEM or MFEM) and Discontinuous Galerkin methods (DG).

MFEM [13, 14] are a family of finite element methods based on a variational principle expressing an equilibrium or saddle point condition that can be satisfied locally on each finite element. MFEM can be regarded as a family of methods that are distinguished mainly by the choice of numerical quadrature applied. They encompass the widely used Cell-Centered Finite Differences (CCFD) on structured rectangular grids and more recent formulations for general corner point or unstructured grids such as mimetic finite differences and multipoint flux approximation methods (MPFA) [15-18]. The advantage of MFEM is that it provides accurate approximation for both the pressure and the velocity, and flux continuity

***Address correspondence to S. G. Thomas:** Institute for Computational Engineering and Sciences, The University of Texas at Austin, Austin, TX, USA; Tel: 1-512-850-2337; Email: sgthomas@ices.utexas.edu.

Fan Zhang, Gour-Tsyh (George) Yeh, Jack C. Parker and Xiaonan Shi (Eds)

is preserved by virtue of the approximation spaces used. MFEM also preserves local mass conservation, which is a desired property in many sub-surface modeling applications considered.

DG schemes are finite element methods using piecewise discontinuous polynomial spaces and specialized bilinear forms to weakly impose boundary conditions and interelement continuities [19-29]. The methods have recently become popular in the scientific and engineering communities due to their many appealing features. In particular, DG methods are element-wise conservative, they support local approximations of high order, they are implementable on unstructured and non-matching meshes, and they are useful for multiscale, adaptive and parallel implementations. We now describe our simulator where MFEM and DG are implemented, among other numerical schemes.

1.2. Overview of IPARS

The Integrated Parallel and Accurate Reservoir Simulator (IPARS) provide a framework and a growing number of physical models suitable for research and practical calculations. Both oil reservoirs and aquifers can be simulated with the program. IPARS runs on parallel and single processor computers and can solve problems involving a million or more grid elements. It can handle multiple fault blocks [30] with unaligned grids and problems that involve different physical models (multimodels/multiphysics) [31, 32] or different numerics (multinumerics) [33-35] in various regions of the reservoir.

IPARS supports three dimensional transient flow of multiple phases containing multiple components plus immobile phases (rock) and adsorbed components. IPARS can handle a variety of physical models such as air-water, oil-water, black-oil (3 phases) and compositional equation of state (EOS) flow models. In addition IPARS can also solve fairly general porescale network models for flow, geomechanics, reactive transport and inverse modeling problems for reservoir characterization and history matching. Most of the flow models as well as geomechanics are available in fully implicit as well as iteratively coupled IMPES/IMPEC (implicit-pressure, explicit saturation/ concentration) formulations. An overview of the features of IPARS can be found in [36].

The spatial discretization of the nonlinear partial differential equations describing flow uses a variety of methods. Cell-centered finite difference methods are available in almost all the models. In addition, continuous and discontinuous Galerkin methods [37] have been implemented for the air-water and single phase flow models as well as for geomechanics [38]. More recently, Mutli-Point Flux Approximation (MPFA) methods [15] have been implemented for single phase and iteratively coupled two-phase flow equations and is being tested on more complex problems, while enhanced velocity mixed FEM [39] has been implemented for solving coupled flow and reactive transport problems. Also, the classical mortar mixed FEM has been extended to include multiscale basis implementations [40].

Phase densities and viscosities may be arbitrary functions of pressure and composition or may be represented by simpler functions (*e.g.* constant compressibility). Actual treatment of phase properties depends on the physical model. Porosity (at standard conditions) and permeability may vary with location in an arbitrary manner. Some physical models treat porosity as a constant and others make it a function of pressure (and volume strain for geomechanics). Permeability is treated as a constant diagonal tensor although full tensor extensions are implemented for MPFA discretizations and for the physical dispersion term [41] in coupled flow-transport problems. Properties such as relative permeability and capillary pressure are functions of saturations and rock type, which is in general a function of location. Several models for three-phase relative permeability are built into the simulator and other models can be added as needed. Any dependence of relative permeability and capillary pressure on composition and pressure is left to the individual physical models.

On multiprocessor machines, the grid system is distributed among the processors such that each processor is assigned a subset of the total grid system. The subgrid assigned to a processor is surrounded by a "communication" layer of grid elements that belong to other processors. The framework provides a routine that updates data in the communication layer. In addition, IPARS supports an arbitrary number of wells each with one or more completion intervals. A well may penetrate more than one fault block but a completion interval must occur in a single fault block. On parallel machines, well grid elements may be assigned to more than one processor. For each well element, the framework provides estimates of the permeability normal to the wellbore, the geometric constant in the productivity index, and the length of the open interval. Other well calculations are left to the individual physical models. Portability of the simulator is emphasized. FORTRAN77/90, classical C and C^{++} code are used in a multilanguage and multicompiler setting.

1.3. Transport-Chemistry Module in IPARS

The TRCHEM (TRansport with general biogeoCHEMistry) module in IPARS represents an accurate, efficient and relatively generic multiphase reactive transport model for phenomena which occur in porous media in a number of scientific and engineering applications including chemical, petroleum and environmental engineering. The overall multiphase multicomponent formulation of the model follows Parker's formulation [42] which we extended to arbitrary number of immiscible compressible flowing phases. The model is implemented as a part of IPARS framework [36, 43-44] and it can be used in principle with any *flow model* in IPARS provided the necessary interface routines have been written. More recently, the IPARS multiblock capability has been extended to TRCHEM in both mortar (classical and multiscale basis implementations) and non-mortar (using an enhanced velocity MFEM implementation). Thus advection, diffusion and reaction can be simulated in a physical domain modeled using sub-domains with non-matching grids.

A number of simplifying assumptions were made in implementing majority of this module. Interphase mass transfer between flowing phases is assumed to be locally equilibrium controlled. In this module we consider only the case of linear partitioning between flowing phases and constant partitioning coefficients, although computational EOS models are provided elsewhere in IPARS. Interphase mass transfer between flowing phases and stationary phases is handled by separate adsorption or general chemistry routines. The chemical species are divided into two groups: the species in flowing phases and the species in stationary phases. If some species exists in a stationary and in a flowing phase, it will be treated as two different species in the input file and in the code. This remark does not apply to linear or non-linear scalar adsorption which can be handled directly in TRCHEM. The density ρ_α and the viscosity μ_α for each phase are currently assumed to be independent of the concentration of species. It is assumed that the rock (solid phase) is not strongly involved in chemical reaction (such as dissolution of the rock by acid), so that the permeability tensor $\mathbf{K}$ and the porosity ϕ are not affected by the reactive transport process. The bulk source-sink term due to mass transfer between phases is assumed to be small so that it can be ignored in the flow equation.

The assumption of local equilibrium is the key assumption in the module and it enables one to obtain the phase-summed transport equations. For many problems, in particular, in environmental subsurface applications, the mass transfer rate between flowing phases is much faster than the rate of change of concentrations in flowing phases with time, justifying the local equilibrium assumption. The kinetic model in which the mass transfer rate between flowing phases is explicitly computed is in theory more generic than the equilibrium model. However, kinetic model is very inefficient when used to simulate fast interphase mass transfer problems. In addition, it is so different from the equilibrium model that it requires separate implementation. Therefore, only the local equilibrium model is used.

The TRCHEM model in IPARS support a number of capabilities: It can handle multiple flowing phases and multiple stationary phases. The number of flowing phases is arbitrary in this model. However, most of the time only one, two, or three flowing phases are used. The number of stationary phases is also arbitrary. The model can handle both molecular diffusion and physical dispersion. Longitudinal dispersivity and transverse dispersivity can be rock-type or phase-type dependent. Each of the flowing phases can be incompressible or compressible. That is, the density ρ_α might vary with position and time. The rock (formation) can be incompressible or can be slightly compressible. If it is slightly compressible, then the porosity ϕ is taken to be a function of pressure only. The TRCHEM model can be coupled easily with any flow model in IPARS framework. Further it has been coupled to the mortar FEM framework in IPARS and hence advection, diffusion reaction problems can be simulated on non-matching multiblock grids. TRCHEM can handle general biogeochemistry including adsorption, ion-exchange, precipitation, dissolution, bioremediation and radionuclide decay. Three types of chemistry are supported. These are: equilibrium controlled reaction, classical mass-action kinetics reaction and Monod type kinetics reaction. The handling of equilibrium controlled reaction is relatively robust, even when mineral (stationary) phases precipitate into existence or dissolve away, because it uses an interior-point algorithm to minimize the Gibbs free energy. Additional features of TRCHEM, especially these associated with DG, are currently under implementation.

The remainder of the article is organized as follows. In the following section, we state the governing equations and numerical algorithms for single-phase and two-phase flow in porous media; in particular the DG and (mortar) mixed FEM methods. Modeling of a multiple component reactive transport system using the DG scheme is given in §3. In §4, we present Godunov-mixed FEM for the reactive transport equations in a time-split scheme. We also comment on iterative coupling schemes at the end of this section. In §5, one of the current development of IPARS, mesh adaptation, is described. Finally, several numerical examples are presented in §6.

2. FLOW IN POROUS MEDIA

In this chapter, we consider coupled flow and reactive transport for a single or multiple phase(s) in porous media. We allow for single or multiple phases, as well as for multiple species in the system. For convenience, we will assume Ω is a polygonal and bounded domain in R^d ($d=1$, 2 or 3) with boundary $\partial\Omega$. We start by describing single-phase flow.

2.1. DG for Single-Phase Flow

We start by first considering a simple model for single-phase flow in porous media with the following equation, although DG can be easily extended to treat more complicated flow models.

$$-\nabla\cdot\mathbf{u}=q, \text{ in } \Omega\times(0,T], \quad (1)$$

$$\mathbf{u}\cdot\mathbf{n}=u_B, \text{ on} \in \partial\Omega\times(0,T], \quad (2)$$

where the unknowns are p (the pressure in the fluid mixture) and $\mathbf{u}$ ($=-\mathbf{K}\nabla p$, the Darcy velocity of the mixture, *i.e.* the volume of fluid flowing cross a unit across-section per unit time). Here, we assume that the conductivity $\mathbf{K}$ is uniformly symmetric positive definite and bounded. The imposed external total flow rate q is a sum of sources (injection) and sinks (extraction) and is assumed to be bounded.

The computed velocity from the flow modeling equation will be needed in the convection part of transport simulation. In addition, it will also affect the mechanical dispersion of species. The dispersion-diffusion tensor $\mathbf{D}(\mathbf{u})$ has contributions from molecular diffusion and mechanical dispersion, and can be calculated by

$$\mathbf{D}(\mathbf{u})=d_m\mathbf{I}+|\mathbf{u}|\left\{d_l\mathbf{E}(\mathbf{u})+d_t\left(\mathbf{I}-\mathbf{E}(\mathbf{u})\right)\right\}, \quad (3)$$

where $\mathbf{E}(\mathbf{u})$ is the tensor that projects onto the $\mathbf{u}$ direction, whose (i,j) component is $\left(\mathbf{E}(\mathbf{u})\right)_{ij}=\frac{u_iu_j}{|\mathbf{u}|^2}$; d_m is the molecular diffusivity and is assumed to be strictly positive; d_l and d_t are the longitudinal and transverse dispersivities, respectively, and are assumed to be nonnegative.

Let E_h be a family of non-degenerate quasi-uniform and possibly non-conforming partitions of Ω composed of triangles or quadrilaterals if $d=2$, or tetrahedra, prisms or hexahedra if $d=3$. The set of all interior edges (for 2 dimensional domain) or faces (for 3 dimensional domain) for E_h are denoted by Γ_h. On each edge or face $\gamma\in\Gamma_h$, a unit normal vector $\mathbf{n}_\gamma$ is chosen. The set of all edges or faces on

Γ_{out} and on Γ_{in} for E_h are denoted by $\Gamma_{h,out}$ and $\Gamma_{h,in}$, respectively, for which the normal vector $\mathbf{n}_\gamma$ coincides with the outward unit normal vector.

The discontinuous finite element space is

$$D_r(E_h) \equiv \left\{\phi \in L^2(\Omega) : \phi|_{E \in E_h} \in P_r(E)\right\}, \quad \textbf{(4)}$$

where $P_r(E)$ denotes the space of polynomials of (total) degree less than or equal to r on E.

We consider NIPG (the Non-Symmetric Interior Penalty Galerkin method [24]), SIPG (the symmetric interior penalty Galerkin method [26, 29, 37]) and IIPG (the incomplete interior penalty Galerkin method [26, 37, 45]) for the single-phase flow. The OBB-DG (the Oden-Babu š ka-Baumann formulation of DG [43]) can be viewed as a special case of NIPG. If we also use them for transport, the three methods for flow and the three schemes for transport lead to nine different combinations for coupled flow and transport problems. However, we note that only IIPG for flow is compatible with primal DG methods for transport in the sense defined in [45].

For flow, we introduce the bilinear form $a(p,\psi)$ and the linear functional $l(\psi)$,

$$a(p,\psi) = \sum_{E \in E_h} \int_E \mathbf{K}\nabla p \cdot \nabla \psi + J_{0,flow}(p,\psi) - \sum_{\gamma \in \Gamma_h} \int_\gamma \{\mathbf{K}\nabla p \cdot \mathbf{n}_\gamma\}[\psi] - S_{flow} \sum_{\gamma \in \Gamma_h} \int_\gamma \{\mathbf{K}\nabla \psi \cdot \mathbf{n}_\gamma\}[p],$$

$$l(\psi) = (q,\psi) - \sum_{\gamma \in \Gamma_{h,in} \cup \Gamma_{h,out}} \int_\gamma \psi u_B,$$

where $S_{flow} = -1$ for NIPG, $S_{flow} = 1$ for SIPG and $S_{flow} = 0$ for IIPG. The interior penalty term $J_{0,flow}(p,\psi)$ for flow is defined as:

$$J_{0,flow}(p,\psi) = \sum_{\gamma \in \Gamma_h} \frac{r_{flow}^2 \sigma_{\gamma,flow}}{h_\gamma} \int_\gamma [p][\psi],$$

where the penalty parameter $\sigma_{\gamma,flow}$ is a constant on each edge or face γ.

The DG velocity $\mathbf{u}^{DG}$, which will be used in the transport simulation, is defined below.

$$2\mathbf{u}^{DG} = -\mathbf{K}\nabla P^{DG} \quad x \in E, E \in E_h,$$

$$\mathbf{u}^{DG} \cdot \mathbf{n} = -\{\mathbf{K}\nabla P^{DG} \cdot \mathbf{n}\} + \frac{r_{flow}^2 \sigma_{\gamma,flow}}{h_\gamma} \int_\gamma \left(P^{DG}\big|_E - P^{DG}\big|_{\Omega \setminus \bar{E}} \right) x \in \gamma = \partial E_i \cap \partial E_j,\ E_i, E_j \in E_h \ \ \mathbf{n} \text{ exterior to } E_i,$$

$$\mathbf{u}^{DG} \cdot \mathbf{n} = u_B \quad x \in \partial\Omega.$$

Here, $\mathbf{u}^{DG}$ is defined at every interior point in each element, but only the normal velocity component $\mathbf{u}^{DG} \cdot \mathbf{n}$ is defined on element interfaces and on domain boundaries, as this is all the information needed in the DG schemes in the transport part.

2.2. Multiblock Mixed Finite Element Methods for Single-phase Flow

We consider the following nonlinear second order parabolic equation, which can be used to model a time-dependent single phase slightly compressible flow in porous media over the interval of time $J = [0,T]$:

$$\frac{\partial(\phi\rho)}{\partial t} - \nabla\cdot(\rho\mathbf{K}(\nabla p - \rho\mathbf{g})) = f \text{ in } \Omega\times J, \qquad \textbf{(5)}$$

$$\rho\mathbf{K}(\nabla p - \rho\mathbf{g})\cdot\mathbf{n} = 0 \text{ on } \partial\Omega\times J, \qquad \textbf{(6)}$$

$$p = p_0 \text{ in } \Omega\times\{0\}. \qquad \textbf{(7)}$$

The above system can also be written in a mixed form:

$$\mathbf{u} = -\rho\mathbf{K}(\nabla p - \rho\mathbf{g}) \quad \text{in } \Omega\times J, \qquad \textbf{(8)}$$

$$\frac{\partial(\phi\rho(p))}{\partial t} + \nabla\cdot\mathbf{u} = f \quad \text{in } \Omega\times J, \qquad \textbf{(9)}$$

$$\mathbf{u}\cdot\mathbf{n} = 0 \quad \text{on } \partial\Omega\times\{0\}, \qquad \textbf{(10)}$$

$$p = p_0 \quad \text{in } \Omega\times\{0\}. \qquad \textbf{(11)}$$

Here, $\Omega\subset\mathbf{R}^d$, $d=2$ or 3, is the flow domain with $\mathbf{n}$ being the unit outward normal vector to $\partial\Omega$; p, $\mathbf{u}$, and $\rho=\rho(p)$ are the fluid pressure, Darcy velocity, and density, respectively; ϕ is the rock porosity, $\mathbf{K}$ is a symmetric, uniformly positive definite tensor representing the rock permeability divided by the fluid viscosity, f is the source term, g is the gravitational constant, and D is the depth. The equation of state is given by:

$$\frac{d\rho}{\rho} = c_f dp,$$

where c_f is the fluid compressibility constant. The homogeneous Neumann boundary conditions are considered merely for simplicity and the results have been generalized to more general boundary conditions.

Let the domain Ω be decomposed into a finite number of non-overlapping subdomain blocks Ω_i so that $\overline{\Omega} = \cup_{i=1}^{n_b}\overline{\Omega}_i$. The blocks may form a geometrically nonconforming partition. Let $\Gamma_{i,j} = \partial\Omega_i\cap\partial\Omega_j$, $\Gamma = \cup_{1\le i<j\le n_b}\Gamma_{i,j}$, and $\Gamma_i = \partial\Omega_i\cap\Gamma = \partial\Omega_i\setminus\partial\Omega$. The unit outer normal vector to $\partial\Omega_i$ is denoted by $\mathbf{n}_i$. Let $T_{h,i}$ be a non- degenerate (quasi-uniform) finite element partition of Ω_i. We allow for $T_{h,i}$ and $T_{h,j}$ to be non-matching on $\Gamma_{i,j}$ [30, 46].

The following functional spaces were used in the weak formulation of the problem:

$$2\mathbf{V}_i := \{\mathbf{v}\in H(\text{div};\Omega_i) : \mathbf{v}\cdot\mathbf{n} = 0 \text{ on } \partial\Omega_i\cap\partial\Omega\},$$
$$W_i := L^2(\Omega_i), \text{ and } \Lambda_{i,j} := H^{\frac{1}{2}}(\Gamma_{i,j}).$$

For the discrete problem, the following finite element space is used let $\mathbf{V}_{h,i}\times W_{h,i}\subset\mathbf{V}_i\times W_i$ be any of the usual mixed finite element spaces defined on $T_{h,i}$ (see [47], Section III.3). The most commonly used mixed spaces are the Raviart-Thomas spaces of order k, RT_k [13, 14]. Our simulator currently uses RT_0 space on rectangular partitions for most models using MFEM (except MPFA which will not be discussed here). On the interface we will use a mortar finite element space to approximate the pressure and impose weakly continuity of flux and pressure. Let $M_{H,i,j} \subset L^2(\Gamma_{i,j})$ be the mortar space on $\Gamma_{i,j}$, containing at least either the continuous or discontinuous piecewise polynomials of degree m on $T_{H,i,j}$. In our numerical experiments we used $m=0$, or $m=1$. Then, the mortar finite element space on Γ is defined as:

$$M_H = \bigoplus_{1 \le i < j \le n_b} M_{H,i,j}$$

Then the fully discrete-in-time mortar MFEM formulation is as follows. Let $0=t_0 < t_1 < t_2 < \dots$, $\Delta t^n = t_n - t_{n-1}$, and $f^n = f(t_n)$. Then, for n=1, 2, 3…, we seek $\left(u_h^n, p_h^n\right)\Big|_{\Omega_i} \in V_{h,i} \times V_{h,i}$ and $\lambda_H^n \in M_H$, for $1 \le i < j \le n_b$ such that:

$$\int_{\Omega_i} \left(\rho_h^n \mathbf{K}\right)^{-1} \mathrm{u}_h^n \cdot \mathrm{v} dx = \int_{\Omega_i} \rho_h^n \nabla \cdot \mathrm{v} dx - \int_{\Gamma_i} \mathrm{v} \cdot \mathrm{n}_i \lambda_H^n ds + \int_{\Omega_i} \rho_h^n g \cdot \mathrm{v} dx, \forall \mathrm{v} \in \mathrm{V}_{h,i}, \quad \textbf{(12)}$$

$$\int_{\Omega_i} \frac{(\phi\rho h)^n - (\phi\rho h)^{n-1}}{\Delta t^n} \omega dx + \int_{\Omega_i} \nabla \cdot \mathrm{u}_h^n \omega dx = \int_{\Omega_i} f^n \omega dx, \forall \omega \in \mathrm{W}_{h,i}, \quad \textbf{(13)}$$

$$\int_{\Omega_i} [\mathrm{u}_h^n \cdot \mathrm{n}]_{i,j} \mu ds = 0, \forall \mu \in M_{H,i,j}. \quad \textbf{(14)}$$

To solve the discrete system (12)–(14) on each time step, we reduce it to an interface problem in the (coarse) mortar space. We define a non-linear interface bivariate form $b_n : M_H \times M_H \to \mathbf{R}$ as follows. For $\psi = \lambda_H^n \in M_H$ and $\mu \in M_H$, let:

$$b_n(\psi, \mu) = \sum_{i=1}^{n_b} \int_{\Gamma_i} \mathrm{u}_h^n\Big|_{\Omega_i} \cdot \mathrm{n}_i \mu ds$$

where $p_h^n = p_h^n(\psi)$, $\mathrm{u}_h^n = \mathrm{u}_h^n(\psi)$ are obtained from the solution of (12)–(13) with Dirichlet boundary data ψ. Now, define a non-linear interface operator

$\mathcal{B}_n : M_H \to M_H$ by

$$\left\langle \mathcal{B}^n \psi, \mu \right\rangle = b^n(\psi, \mu), \qquad \forall \mu \in M_H,$$

where $\langle \cdot, \cdot \rangle$ is the L^2-inner product in M_H. Then, it follows that $\left(\psi, p_h^n(\psi), \mathrm{u}_h^n(\psi)\right)$ is the solution to (12)–(14), when $\psi \in M_H$ solves:

$$\mathcal{B}^n(\psi) = 0. \quad \textbf{(15)}$$

We solve the system of nonlinear equations (25) on the interface by an inexact Newton method. Each Newton step s is computed by a forward difference GMRES iteration for solving $\left(\mathcal{B}^n\right)'(\psi) s = -\mathcal{B}^n(\psi)$. On each GMRES iteration the action of the Jacobian $\left(\mathcal{B}^n\right)'(\psi)$ on a vector μ is approximated by a forward difference which requires only one evaluation of the nonlinear operator $\mathcal{B}^n$. The evaluation of $\mathcal{B}^n$ involves solving subdomain problems (12)–(13) in parallel and two inexpensive projection steps - from the mortar grid onto the local subdomain grids and from the local grids.

2.3. Mixed Finite Elements for Two-Phase Flow in Porous Media

Next, we formulate the mortar mixed finite element method for the two-phase Darcy flow system. The governing mass conservation equations are imposed on each subdomain Ω_i

$$\frac{\partial(\phi\rho_\alpha S_\alpha)}{\partial t}+\nabla \mathrm{u}_\alpha = q_\alpha, \tag{16}$$

where $\alpha=\omega$ (wetting), n (non-wetting) denotes the phase, S_α is the phase saturation, $\rho_\alpha=\rho_\alpha(p_\alpha)$ is the phase density, ϕ is the porosity, q_α is the source term, and

$$\mathrm{u}_\alpha=-\frac{\kappa_{r\alpha}(S_\alpha)\mathbf{K}}{\mu_\alpha}\rho_\alpha(\nabla p_\alpha-\rho_\alpha g) \tag{17}$$

is the α -phase Darcy velocity. Here p_α is the phase pressure, $\kappa_{r\alpha}(S_\alpha)$ is the phase relative permeability, μ_α is the phase viscosity, $\mathbf{K}$ is the rock permeability tensor, g is the gravity vector. On each interface $\Gamma_{i,j}$ the following physically meaningful continuity conditions are imposed:

$$p_\alpha|_{\Omega_i}=p_\alpha|_{\Omega_j}, \tag{18}$$

$$[\mathrm{u}_\alpha\cdot \mathrm{n}]_{i,j}\equiv \mathrm{u}_\alpha|_{\Omega_i}\cdot \mathrm{n}_i+\mathrm{u}_\alpha|_{\Omega_j}\cdot \mathrm{n}_j=0, \tag{19}$$

The above system of equations are closed *via* the volume balance and capillary pressure relationships

$$S_\omega+S_n=1, \qquad p_c(S_\omega)=P_n-P_\omega, \tag{20}$$

which are imposed on each Ω_i and $\Gamma_{i,j}$. We assume that no flow $\mathrm{u}_\alpha\cdot\mathrm{n}$ is imposed on $\partial\Omega$, although more general types of boundary conditions can be treated in practice.

Using the expanded mixed finite element formulation [48], let, for $\alpha=\omega, n$,

$$\tilde{\mathrm{u}}_\alpha=-\mathrm{K}(\nabla p\alpha-\rho\alpha g)$$

Then

$$\mathrm{u}_\alpha=\frac{\kappa_{r\alpha}(S_\alpha)}{\mu_\alpha}\rho_\alpha\tilde{\mathrm{u}}_\alpha$$

The motivation for introducing the new variable $\tilde{\mathrm{u}}_\alpha$ is to avoid inverting $\kappa_{r\alpha}(S_\alpha)$, which can be zero if phase α is immobile. The gradient $\tilde{\mathrm{u}}_\alpha$ is discretized in the space $\tilde{\mathrm{V}}_{h,i}$, which is the space $\mathrm{V}_{h,i}$ without imposing the no-flow boundary condition. This choice, combined with appropriate quadrature rules for the mass matrices, allows for local elimination of both $\tilde{\mathrm{u}}_\alpha$ and u_α, reducing the method to cell-centered finite differences for the subdomain primary variables p_h and S_h, coupled with the mortar primary variables p_H^M and S_H^M (α dropped here as it refers to a fixed reference phase), see [48] for details.

Let $0=t_0<t_1<t_2<\ldots$, let $\Delta t^n=t_n-t_{n-1}$, and let $f^n=f(t_n)$. In the backward Euler multiblock expanded mixed finite element approximation of (16)-(20) we seek, for $1 \le i < j \le n_b$ and n=1, 2, 3…, $\mathrm{u}_{h,\alpha}^n\big|_{\Omega_i} \in V_{h,i}$, $\tilde{\mathrm{u}}_{h,\alpha}^n\big|_{\Omega_i} \in \tilde{V}_{h,i}$, $p_h^n\big|_{\Omega_i} \in W_{h,i}$, $S_h^n\big|_{\Omega_i} \in W_{h,i}$, $p_H^{M,n}\big|_{\Gamma_{i,j}} \in M_{H,i,j}$, and $S_H^{M,n}\big|_{\Gamma_{i,j}} \in M_{H,i,j}$, such that, for $\alpha = \omega$ and n,

$$\int_{\Omega_i} \frac{S_{h,\alpha}^n - S_{h,\alpha}^{n-1}}{\Delta t^n} \omega dx + \int_{\Omega_i} \nabla \cdot \mathrm{u}_{h,\alpha}^n \omega dx = \int_{\Omega_i} q_\alpha \omega dx, \forall \omega \in \mathrm{W}_{h,i}, \tag{21}$$

$$\int_{\Omega_i} \mathbf{K}^{-1} \tilde{\mathrm{u}}_{h,\alpha}^n \cdot \mathrm{v}dx = \int_{\Omega_i} p_{h,\alpha}^n \nabla \cdot \mathrm{v}dx - \int_{\partial\Omega_i \backslash \partial\Omega} p_{H,\alpha}^{M,n} \mathrm{v} \cdot \mathrm{n}_i ds + \int_{\Omega} \rho_{h,\alpha}^n g \cdot \mathrm{v}dx, \mathrm{v} \in \mathrm{V}_{h,i}, \tag{22}$$

$$\int_{\Omega_i} \mathrm{u}_{h,\alpha}^n \cdot \tilde{\mathrm{v}}dx = \int_{\Omega_i} \frac{k_{h,r\alpha}^n}{\mu_{h,\alpha}} p_{h,\alpha}^n \tilde{\mathrm{u}}_{h,\alpha}^n \cdot \tilde{\mathrm{v}}dx, \quad \tilde{\mathrm{v}} \in \mathrm{V}_{h,i}, \tag{23}$$

$$\int_{\Gamma_{i,j}} [\mathrm{u}_{h,\alpha}^n \cdot \mathrm{n}]_{i,j} \mu ds = 0, \mu \in M_{H,i,j}. \tag{24}$$

Here $k_{h,\alpha}^n$ and $\rho_{h,\alpha}^n \in W_{h,i}$ are given functions of the subdomain primary variables p_h^n and S_h^n (in a prescribed reference phase variable). The mortar functions $p_{H,\alpha}^{M,n}$ can be computed using (20), given the mortar primary variables $p_H^{M,n}$ and $S_H^{M,n}$. It is to be noted that when the mortar separates two different rock types (*e.g.* faults), the saturations can be discontinuous across the mortar even in the absence of moving fronts. But the phase and capillary pressures must be continuous. Hence, in such situations, the proper choice of mortar primary variables is either the phase variables $p_{\alpha,H}^{M,n}$ ($\alpha = \omega, n$) or a phase pressure and the capillary pressure, $p_H^{M,n}$ and $p_{cn\omega,H}^{M,n}$.

To solve the discrete system (21)–(24) on each time step, we reduce it to an interface problem in the coarse mortar space, see also [49]. Here we assume that two phases exist (no single-phase degeneracy) on both sides of the interface.

Let $\mathcal{M}_H = M_H \times M_H$ be the space of mortar primary variables. We define a non-linear interface bivariate form $b^n : \mathcal{M}_H \times \mathcal{M}_H \to \mathbf{R}$ as follows. For $\left(p_H^{M,n}, S_H^{M,n}\right)^T \in \mathcal{M}_H$ and $\mu = (\mu_\omega, \mu_n) \in \mathcal{M}_H$, let

$$b_n(\psi, \mu) = \sum_{1 \le i < j \le n_b} \int_{\Gamma_{i,j}} \sum_\alpha \left[\mathrm{u}_{h,\alpha}^n(\psi) \cdot \mathrm{n}\right]_{i,j} \mu_\alpha ds$$

where $\left(S_h^n(\psi), \mathrm{u}_{h,\alpha}^n(\psi)\right)$ are solutions to the series of subdomain problems (21)–(23) with Dirichlet boundary data $p_{H,\alpha}^{M,n}(\psi)$.

Define a non-linear interface operator $\mathcal{B}_n : \mathcal{M}_H \to \mathcal{M}_H$ by

$$\left\langle \mathcal{B}^n \psi, \mu \right\rangle = b^n(\psi, \mu), \quad \forall \mu \in \mathcal{M}_H,$$

where $\langle \cdot, \cdot \rangle$ is the L^2-inner product in $\mathcal{M}_H$. It is easy to see that $\left(\psi, S_h^n(\psi), \mathrm{U}_{h,\alpha}^n(\psi)\right)$ is the solution to (21)–(24), where $\psi \in \mathcal{M}_H$ solves

$$\mathcal{B}^n(\psi) = 0. \tag{25}$$

We solve the system of nonlinear equations on the interface (25) by an inexact Newton method. Each Newton step s is computed by a forward difference GMRES iteration for solving $(\mathcal{B}^n)'(\psi)s = -\mathcal{B}^n(\psi)$. On each GMRES iteration the action of the Jacobian $(\mathcal{B}^n)'(\psi)$ on a vector μ is approximated by a forward difference which requires only one evaluation of the nonlinear operator $\mathcal{B}^n$. The evaluation of $\mathcal{B}^n$ involves solving subdomain problems (21)–(23) in parallel and two inexpensive projection steps - from the mortar grid onto the local subdomain grids and from the local grids onto the mortar grid. Since each block can be distributed among a number of processors, the subdomain solvers are parallel themselves. The subdomain problems are also nonlinear and are solved by a preconditioned Newton- Krylov solver [50, 51].

3. DISCONTINUOUS GALERKIN METHODS FOR REACTIVE TRANSPORT IN POROUS MEDIA

We consider the multiple component reactive transport system for a single or multiple flowing phase (and possibly multiple stationary phases) in porous media. The chemical system is comprised of N_S species. We allow for possible linear equilibrium type adsorptions, which are treated by using retardation factors. Except for these adsorptions, all species considered are either in a flowing phase (mobile species) or in a stationary phase (immobile species or minerals). If the same physical species occurs in both flowing and stationary phases, we treat it as two distinct species, one in each phase, in the modeling equations. In this case, the interphase mass transfer of the physical species is incorporated into the modeling system as a chemical reaction. Let N_M denote the number of immobile species and I_M the corresponding index set. Similarly, let N_F denote the number of mobile species and I_F the corresponding index set. Obviously, $N_{S=} N_M+N_F$, $I_M \cup I_F = \varnothing$ and $I_M \cup I_F = \{1, 2, \ldots, N_s\}$.

3.1. The Advection-Diffusion-Reaction Equations

The classical advection-diffusion-reaction equations governing the reactive transport process for a single flowing phase in porous media are as follows:

$$\frac{\partial(\phi_i c_i)}{\partial t} + \nabla\cdot(\mathrm{u}c_i - \mathrm{D}_i(\mathrm{u})\nabla c_i) = s_i + r_i, \quad in \quad \Omega\times(0,T], \quad i\in I_F, \tag{26}$$

$$\frac{\partial(\phi_i c_i)}{\partial t} = s_i + r_i, \quad in \quad \Omega\times(0,T], \quad i\in I_M, \tag{27}$$

where the unknown variables are c_i (i=1,…,N_S), the concentrations of all species (amount per volume). Here T is the final simulation time. ϕ_i is the effective porosity (the product of porosity and the retardation factor) for the species i and it is assumed to be time-independent and uniformly bounded from above and below. $\mathbf{D}_i(\mathbf{u})$ is the dispersion-diffusion tensor for the species i, and it is assumed to be uniformly symmetric positive definite and bounded from above. $r_i = r_i(c_1, c_2, \ldots, c_{Ns})$ is the reaction term for the species i which will be described in the sequel. In (26), $s_i = qc_i^* + s_{i,pure}$ is the source term. The imposed external total flow rate q is a sum of sources (injection) and sinks (extraction). c_i^* is the injected concentration $c_{i,inj}$ if $q > 0$ and is the resident concentration c_i if $q \le 0$. In (27), the source term s_i is simply the pure transport source $s_{i,pure}$.

To close the system of advection-diffusion-reaction, we consider the following boundary conditions for this problem:

$$(\mathrm{u}c_i - \mathrm{D}_i\nabla c_i)\cdot \mathrm{n}\partial\Omega = cB,i\mathrm{u}\cdot\mathrm{n}\partial\Omega, \quad \text{on } \Gamma_{\text{in}}\times(0,T], \tag{28}$$

$$-\mathrm{D}_i\nabla c_i\cdot\mathrm{n}\partial\Omega = 0 \qquad \text{on } \Gamma_{\text{out}}\times(0,T], \tag{29}$$

where $i\in I_F$ and $c_{B,i}$ is the inflow concentration for the species i. The initial concentrations are specified in the following way:

$$c_i(x,0) = c_{0,i}(x), \quad x \in \Omega, i \in I_M \cup I_F \tag{30}$$

Here the domain boundary $\partial\Omega = \overline{\Gamma}_{\text{in}} \cup \overline{\Gamma}_{\text{out}}$ contains the inflow boundary $\overline{\Gamma}_{\text{in}}$ and the outflow/noflow boundary $\overline{\Gamma}_{\text{out}}$, defined by:

$$\Gamma_{\text{in}} = \{x \in \partial\Omega: \mathbf{u} \cdot \mathbf{n}_{\partial\Omega} < 0\}$$
$$\Gamma_{\text{out}} = \{x \in \partial\Omega: \mathbf{u} \cdot \mathbf{n}_{\partial\Omega} \geq 0\}$$

and $n_{\partial\Omega}$ denotes the unit outward normal vector to $\partial\Omega$.

3.2. General Reaction Terms

Let N_R be the number of total reactions. The collection of N_R reactions may or may not form a linearly independent set. Denote by $\hat{r}_\beta$ $(\beta = 1,2,...,N_R)$ the extent of reaction (or reaction coordinate) for the reaction β. Let **V** be the associated stoichiometric matrix for the reaction system, *i.e.* $(V)_{ij}$ is the stoichiometric coefficient of the *i*-th species in the *j* -th reaction. Then, we have:

$$\text{r}=\text{V}\hat{\text{r}},$$

where

$$\text{r}=\left(r_1, r_2, \cdots, r_{N_S}\right)^T, \hat{\text{r}}=\left(\hat{r}_1, \hat{r}_2, \cdots \hat{r}_{N_R}\right)^T$$

To illustrate the above notations, we provide a simple example for the system (multiphase) consisting of stationary and flowing toluene (C_7H_8), denoted by symbols $C_7H_8(s)$ and $C_7H_8(f)$ respectively, oxygen (O_2), nitrogen (N_2), aerobic microbes denoted by M_l (living) and M_d (dead), complex organic (biodegraded) by-products denoted simply by symbols P_1 and P_1 and possibly, some non-reactive tracers. Also, let us denote the indices of the species: $C_7H_8(s)$, $C_7H_8(f)$, O_2, N_2, M_l, P_1, P_2, and M_d by the integers 1 through 8 respectively. The species undergo the following chemical reactions.

$$\begin{array}{ccccc} C_7H_8(f) & \rightleftharpoons & C_7H_8(s) & \hat{r}_1 \\ 10.87\ C_7H_8(f) + 62.5\ O_2 & \rightleftharpoons & 0.5\ M_l + P_1 & \hat{r}_2 \\ 10.87\ C_7H_8(f) + 92.86\ N_2 & \rightleftharpoons & 0.5\ M_l + P_2 & \hat{r}_3 \\ M_l & \rightleftharpoons & M_d & \hat{r}_4 \end{array}$$

We then have, N_R=4, N_S=8, N_F=7, N_M=1, I_F={2, …, 8}, I_M={1}, and r=Vr̂ is given by:

$$\text{r} = \begin{bmatrix} r_1 \\ r_2 \\ r_3 \\ r_4 \\ r_5 \\ r_6 \\ r_7 \\ r_8 \end{bmatrix} = \begin{bmatrix} 1 & 0 & 0 & 0 \\ -1 & -10.87 & -10.87 & 0 \\ 0 & -62.5 & 0 & 0 \\ 0 & 0 & -92.86 & 0 \\ 0 & 0.5 & 0.5 & -1 \\ 0 & 1 & 0 & 0 \\ 0 & 0 & 1 & 0 \\ 0 & 0 & 0 & 1 \end{bmatrix} \begin{bmatrix} \hat{r}_1 \\ \hat{r}_2 \\ \hat{r}_3 \\ \hat{r}_4 \end{bmatrix}$$

3.3. DG for Reactive Transport

We denote the upwind value of a concentration *c* as follows:

$$c^*|_\gamma := \begin{Bmatrix} c|_{E_i}, & \textbf{if } \mathbf{u} \cdot \mathbf{n}_\gamma \geq 0 \\ c|_{E_i}, & \textbf{if } \mathbf{u} \cdot \mathbf{n}_\gamma < 0 \end{Bmatrix}$$

For convenience of presentation, we write c $=(c_1, c_2,\ldots,c_{Ns})^T$, and $\mathrm{w}=\left(\omega_1,\omega_2,\ldots,\omega_{N_S}\right)^T$. We introduce the bilinear form B(c, w; u) as:

$$B(c,\mathrm{w};\mathrm{u}):=\sum_{i\in I_F}\left\{\sum_{E\in\varepsilon_h}\int_E\left(\mathrm{D}_i(\mathrm{u})\nabla c_i-c_i\mathrm{u}\right)\cdot\nabla\omega_i-\int_\Omega c_i q^-\omega_i-\sum_{\gamma\in\Gamma_h}\int_\gamma\{\mathrm{D}_i(\mathrm{u})\nabla c_i\cdot\mathrm{n}_\gamma\}[\omega_i]\right.$$

$$\left.-s_{\mathrm{form}}\sum_{\gamma\in\Gamma_h}\int_\gamma\{\mathrm{D}_i(\mathrm{u})\nabla\omega_i\cdot\mathrm{n}_\gamma\}[c_i]+\sum_{\gamma\in\Gamma_h}\int_\gamma c_i^*\mathrm{u}\cdot\mathrm{n}_\gamma[\omega_i]+\sum_{\gamma\in\Gamma_{h,\mathrm{out}}}\int_\gamma c_i\mathrm{u}\cdot\mathrm{n}_\gamma\omega_i+J_0^\sigma(c_i,\omega_i)\right\}$$

Here, c_{form}=-1, 0, or 1. The injection source term q^+ and the extraction part q^+ are defined as usual:

$$q^+ := \max(q,0), \qquad q^- := \min(q,0).$$

The interior penalty term $J_0^\sigma(c_i,\omega_i)$ is defined by:

$$J_0^\sigma(c_i,\omega_i):=\sum_{\gamma\in\Gamma_h}\frac{r^2\sigma_{\gamma,i}}{h_\gamma}\int_\gamma[c_i][\omega_i]$$

where σ is a discrete positive function that takes a constant value $\sigma_{\gamma,i}$ on the face γ for the species *i*. Let $I_{FM}=I_F\cup I_M$. Then, the linear functional L(w; u, c) is defined as:

$$L(\mathrm{w};\mathrm{u},c):=\sum_{i\in I_{FM}}\int_\Omega\left(s_{\mathrm{pure},i}+r_i(c)\right)\omega_i+\sum_{i\in I_F}\int_\Omega c_{\omega,\mathrm{inj}}q^+\omega_i-\sum_{i\in I_F}\sum_{\gamma\in\Gamma_{h,\mathrm{in}}}\int_\gamma c_{B,i}\mathrm{u}\cdot\mathrm{n}_\gamma\omega_i$$

The continuous-in-time DG approximation $\mathrm{C}^{DG}(\cdot,t)\in\left(\mathcal{D}_r(\varepsilon_h)\right)^{N_S}$ of (26)–(30) is defined by:

$$\sum_{i\in I_{FM}}\left(\frac{\partial\phi_i C_i^{DG}}{\partial t},\omega_i\right)+B\left(\mathrm{C}^{DG},\mathrm{w};\mathrm{u}\right)=L\left(\mathrm{w};\mathrm{u},\mathrm{C}^{DG}\right),\quad\forall w\in\left(\mathcal{D}_r(\varepsilon_h)\right)^{N_S},\quad\forall t\in(0,T] \tag{31}$$

$$\sum_{i\in I_{FM}}\left(\phi_i C_i^{DG},\omega_i\right)=\sum_{i\in I_{FM}}\left(\phi_i c_{0,i},\omega_i\right),\quad\forall w\in\left(\mathcal{D}_r(\varepsilon_h)\right)^{N_S},\quad t=0 \tag{32}$$

We note that OBB-DG has s_{form}=-1 and $\sigma_\gamma=0$ in the bilinear form; NIPG has s_{form}=-1 and $\sigma_\gamma\geq 0$; SIPG has s_{form}=1 and $\sigma_\gamma>0$; and IIPG has s_{form}=0 and $\sigma_\gamma>0$.

We emphasize the important feature of local mass conservation that holds for the four primal DG schemes: The DG approximation of a concentration satisfies on each element E the following local mass balance property:

$$\int_E\frac{\partial\phi_i C_i^{DG}}{\partial t}-\int_{\partial E\backslash\partial\Omega}\{\mathrm{D}_i(u)\nabla C_i^{DG}\cdot\mathrm{n}_{\partial E}\}+\int_{\partial E\cap\Gamma_{h,\mathrm{in}}}c_{B,i}\mathrm{u}\cdot\mathrm{n}_{\partial E}+\int_{\partial E\cap\Gamma_h}C_i^{DG,*}\mathrm{u}\cdot\mathrm{n}_{\partial E}$$

$$+\int_{\partial E\cap\Gamma_{h,\mathrm{out}}}C_i^{DG}\mathrm{u}\cdot\mathrm{n}_{\partial E}+\sum_{\gamma\in\partial E\backslash\partial\Omega}\frac{r^2\sigma_{\gamma,i}}{h_\gamma}\int_\gamma\left(C_i^{DG}\big|_E-C_i^{DG}\big|_{\Omega\backslash\bar{E}}\right)$$

$$=\int_E\left(C_i^{DG}q^-+c_{\omega,\mathrm{inj}}q^+\right)+\int_E\left(s_{\mathrm{pure},i}+r_i\left(\mathcal{M}\left(\mathrm{C}^{DG}\right)\right)\right)$$

for each $i \in I_F$; and

$$\int_E \frac{\partial \phi_i C_i^{DG}}{\partial t} = \int_\Omega \left(s_{\text{pure},i} + r_i \left(\mathcal{M}\left(C^{DG} \right) \right) \right)$$

for each $i \in I_M$.

3.4. DG for A Special Case

In this subsection, we assume that no immobile species presents in the system, and that the effective porosities and the dispersion-diffusion tensors are identical for all species. Under these assumptions, the system of multicomponent transport equations can be significantly simplified by using chemical stoichiometry. The transport equations now become:

$$\frac{\partial (\phi_i c_i)}{\partial t} + \nabla \cdot \left(\mathrm{u} c_i - \mathrm{D(u)} \nabla c_i \right) = s_i + r_i, i \in I_F$$

Writing $\mathrm{c} = (c_1, c_2, \ldots, c_{Ns})^T$, $\mathrm{s} = (s_1, s_2, \ldots, s_{Ns})^T$, and using the standard notation of ∇ applied to c, we have

$$\frac{\partial (\phi \mathrm{c})}{\partial t} + \nabla \cdot \left(\mathrm{uc} - \mathrm{D(u)} \nabla \mathrm{c} \right) = \mathrm{s} + \mathrm{r} \tag{33}$$

It is always possible to perform a Gauss-Jordan reduction on the stoichiometric matrix **V** to obtain a reduced form $\mathbf{V}_r$, *i.e.*

$$\mathbf{V}_r = \begin{pmatrix} \mathbf{I}_{N_I \times N_I} & \hat{\mathbf{V}} \\ 0 & 0 \end{pmatrix}$$

and

$$\mathbf{V} = \mathbf{P}^{-1} \mathbf{V}_r,$$

where **P** is a non-singular $N_S \times N_S$ matrix, $\hat{\mathrm{V}}$ is an $N_I \times (N_R - N_I)$ matrix, and N_I is the number of independent reactions. Left-multiplying the matrix **P** on both sides of (33), we have,

$$\frac{\partial \phi \mathbf{Pc}}{\partial t} + \nabla \cdot (\mathbf{u}(\mathbf{Pc}) - \mathbf{D}(\mathbf{u}) \nabla (\mathbf{Pc})) = \mathbf{Ps} + \mathbf{Pr}.$$

Let us denote $\mathrm{c}^r = \mathrm{Pc}$, $\mathrm{s}^r = \mathrm{Ps}$. We note that:

$$\mathbf{Pr} = \mathbf{PV}\hat{\mathbf{r}} = \mathbf{V}_r \hat{\mathbf{r}} = \begin{pmatrix} \mathbf{I}_{N_I \times N_I} & \hat{\mathbf{V}} \\ 0 & 0 \end{pmatrix} \hat{\mathbf{r}}.$$

Consequently, we have the following reduced form of the reactive transport equations.

$$\begin{aligned} &\frac{\partial \phi \mathbf{c}^r}{\partial t} + \nabla \cdot (\mathbf{u}\mathbf{c}^r - \mathbf{D}(\mathbf{u}) \nabla \mathbf{c}^r) \\ &= \mathbf{s}^r + \begin{pmatrix} \mathbf{I}_{N_I \times N_I} & \hat{\mathbf{V}} \\ 0 & 0 \end{pmatrix} \hat{\mathbf{r}}, \end{aligned} \tag{34}$$

A DG scheme to the reduced equation (34) is an efficient algorithm for the multicomponent reactive transport system especially for the case where the number of species is much greater than the number of independent kinetic reactions. We note that in the reduced system only N_I equations are coupled due to reactions, the remaining (N_S-N_I) equations do not contain chemistry and can be solved independently. On the other hand, DG applied to the original equations needs a solution of N_S coupled equations.

4. GODUNOV-MIXED METHODS FOR REACTIVE TRANSPORT

Consider the reactive transport problem described by:

$$\frac{\partial(\phi c_{i\alpha} S_\alpha)}{\partial t}+\nabla\cdot(c_{i\alpha}u_\alpha-\phi S_\alpha \mathrm{D}_{i\alpha}\nabla c_{i\alpha})=r(c_{i\alpha}) \quad \textbf{(35)}$$

$$\mathrm{D}_{i\alpha}\nabla c_{i\alpha}\cdot \mathrm{n}=0 \quad \textbf{(36)}$$

where $\mathrm{D}_{i\alpha}=\mathrm{D}_{i\alpha}^{\mathrm{diff}}+\mathrm{D}_{i\alpha}^{\mathrm{hyd}}$ is the sum of *molecular diffusion* and *hydrodynamic dispersion*, $\mathrm{D}_{i\alpha}^{\mathrm{diff}}=\tau_\alpha d_{m,i\alpha}\mathrm{I}$,

$$\phi S_\alpha \mathrm{D}_{i\alpha}^{\mathrm{hyd}}=d_{t,\alpha}|\mathrm{u}_\alpha|\mathrm{I}+(\mathrm{d}_{1,\alpha}-\mathrm{d}_{t,\alpha})/|\mathrm{u}_\alpha|\mathrm{u}_\alpha\mathrm{u}_\alpha^{\mathrm{T}}.$$

Here τ_α is the "tortuousity" of flow of phase α, $d_{m,i\alpha}$, $\mathrm{d}_{1,\alpha}$, $\mathrm{d}_{t,\alpha}$ are the *molecular diffusion*, *longitudinal*, and *transverse dispersion coefficients*, respectively. The source takes the general form, $r(c_{i\alpha})=r_{i\alpha}^I+\phi S_\alpha r_{i\alpha}^C+q_{i\alpha}$ where the terms $r_{i\alpha}^I$ and $r_{i\alpha}^C$ model the influx (or efflux) from other phases and the chemical rate of decay (or formation) of species *i* in phase α, respectively. The term $q_{i\alpha}$ models a source (or sink) for species *i* in phase α. Further, note that the net interchange of species between phases is zero; *i.e.*, $\sum_\alpha r_{i\alpha}+r_{iR}=0$. Here, r_{iR} is the influx (efflux) of species *i* into the stationary phases (for *e.g.*, the rock matrix). For simplicity, assume there is no adsorption, hence $r_{iR}\equiv 0$.

4.1. An Operator-Split (Time-Split) Scheme

We first present a "phase-summed" formulation of (35)–(36). An equilibrium partitioning of the species among the phases is assumed, given by constants $\theta_{i\alpha}$ so that $c_{i\alpha}=\theta_{i\alpha}c_{i\alpha_0}$, where α_0 is a reference phase, say, the water phase. Then summing the equations (35) and (36) over α, for a given species *i*, reduces it to,

$$\frac{\partial(\phi_i^* c_{i\omega})}{\partial t}+\nabla\cdot(c_{i\omega}\mathrm{u}_i^*-\mathrm{D}_i^*\nabla c_{i\omega})=r^*(c_{i\omega}) \quad \textbf{(37)}$$

$$\mathrm{D}_{i\omega}\nabla c_{i\omega}\cdot\mathrm{n}=0 \quad \textbf{(38)}$$

Note that in (37) the reference phase is chosen to be the water phase. The phase-summed asterisked (*) terms are defined as follows:

$$\phi_i^*=\phi\sum_\alpha \theta_{i\alpha}S_\alpha,$$

$$u_i^*=\phi\sum_\alpha \theta_{i\alpha}\mathrm{u}_\alpha,$$

$$\mathrm{D}_i^*=\phi\sum_\alpha S_\alpha\theta_{i\alpha}\mathrm{D}_{i\alpha},$$

$$r^*(c_{i\omega})=\phi\sum_\alpha r_{i\alpha}^C+r_{iR}S_\alpha+\sum_\alpha q_{i\alpha}.$$

Assume that at time $t=\tau_m$, the concentrations of all species are known. Assume also that $(\tau_m,\tau_{m+1})\subset(t_n,t_{n+1})$ and that the values of u_i^* and ϕ_i^* are known at the old and new flow time-steps, *i.e.*, t_n and t_{n+1}. A direct discretization of the equation (37) yields,

$$\frac{T_i^{m+1}-T_i^m}{\Delta\tau^{m+1}}+\nabla\cdot\left(c_{i\omega}^m\mathrm{u}_i^{*,m+1/2}-\mathrm{D}_i^{*,m}\nabla c_{i\omega}^{m+1}\right)=r^*\left(c_{i\omega}^{m+1/2}\right) \quad \textbf{(39)}$$

Here, $\Delta\tau^{m+1}=\tau_{m+1}-\tau_m$ and $T_i=\phi_i^*c_{i\omega}$. Note also that ϕ_i^* and u_i^* are evaluated at time $t\in(\tau_m,\tau_{m+1})$ by linear interpolation between the known values at t_n and t_n. Direct solution of equation (39) is impractical. Hence, a time-split algorithm is employed in which the advection, chemical reaction and diffusion-dispersion components are solved "independently" of each other. Each component delivers intermediate values for T_i, labeled $\overline{T}_i$, $\hat{T}_i$ and T_i^{m+1}. The individual steps of this algorithm are described below.

4.2. Advection: Godunov Methods

The basic idea of the Godunov method [52, 53] is to construct a numerical flux approximation on the boundary of each element. This consists of calculating left and right states, and solving a numerical approximation to a 1-D Riemann problem (normal to the boundary of elements). Given left and right states c^L and c^R, the Riemann solution is determined by the Godunov flux [54]. For a given flux function $\omega(c)$, the Godunov flux $H_\omega(c^L,c^R)$ is given by:

$$\boldsymbol{H_\omega(c^L,c^R)}=\begin{cases}\mathbf{min}_{c^L\le c\le c^R}\ \boldsymbol{\omega(c)}\ \ \mathbf{if}\ \boldsymbol{c^L\le c^R}\\ \mathbf{max}_{c^R\le c\le c^L}\ \boldsymbol{\omega(c)}\ \ \mathbf{otherwise}\end{cases} \quad \textbf{(40)}$$

It can also be shown that H_ω is Lipschitz in its arguments if ω is Lipschitz in *c*, and H_ω is consistent, that is, $H_\omega(c,c)=\omega(c)$. In the case of a nondecreasing (resp. nonincreasing) functions $\omega(c)$, the Godunov monotone flux scheme reduces to $H_\omega(c^L,c^R)=\omega(c^L)$ (resp. $\omega(c^L)$). Then, in the case of a non-decreasing function $\omega(c)$, the advection step reduces to:

$$\left(\frac{\overline{T}_i-T_i^m}{\Delta\tau^{m+1}},\omega\right)_{\Omega_j}+\sum_{E\in\Gamma_{h,j}}\left\langle c_{h,i\omega}^{m,\mathrm{upw}}\mathrm{u}_{h,i}^{*,m+1/2}\cdot\mathrm{n}_E,\omega\right\rangle_{\partial E}=\left(\sum_\alpha q_{i\alpha}^{m+1/2},\omega\right)_{\Omega_j},\omega\in W_{h,j} \quad \textbf{(41)}$$

where $c_{h,i\omega}^{m,\mathrm{upw}}$ is the "upwind" value of the species concentration in the reference phase. The scheme is sometimes also called "upstream finite difference scheme". The updated concentration is then calculated from $\overline{c}_{h,i\omega}=\dfrac{\overline{T}_i}{\phi^{*,m+1}}$.

But the standard Godunov scheme while consistent, conservative and stable, can be very diffusive on coarser grids. To address this problem, higher-order Godunov schemes can be applied. A classical way to do this is to introduce slopes s_i for the scalar variable using the "Monotone Upwind Scheme for Conservative Laws (MUSCL)" due to Van Leer [55]. Then the left and right states may be calculated from a Taylor expansion about the element centers in each direction and using the element slopes, as presented in [52]. The slopes may have to be limited in order to prevent overshoot and undershoot of the solution. This is achieved by solving a maximization problem for the flux limiter $\gamma\in[0,1]$ subject to the constraint that a slope of γs_i renders the left and right states of a face shared by elements E_i and E_j to lie within the interval $\left[c|_{E_i},c|_{E_j}\right]$. There are many element slope limiter calculation methods in the literature [53]. Some of these have been incorporated into IPARS.

4.3. Chemical Reaction: Higher Order ODE Integration

After the advection step is completed, we solve the chemical reaction component of (39) given by:

$$\frac{\partial T_i}{\partial t} = \phi \sum_{\alpha} r_{i\alpha}^C \tag{42}$$

with an initial condition given by the concentration obtained from the advections step, viz., $T_i(0)=\overline{T}_i$. Explicit ODE integration can be used to solve (42), even exactly in some cases (depending on the right hand side). Approximations can be obtained by numerical integration; for *e.g.*, with an explicit second-order Runge-Kutta scheme, the combined reaction ODE system corresponding to (42),

$$\frac{\mathrm{dT}}{\mathrm{d}t} = \mathrm{r}\left(\mathrm{T}\right). \tag{43}$$

yields an update to the concentrations of species i=1, …, N_C, (N_C being the number of species entering into chemical reactions) from the value **T** before the chemistry to $\hat{\mathrm{T}}$ after the chemistry step, given by:

$$\begin{aligned} \mathbf{k}_{1,i} &= \Delta\tau^l \mathbf{r}_i(\mathbf{T}) \\ \mathbf{k}_{2,i} &= \Delta\tau^l \mathbf{r}_i\left(\mathbf{T} + \tfrac{1}{2}\mathbf{k}_1\right) \\ \hat{\mathbf{T}} &= \mathbf{T} + \mathbf{k}_2 \end{aligned} \tag{44}$$

Then, the new values of the concentration after the chemistry step are given by $\hat{c}_{h,i\omega} = \frac{\hat{\mathrm{T}}_i}{\phi^{*,\mathrm{m}+1}}$.

4.4. Diffusion-Dispersion: Backward Euler Mixed FEM

Next, we solve the diffusion-dispersion equation. This takes the form,

$$\frac{\partial(\phi_i^* c_{i\omega})}{\partial t} - \nabla \cdot \mathbf{D}_i^* \nabla c_{i\omega} = 0 \tag{45}$$

with initial conditions $c_{i\omega}(0)=\hat{c}_{h,i\omega}$ following the chemistry step. This is solved using a backward Euler in time discretization and an expanded MFEM with a full diffusion tensor D_i^* as discussed in Section 2. In the discretized weak form of (45), introducing $\tilde{\mathrm{z}} = -\nabla c$ and $\mathrm{z}=\mathrm{D}_i^*\tilde{\mathrm{z}}$, an ExpMFEM seeks $\tilde{\mathrm{z}}_{h,i\omega}^{m+1}\big|_{\Omega_j} \in \tilde{\mathrm{V}}_{h,j}$, $\mathrm{z}_{h,i\omega}^{m+1}\big|_{\Omega_j} \in \mathrm{V}_{h,j}$, $c_{h,i\omega}^{m+1}\big|_{\Omega_j} \in \mathrm{W}_{h,j}$, such that, for $1 \le j \le n_b$,

$$\left(\frac{\phi_i^{*,m+1} c_{h,i\omega}^{m+1} \; \phi_i^{*,m} c_{h,i\omega}^{m}}{\Delta\tau^{m+1}}, \omega\right)_{\Omega_j} + \left(\nabla \cdot \mathbf{z}_{h,i\omega}^{m+1}, \omega\right)_{\Omega_j} = 0, \forall \omega \in W_{h,j} \tag{46}$$

$$\left(\tilde{\mathbf{z}}_{h,i\omega}^{m+1}, \mathbf{v}\right)_{\Omega_j} = \left(c_{h,i\omega}^{m+1}, \nabla \cdot \mathbf{v}\right)_{\Omega_j} - \langle \mathcal{P}_j c_{h,i\omega}^{m}, \mathbf{v} \cdot \mathbf{n}_j \rangle_{\Gamma_j}, \forall \mathbf{v} \in \mathbf{V}_{h,j} \tag{47}$$

$$\left\langle \phi - \mathcal{P}_j \phi, \mathrm{v} \cdot \mathrm{n}_j \right\rangle_{\Gamma_{k,j}} = 0, \quad \forall \mathrm{v} \in \mathrm{V}_{h,i}, \quad \forall k \text{ such that } \Omega_k \cap \Omega_j \ne \varnothing \tag{48}$$

Here, $\mathcal{P}_j : L^2(\Gamma_j) \to L^2(\Gamma_k)$ is an L^2-orthogonal projection satisfying $\forall \phi \in L^2(\Gamma_j)$

$$\langle \boldsymbol{\phi} - \mathcal{P}_j\boldsymbol{\phi}, \mathbf{v}\cdot\mathbf{n}_j\rangle_{\Gamma_{k,j}} = 0, \qquad \forall \mathbf{v} \in \mathbf{V}_{h,i}, \ \forall k \text{ such that } \bar{\Omega}_k \cap \bar{\Omega}_j \neq \emptyset$$

Following [48], suitable quadrature rules can be defined to approximate the integrals appearing in (46)–(48), thereby eliminating $\tilde{z}_{h,i\omega}$ and $z_{h,i\omega}$ in terms of $c_{h,i\omega}$.

4.5. Iterative Coupling for Flow and Reactive Transport

To conclude this section, we remark that the optimal implementation of a multiphase flow coupled to reactive transport models is obtained using an iterative coupling technique. The underlying concept of iterative coupling as applied to the flow problem can be found in [56, 57] wherein the fully coupled system is split into pressure and saturation equations (or pressure and concentration equations in the case of compositional flow) which are then iteratively solved using most recent updates until convergence. It may be regarded as a Picard method of successive approximation for the coupled system of PDEs. For the fully coupled flow-transport problem, the successive flow and transport steps are repeated iteratively, each step using the most recent update to the solution from the other until some global convergence criterion (*e.g.*, a residual which is a function of the solutions from the flow and transport steps, such as the mass balance of phase and species) is satisfied or a prescribed maximum number of iterations is exceeded.

5. CURRENT DEVELOPMENT – MESH ADAPTATION

5.1. A Priori Mesh Adaptation

For simplicity, let us restrict ourselves to single-species transport in this section. Let c be the exact solution of the transport equation and C^{DG} be its approximation from SIPG, NIPG or IIPG. Applying a similar argument as the one used in [26], we can show that, under a certain regularity assumptions on the solution and on the mesh, there exists a constant, K, independent of mesh:

$$\begin{aligned} &\left\|C^{DG} - c\right\|_{L^\infty(J;L^2)} + \left\|\nabla\left(C^{DG} - c\right)\right\|_{L^2(J;L^2)} \\ &\leq K \sum_{E\in\mathcal{E}_h} \frac{h_E^{\mu-1}}{r_E^{s-1-\delta}}\left\|c\right\|_{L^2(J;H^s(E))} + \frac{h_E^{\mu-1}}{r_E^{s-1}}\left\|\frac{\partial c}{\partial t}\right\|_{L^2(J;H^{s-1}(E))} + \frac{h_E^{\mu-1}}{r_E^{s-1}}\left\|c_0\right\|_{H^{s-1}(E)} \end{aligned} \tag{49}$$

Here, $r_E \geq 1$ is the local approximation degree of polynomials in element E; $s_E \geq 2$ is the local regularity of the solution in element E; $\mu = \min(r_E + 1, s_E)$; and $\delta = 0$ for conforming meshes with triangles or tetrahedra, and $\delta = \frac{1}{2}$ in general. A similar error estimate holds for an OBB-DG solution C^{DG}, optimal in mesh sizes but with a larger loss of optimality in polynomial degrees.

For SIPG, one can apply the parabolic lift technique [26] to show that, under certain regularity assumptions on the solution and on the mesh, there exists a constant, K, independent of mesh sizes and approximation degrees, such that:

$$\left\|C^{DG} - c\right\|_{L^2(J;L^2)} \leq K \sum_{E\in\mathcal{E}_h} \frac{h_E^{\mu}}{r_E^{s-1-\delta}}\left\|c\right\|_{L^2(J;H^s(E))} + \frac{h_E^{\mu}}{r_E^{s-\delta}}\left\|\frac{\partial c}{\partial t}\right\|_{L^2(J;H^{s-1}(E))} + \frac{h_E^{\mu}}{r_E^{s-\frac{1}{2}}}\left\|c_0\right\|_{H^{s-1}(E)} \tag{50}$$

Here, again, $r_E \geq 1$ is the local approximation degree of polynomials in element E; $s_E \geq 2$ is the local regularity of the solution in element E; $\mu = \min(r_E + 1, s_E)$; and $\delta = 0$ for conforming meshes with triangles or tetrahedra, and $\delta = \frac{1}{2}$ in general.

We remark that the error estimation of continuous finite element methods usually requires a global regularity of the solutions and their error estimates are dependent on certain global norms of the solutions. On the contrary, the error estimation of DG allows the solution singularities on element interfaces. In addition, the DG errors can be estimated in terms of element-wise norms of the solutions as shown in (49) and (50).

The dependence of errors on the local regularity of a DG solution can be utilized to design *a priori* adaptation of mesh and approximation degrees. Sometimes it is possible to analyze the regularity of the solution by PDE theories or by insights from physical processes. If we know *a priori* the local regularity of the solution and a rough estimate on the local norms of the solution, we can choose an adaptive distribution of mesh sizes and polynomial degrees such that the right-hand side of (49) or (50) is minimized assuming the constant K is a uniform across the entire domain. However, *a priori* adaptivity based on the solution properties is usually less effective than the *a posteriori* adaptivity due to inaccurate knowledge on the exact solution.

5.2. Physics-Based Mesh Adaptation

Mesh adaptation can be achieved using heuristic criteria based on physics. There are four sub-processes, *i.e.* convection, diffusion/dispersion, reaction, and injection/ extraction (or source-sink), occur in our modeling system. We can define the following indices to represent the size of each process in a given element $E \in \varepsilon_h$:

$$Q_{\text{conv},E}(c) := \sum_{\gamma \subset \partial E} \|\mathbf{u}c\|_{L^2(\gamma)}$$

$$Q_{\text{diff},E}(c) := \sum_{\gamma \subset \partial E} \|\mathbf{D}(\mathbf{u})\nabla c\|_{L^2(\gamma)}$$

$$Q_{\text{rxn},E}(c) := \|r(c)\|_{L^2(E)}$$

$$Q_{\text{s},E}(c) := \|qc^*\|_{L^2(E)}$$

where $Q_{\text{conv},E(c)}$, $Q_{\text{diff},E(c)}$, $Q_{\text{rxn},E(c)}$ and $Q_{\text{s},E(c)}$ are the magnitudes of local convection, diffusion/dispersion, reaction, and injection/ extraction, respectively, occurring in element E. Obviously, these four quantities are functions of the mesh sizes as they intend to decrease as the mesh size decreases. If we have a rough *a priori* estimate of the exact solution in terms of $Q_{\text{conc},E(c)}$, $Q_{\text{diff},E(c)}$, $Q_{\text{rxn},E(c)}$ and $Q_{\text{s},E(c)}$, we can design an *a priori* adaptive mesh such that:

$$Q_E(c) := \omega_{\text{conv}} Q_{\text{conc},E}(c) + \omega_{\text{diff}} Q_{\text{diff},E}(c) + \omega_{\text{rxn}} Q_{\text{rxn},E}(c) + \omega_{\text{s}} Q_{\text{s},E}(c)$$

is balanced among all elements $E \in \varepsilon_h$. Here, ω_{conc}, ω_{diff}, ω_{rxn} and ω_{s} are the weight coefficients for each sub-process, which are constant across the entire domain. Depending on the relative importance of sub-processes in the system, we can put different weight for different sub-process. Without specific preference, we can also simply let $\omega_{\text{conv}}=\omega_{\text{diff}}=\omega_{\text{rxn}}=\omega_{\text{s}}=1$.

If a rough *a priori* estimate of the exact solution in terms of $Q_E(c)$ is not available, we can approximate this quantity by $Q_E\left(C^{DG}\right)$. Mesh redistribution making $Q_E\left(C^{DG}\right)$ approximately equal size among all elements yields *a posteriori* mesh adaptation based on physics.

Physics-based adaptation is conceptually intuitive and computationally simple and it is useful especially for the simulations of complex systems where mathematically sound adaptive approach is not available. However, it is heuristic-based and is not mathematically rigorous, and hence is less reliable. In particular, it might lose effectively in special scenarios. For example, if the concentration solution is a constant with a large value in an area of the domain. The above adaptation scheme will yield refined mesh in the domain, which is not necessary.

5.3. Residual-Based Explicit A Posteriori Error Estimators

A posteriori error estimators or indicators can be used to signify where modifications in discretization parameters need to be made, thus achieving adaptivity. We first introduce two computationally efficient explicit estimators. These estimators are computed in terms of residual quantities that only depend on the approximate solution and the data. The residuals consist of the interior residual R_I, the zeroth order boundary residual R_{B0} given by:

$$\boldsymbol{R_I := qC^{DG*} + r(C^{DG}) - \phi\frac{\partial C^{DG}}{\partial t} - \nabla\cdot(C^{DG}\mathbf{u} - \mathrm{D}(\mathbf{u})\nabla C^{DG})} \quad (51)$$

$$R_{B0} := \begin{cases} \left[C^{DG}\right], & \text{on } \Gamma_h, \\ 0, & \text{on } \partial\Omega, \end{cases} \quad (52)$$

and the first order boundary residual R_{B1}, defined as [58, 59].

$$R_{B1} := \begin{cases} \left\lfloor \mathrm{D}(\mathrm{u})\nabla C^{DG}\cdot \mathrm{n}_\gamma \right\rfloor, & \text{on } \Gamma_h \\ \left((c_B - C^{DG})\mathrm{u} + \mathrm{D}(\mathrm{u})\nabla C^{DG}\right)\cdot \mathrm{n}_\gamma, & \text{on } \Gamma_{h,\text{in}} \\ \mathrm{D}(\mathrm{u})\nabla C^{DG}\cdot \mathrm{n}_\gamma, & \text{on } \Gamma_{h,\text{out}} \end{cases}$$

where, in element interior, $C^{DG*}=C^{DG}$ if $q<0$ and $C^{DG*}=c_\omega$ if $q\geq 0$.

Residual-based explicit estimator in $L^2(L^2)$.

The estimator η_E is defined by:

$$\begin{aligned} \eta_E^2 &= \frac{h_E^4}{r^4}\|R_I\|^2_{L^2(J;L^2(E))} \\ &+ \frac{1}{2}\sum_{\gamma\in\partial E\backslash\partial\Omega}\left(\frac{h_\gamma}{r} + \delta r h_\gamma\right)\|R_{B0}\|^2_{L^2(J;L^2(\gamma))} \\ &+ \frac{1}{2}\sum_{\gamma\in\partial E\backslash\partial\Omega}\frac{h_\gamma^3}{r^3}\|R_{B1}\|^2_{L^2(J;L^2(\gamma))} \\ &+ \sum_{\gamma\in\partial E\cap\partial\Omega}\frac{h_\gamma^3}{r^3}\|R_{B1}\|^2_{L^2(J;L^2(\gamma))}, \end{aligned}$$

where $\delta=0$ in the case of conforming meshes with triangles or tetrahedra. In general cases, $\delta=1$. It has been shown [59] that, under certain conditions, there exists a constant K, independent of h and r, such that the error in the $L^2(L^2)$ norm for SIPG can be estimated by η_E in the following sense:

$$\left\|C^{DG} - c\right\|_{L^2(0,T;L^2)} \leq K\left(\sum_{E\in\varepsilon_h}\eta_E^2\right)^{1/2} \quad (53)$$

We remark that the *a posteriori* error estimator in the $L^2(L^2)$ norm has been shown to be effective, particularly for the problems concerning concentration itself rather than transport flux.

Residual-based explicit estimator in $L^2(H^1)$

One drawback of the residual-based explicit estimator in $L^2(L^2)$ is that it only applies to SIPG. On the contrary, the optimal *a posteriori* error estimator in the energy norm is applicable to all primal DG methods, *i.e.* OBB-DG, NIPG, SIPG and IIPG. Letting

$\partial\Omega_E = \partial\Omega \cap \partial E$ and $\partial\hat{\Omega}_E = \partial E \setminus \partial\Omega$, this estimator χ_E is defined by:

$$\chi_E^2 = h_E^2 \|R_I\|_{L^2(J;L^2(E))}^2 + \sum_{\gamma\in\partial\Omega_E} h_\gamma \|R_{B1}\|_{L^2(J;L^2(\gamma))}^2$$
$$+\frac{1}{2}\sum_{\gamma\in\partial\hat{\Omega}_E}\Big(h_\gamma \|R_{B1}\|_{L^2(J;L^2(\gamma))}^2 + \frac{1}{h_\gamma}\|R_{B0}\|_{L^2(J;L^2(\gamma))}^2$$
$$+h_\gamma \|R_{B0}\|_{L^\infty(J;L^2(\gamma))}^2 + h_\gamma \|\partial R_{B0}/\partial t\|_{L^2(J;L^2(\gamma))}^2\Big).$$

It has been proved [58] that, under certain conditions, there exists a constant *K*, independent of *h*, such that

$$\|C^{DG} - c\|_{L\infty(0,T;L^2)} + |||\nabla(C^{DG} - c)|||_{L^2(0,T;L^2)} \le\le K\left(\sum_{E\in\varepsilon_h} x_E^2\right)^{1/2} \quad \textbf{(54)}$$

We note that the *a posteriori* error estimator in the energy norm is flexible. In particular, general boundary conditions may easily be taken into consideration for this error estimator. Both estimators in the $L^2(L^2)$ and the energy norms are explicit and residual-based, thus computationally efficient. In addition, no saturation or regularity assumption for the dual problem is required for these estimators.

5.4. A Posteriori Error Estimators Using Hierarchic Bases

Residual-based explicit error estimators are efficient to compute and may be used to indicate the subset of elements that need to be refined or coarsened to guide adaptivity. However, these residual-based estimators yield only one piece of information for each element, so they do not provide guidance on anisotropic refinements. In this section we present two error estimators using hierarchic bases. These hierarchic estimators provide point-wise information on the error and can be used to guide fully anisotropic *hp*-adaptation. In addition, unlike residual-based estimators, the hierarchic estimators do not contain a generic constant, and thus can also be used as a stopping criterion.

Brute-force hierarchic error estimator,

We first consider the error estimator computed from the difference between two approximations on discretizations of different accuracy [60-63]. The approach applies to many classes of problems and is simple to implement. However, it can be computational expensive since it requires a solution on a finer mesh.

For a given mesh ε_h, we construct the mesh $\varepsilon_{h/2}$ by isotropically refining each element in ε_h. We denote by C^{DG} the DG solution in the original space $\mathcal{D}_r(\varepsilon_h)$. We denote by $r'(r \le r' \le r + C)$ the improved approximation degree and by $h'(h/C \le h' \le h)$ the refined mesh size. Often we use $r' = r+1$ and/or $h' = h/2$. We now define C^{DG} as the DG solution in the fine space $\mathcal{D}_{r'}(\varepsilon_{h'})$. The brute-force hierarchic error estimator is defined as:

$$\varsigma := C^{DG} - C^{DG,F}. \tag{55}$$

If we make the saturation assumption of $\left\|C^{DG,F} - c\right\|_X \leq \beta_X \left\|C^{DG} - c\right\|_X$, $0 \leq \beta_X < 1$ where X could be the $L^\infty(L^2)$, $L^2(L^2)$ or $L^2(H^1)$ norm, then the hierarchic error estimator ς is close to the true error in the following sense:

$$\frac{\|\varsigma\|_X}{1+\beta_X} \leq \left\|C^{DG} - c\right\|_X \leq \frac{\|\varsigma\|_X}{1-\beta_X}. \tag{56}$$

We note that the saturation assumption is often satisfied for DG.

Numerical experiments [64] have indicated that this hierarchic error estimator ς is very effective, in particular for guiding anisotropic mesh adaptation [64]. However, the expensive computational cost of ς motivates us to consider an estimator that is close to ς but only involves solutions of local problems.

Error estimator based on solving local problems,

We now define a computationally efficient error estimator ς to approximate ς. For each $E \in \varepsilon_h$, we use the local space $\mathcal{D}_{r'}(\varepsilon_{h'} \cap E)$ obtained by restricting $\mathcal{D}_{r'}(\varepsilon_{h'})$ to E. We now define $\zeta_E \in \mathcal{D}_{r'}(\varepsilon_{h'} \cap E)$ as the solution of the following local problem for $\omega_E^F \in \mathcal{D}_{r'}(\varepsilon_{h'} \cap E)$,

$$\left(\frac{\partial \phi \xi_E}{\partial t}, \omega_E^F\right) + B(\xi_E, \omega_E^F; \mathbf{u}) = \left(\frac{\partial \phi C^{DG}}{\partial t}, \omega_E^F\right) + B\left(C^{DG}, \omega_E^F; \mathbf{u}\right) - L\left(\omega_E^F; \mathbf{u}, C^{DG} - \xi_E\right) \tag{57}$$

$$\left(\phi \xi_E - \phi C^{DG}, \omega_E^F\right)(0) = -\left(\phi c_0, \omega_E^F\right) \tag{58}$$

The function ζ_E involves only local computation in the element E. The estimator ζ_E for all elements $E \in \varepsilon_h$ may be computed independently of each other and in parallel. Moreover, the function $\zeta := \sum_{E\in\varepsilon_h} \zeta_E$ is a good approximate to ς in the following senses. Let $\hat{\Omega}_E = \Omega \backslash E$ and $E_{\varepsilon'_h} = \varepsilon_{h'} \cap E$ for brevity. Then, it has been shown in [65] that under certain conditions ζ is a lower bound estimate of ς:

$$\begin{aligned}
&\left\|\sqrt{\phi}\xi\right\|_{L\infty(J;L^2(E))} + \left\|D^{1/2}\nabla\xi\right\|_{L^2\left(J;L^2\left(E_{\varepsilon'_h}\right)\right)} \leq \left\|\sqrt{\phi}\varsigma\right\|_{L\infty(J;L^2(E))} + \left\|D^{1/2}\nabla\varsigma\right\|_{L^2\left(J;L^2\left(E_{\varepsilon'_h}\right)\right)} \\
&+K\frac{r_E}{h_E^{1/2}}\left\|\varsigma\big|_{\Omega\backslash E}\right\|_{L^2(J;L^2(\partial\hat{\Omega}_E))} + K\left(\frac{h_E^{1/2}}{r_E} + \delta_{\mathrm{OBB}}\frac{r_E}{h_E^{1/2}}\right)\left\|\nabla\varsigma\big|_{\hat{\Omega}_E}\cdot \mathrm{n}_{\partial E}\right\|_{L^2(J;L^2(\partial\hat{\Omega}_E))}, \quad \forall E \in \varepsilon_h
\end{aligned} \tag{59}$$

and

$$\begin{aligned}
&\left\|\sqrt{\phi}\xi\right\|_{L\infty(J;L^2(\Omega))} + \left\|D^{1/2}\nabla\xi\right\|_{L^2\left(J;L^2\left(E_{\varepsilon'_h}\right)\right)} \leq \left\|\sqrt{\phi}\varsigma\right\|_{L\infty(J;L^2(\Omega))} + \left\|D^{1/2}\nabla\varsigma\right\|_{L^2\left(J;L^2\left(E_{\varepsilon'_h}\right)\right)} \\
&+K\sum_{E\in\varepsilon_h}\frac{r_E}{h_E^{1/2}}\left\|\varsigma\right\|_{L^2(J;L^2(\partial\hat{\Omega}_E))} + K\sum_{E\in\varepsilon_h}\left(\frac{h_E^{1/2}}{r_E} + \delta_{\mathrm{OBB}}\frac{r_E}{h_E^{1/2}}\right)\left\|\nabla\varsigma\cdot \mathrm{n}_{\partial E}\right\|_{L^2(J;L^2(\partial\hat{\Omega}_E))}
\end{aligned} \tag{60}$$

where K is constant independent of the mesh sizes and polynomial degrees, $\delta_{OBB} = 1$ for OBB-DG or NIPG with arbitrary penalty parameters, and $\delta_{OBB} = 0$ for SIPG, IIPG or NIPG with sufficiently large penalty parameters.

In addition, it has also been shown in [65] that under certain conditions ζ is a upper bound estimate of ς:

$$\left\|\sqrt{\phi}\varsigma\right\|_{L\infty\left(J;L^2(\Omega)\right)}+\left\|D^{1/2}\nabla\varsigma\right\|_{L^2\left(J;L^2\left(E_{\varepsilon_h}\right)\right)}\leq\left\|\sqrt{\phi}\xi\right\|_{L\infty\left(J;L^2(\Omega)\right)}+\left\|D^{1/2}\nabla\xi\right\|_{L^2\left(J;L^2\left(E_{\varepsilon_h}\right)\right)}$$
$$+K\sum_{E\in\varepsilon_h}\frac{r_E}{h_E^{1/2}}\left\|\nabla\xi\cdot\mathrm{n}_{\partial E}\right\|_{L^2\left(J;L^2\left(\partial\Omega_E\right)\right)}+K\sum_{E\in\varepsilon_h}\left(\frac{h_E^{1/2}}{r_E}+\left(1-\delta_{\mathrm{OBB}}\right)\frac{r_E^3}{h_E^{3/2}}\right)\left\|\xi\right\|_{L^2\left(J;L^2\left(\partial\Omega_E\right)\right)} \quad \textbf{(61)}$$

where K is constant independent of the mesh sizes and polynomial degrees, $\delta_{OBB}=1$ for OBB-DG, and $\delta_{OBB}=0$ for SIPG, IIPG or NIPG.

5.5. Adaptive Strategy Based on Error Estimators

Reactive transport in subsurface media usually exhibit localized behaviors, such as concentrated plumes, sharp fronts, shocks and layers, which may also change with time. Consequently, efficient simulations of these phenomena require effective, dynamic, and self-adaptive local grid refinement and coarsening. We now describe a technique that utilizes *a posteriori* error estimators and indicators to lead to adaptive local grid modification dynamically with time. We divide the entire simulation period into a collection of time slices. Each time slice contains a number of time steps. Between each time slices, the mesh is adaptively adjusted while the number of elements remains constant. The concentration is transferred between meshes by the L^2 projection. The initial mesh is usually chosen to be a uniform fine grid. We denote by #(S) the number of elements in a set S. The adaptive strategy is given in Algorithm 1 below.

Algorithm 1. (dynamic mesh adaptation),

Given an initial mesh ε_0, a modification factor $\alpha\in(0,1)$, time slices $\{(T_0, T_1), (T_1, T_2), \ldots, (T_{N-1}, T_N)\}$, and iteration numbers $\{M_1, M_2, \ldots, M_N\}$:

For n=1 to N do
  For m=1 to M_n do
    If (n==1) then
      compute C_{init} for time slice
          (T_{n-1}, T_n) using I.C.
        If m=1, Let $\varepsilon_{m.n}=\varepsilon_0$
    else
        compute C_{init} for time slice
          (T_{n-1}, T_n) by L^2 projn.
        If m=1, Let
          $\varepsilon_{m.n}=\varepsilon_{M_{n-1}+1,n-1}$
    end
    solve DG on (T_{n-1}, T_n)
          with mesh $\varepsilon_{m.n}$
    compute error indicator
          for each $E\in\varepsilon_{m.n}$
    select $\varepsilon_r\in\varepsilon_{m.n}$
        s.t. $\#(\varepsilon_r)\in\lceil\alpha\#(\varepsilon_{m,n})\rceil$
        and $\{\eta_E{:}E\in\varepsilon_r\}\geq$
         $\max\{\eta_E{:}E\in\varepsilon_{m.n}\setminus\varepsilon_r\}$
    select $\varepsilon_c\in\varepsilon_{m,n}$
        to minimize

$\max\{\eta_E : E \in \varepsilon_c\}$ s.t.

ε_c compatible with $(\varepsilon_{m.n}, \varepsilon_r)$

$\#(\varepsilon_c) \in \lceil \alpha \#(\varepsilon_{m,n}) \rceil$

refine all $E \in \varepsilon_r$

and coarsen all $E \in \varepsilon_c$

to form a new mesh $\varepsilon_{m+1,n}$

end

end

Report the solution and stop.

5.6. Other Error Estimation and Adaptive Approaches

We would like to mention a few other approaches of error estimation and adaptivity that are useful to discontinuous Galerkin solutions to reactive transport problems.

Quite often we are not interested in the entire solution in the whole domain; instead, we might only look for a particular quantity of interest, which is a functional of the solution. For example, we might want to know the total convective and diffusive flux across certain part of domain boundary as an indication of leakage level; or we might want to estimate a certain weighted average of chemical concentration around a certain area to reflect the influence of this chemical on the area. In these situations, the numerical error is estimated and controlled in terms of quantities of interest rather than the convectional L^2 or energy norms. Usually the quantity of interest can be characterized by a linear functional on the space of functions to where the solution belongs, and estimation of such quantity involves solving the dual problem of the original weak formulation numerically [66, 67]. This approach is sometimes referred goal-oriented error estimation, and it can be directly applied to SIPG solutions of reactive transport problems. A main advantage of adaptivity guided by this estimation is improved accuracy for resolving the quantity of interest. The dual problem only needs to be solved approximately to a certain level lower than that for the original problem. In addition, the dual problem is always linear even for a nonlinear original problem.

Instead of achieving mesh adaptation by refinement and coarsening of selected elements, we could also move the nodes in the mesh to better accommodate dynamically changed physical processes. The mesh redistribution can be implemented by using harmonic maps between the physical space and the parameter space by an iteration procedure [68-70], where the topology of the mesh is not modified during the mesh adaptation process. An alternative strategy is to generate conforming centroidal Voronoi Delaunay triangulations (in a two dimensional domain) with nodes redistributed dynamically using a density function built with *a posteriori* error estimates and local mesh sizes [71, 72]. This approach allows the topology of the mesh being changed during mesh adaptation; however it is quite complicated to extend into three spatial dimensions.

6. NUMERICAL RESULTS

6.1. A Numerical Example of Bioremediation Processes

We consider the problem (21)–(24) and (41)–(48) for the case of multicomponent reactive transport in a two-phase immiscible flow setting. The components are toluene (flowing and stationary), microbes (living and dead), oxygen (O_2), nitrogen (N_2), bio-degraded by-products (CO_2 and water) and a non-reactive tracer. The chemical reactions are as described in Section 3.2.

The physical domain has dimensions 24 ft×400 ft×400 ft (*x*-direction being aligned with gravity). Two wells, injecting one phase and producing the other at opposite corners of the domain (areal), drives the flow. For simplicity no-flow boundary conditions for the flow and zero (advective and) diffusive flux of the species is assumed. The domain is split into 4 blocks with non-matching grids. The grids, permeability field, are shown in Fig. **1**. The block containing the injection well is finest with a discretization of 12×16×16, while the block containing the production well is the coarsest at 8×12×12. The other two blocks have a discretization of 10×14×14. The permeability field has been derived from the 6th SPE comparative project. The heterogenous flow features makes this an interesting and realistic problem to test.

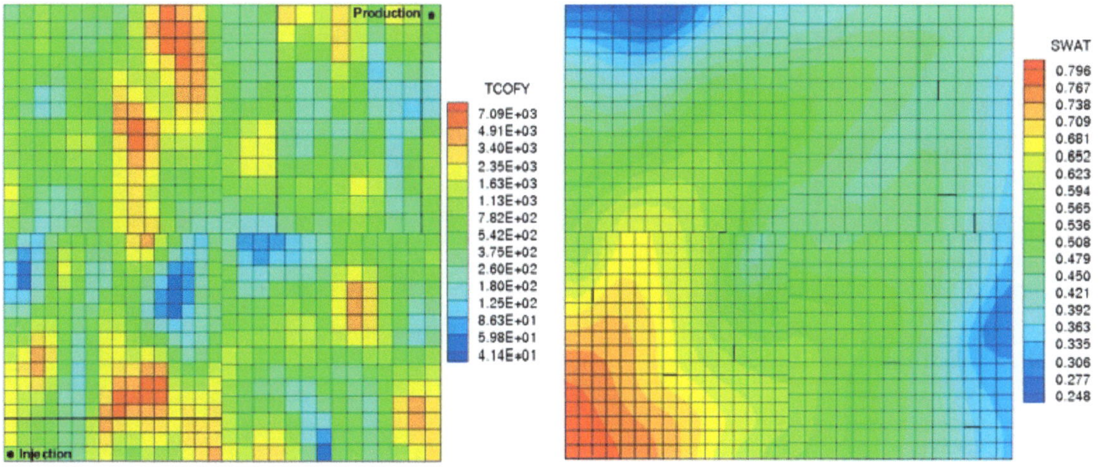

(a) Permeability **(b) Water content, 120d_**

Figure 1: Permeability, grid and flow solution.

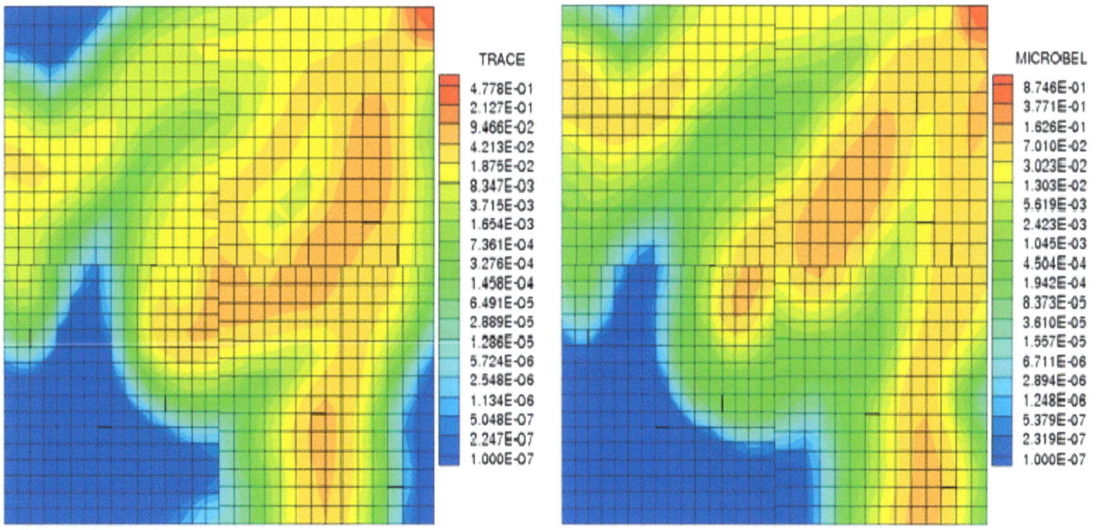

(a) Tracer **(b) Microbe**

Figure 2: Component concentration, 120d.

At initial time, all components except N_2 and O_2 occupy the square column of width 50 ft×50 ft at the lower left region of the domain (near the injection well). N_2 and O_2 initially occupy the remainder of the domain. Molecular diffusivity of 1.0 sq-ft/day and a longitudinal and transverse dispersivity of 1.0 ft and 0.2 ft respectively have been assumed. The injection and production wells are assumed to be pressure specified with the bottom hole pressure of the injection well increasing gradually from 550 to 1000 psi and that at the production well decreasing from 500 to 450 psi over a period of the first 50 days. All solutions are shown in a top view for clarity. Fig. **2** shows the concentration of tracer and microbes at t=120 days. Fig. **3** shows the concentration of the contaminant and the byproducts and t=120 days. From the result it is clear that the propagation of the contaminant into the domain under consideration is reduced significantly when compared to concentrations of tracer and microbes at the same time and extent of their spread.

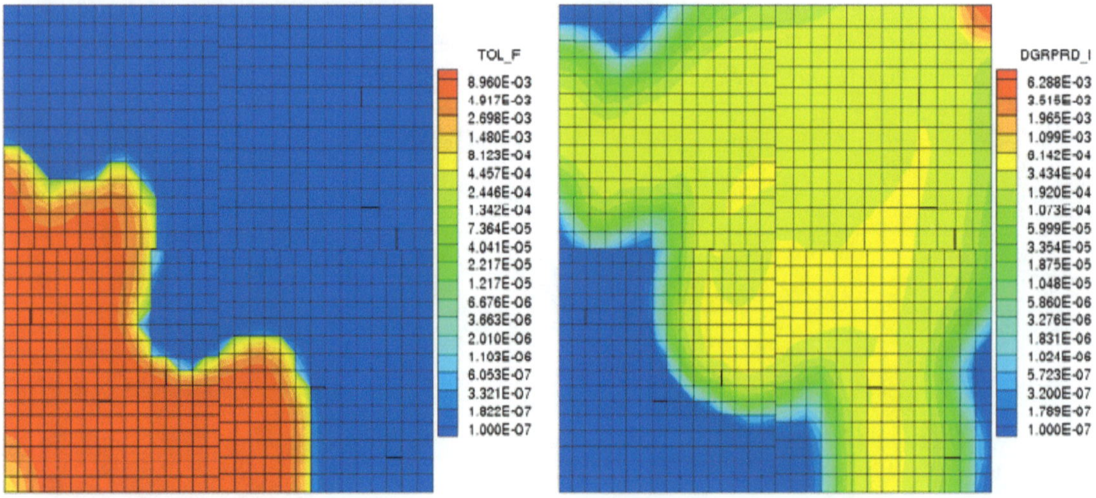

(a) Contaminant **(b) By-product**

Figure 3: Contaminant, product conc., 120d.

The purpose of this example is to motivate the use of the simple, yet natural mechanism of respiration by which aerobic microbes, in the presence of O_2 and N_2, can break down some harmful contaminants that may find their way into sources of drinking water. The by-products of such chemical reactions are relatively harmless. This phenomenon can be used as a safe means of treating contaminated water or spills of harmful chemicals such as heavy oils into lakes and rivers that provide potable water to humans. In addition it is found that the use of variably refined subdomains (coarser away from the injection well in this particular example) saved more than 25% of the computational time compared to running the same problem using a fine grid everywhere, while preserving the accuracy of the solution. Naturally, it is to be expected that the solutions will not exactly agree with the single-block fine-grid everywhere solution since permeabilities on the two grids differ, as well as the effect this has on the phase flow and concentration solution (latter due to the dependence of advection and physical dispersion on phase velocities). In addition coarser grids contribute to more numerical diffusion.

6.2. An Adaptive Example on A Two- Dimensional Mesh

We consider the problem (26) over the domain $\Omega=(0,10)^2$ with a single species in the flowing phase. The domain is divided into two parts, *i.e.* the lower half $\Omega_l=(0,10)\times(0,5)$ and the upper half $\Omega_u=(0,10)\times(5,10)$. We model adsorption by using effective porosity.

Adsorption occurs only in the lower part of the domain, resulting in an effective porosity ϕ of 0.2 in Ω_l. The effective porosity ϕ in Ω_u is 0.1. The diffusion-dispersion tensor D is a diagonal matrix with D_{ii}=0.01, and the Darcy velocity is u= (-0.2, 0). We impose no injection or extraction, and we ignore kinetic reactions; that is, we assume that q=r=0. Due to contamination, the initial total concentration ϕc_0 is 0.1 inside the square centered at (5, 5) with a side of length 0.3125, and 0.0 elsewhere. The velocity field and initial concentration are shown in Fig. **4** (left). The inflow concentration c_B is zero.

SIPG with dynamic mesh modification is employed to solve this problem. The penalty parameter is chosen to be 0.1, based on an error bound established in [26]. For simplicity, each time slice contains only one time step and the number of iterations is chosen to be 2 for all time slices. We set the modification factor α=0.05. The implicit Euler method is employed for time integration. The initial mesh is a structured triangular grid as shown in Fig. **4** (right). The simulation time interval is 0.2 with a uniform time step size $\Delta t = 0.01$. The complete quadratic basis function is used for each element. The residual-based explicit error indicator η in the $L^2(L^2)$ norm is computed for each element E at each time slice. Refinement for each element is isotropic, *i.e.* each element is refined into four congruent sub-elements. The concentration is projected between meshes in a locally conservative manner using the L^2 projection.

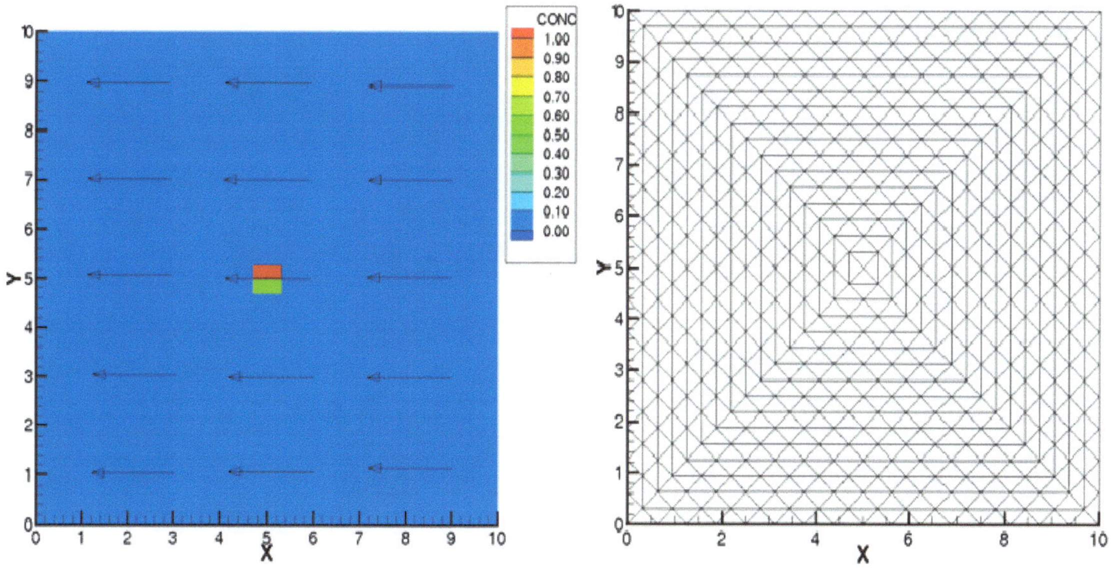

Figure 4: Input for the 2D example (left: initial fluid concentration and velocity; right: initial mash).

Results on the concentration profiles and the mesh structures as functions of time are shown in Fig. **5**. Simulation results show that a contaminant plume moves to the left due to convection, expanding its area with time due to diffusion. Clearly, the resultant mesh in the simulation follows the movement of the plume. Due to the retardation effect arising from adsorption, the convection of the contaminant is slower in the lower part of the domain. Nevertheless, a continuous concentration profile is formed because of non-zero diffusion-dispersion over the entire domain. Compared with non-adaptive DG or mixed finite element methods (results not shown here), adaptive DG has much less numerical diffusion. This is a particularly appealing feature for convection-dominated problems. We also remark, DG on uniform meshes usually requires slope limiting to reduce concentration oscillations. Here, dynamic mesh modification eliminates the need of slope limiting for DG.

The mesh modification for DG is much simpler than classic continuous Galerkin methods as mesh conformity needs not be maintained. In addition, unlike adaptivity in continuous Galerkin, all daughter elements in DG's meshes are similar to parent elements, resulting in high-quality element aspect ratios during mesh refinements. We refer the reader to [59, 64, 73] for adaptive results of DG with quadrilateral elements.

6.3. An Adaptive Example on A Three- Dimensional Mesh

We now consider the problem (26) over the three-dimensional domain $\Omega=(0,10)^3$ with a single species in the flowing phase. Similar to the previous example, the domain is divided into two parts, *i.e.* the lower half $\Omega_l=(0,10)\times(0,10)\times(0,5)$ and the upper half $\Omega_u=(0,10)\times(0,10)\times(5,10)$. Again, the effective porosity ϕ is 0.2 in Ω_l due to adsorption, and 0.1 in Ω_u without adsorption. The diffusion-dispersion tensor D is a diagonal matrix with D_{ii}=0.01, and the velocity is u= (-0.2, 0.0). The initial total concentration ϕc_0 is 0.1 inside the cube centered at (5, 5, 5) with a side of length 0.3125, and 0.0 elsewhere. All other problem parameters are the same as in the example in 2D.

Once again, SIPG with dynamic mesh modification is applied with the initial mesh being a uniform 8×8×8 rectangular grid. All parameters are the same as in the previous example except now the number of iterations is chosen to be 5 initially and 2 for all other time slices, and each time slice contains 10 time steps. Refinement and coarsening are isotropic. For example, a cubic element will be refined into eight congruent small cubes.

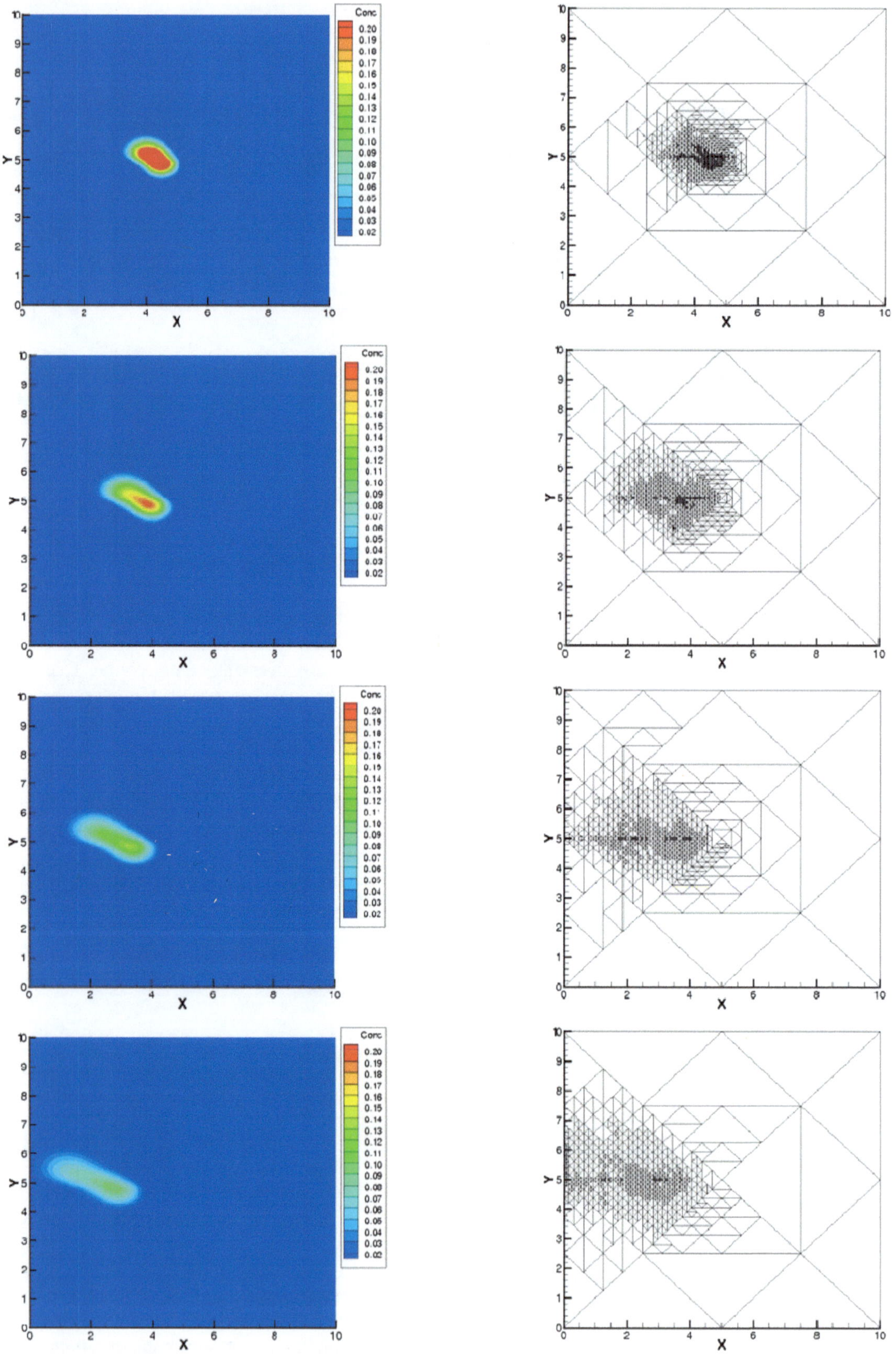

Figure 5: DG results for the 2D example (left column: the concentration in fluid; right column: the mesh structure; from top row to bottom: t=0.5, 1, 1.5 and 2, respectively).

Simulation results are shown in Fig. **6** for concentrations and meshes, where the volume (2.5, 10) × (5, 10) × (5, 10) is removed for better visualization. From the concentration profiles, one can observe varied convection speeds in upper and lower parts of the domain due to the retardation effect. Clearly, resultant meshes are adapted to the time-evolving transport phenomena. The adaptive approach does not only provide physics-driven density distribution of the mesh, but it also yields small numerical diffusion and eliminates the needs of slope limiting.

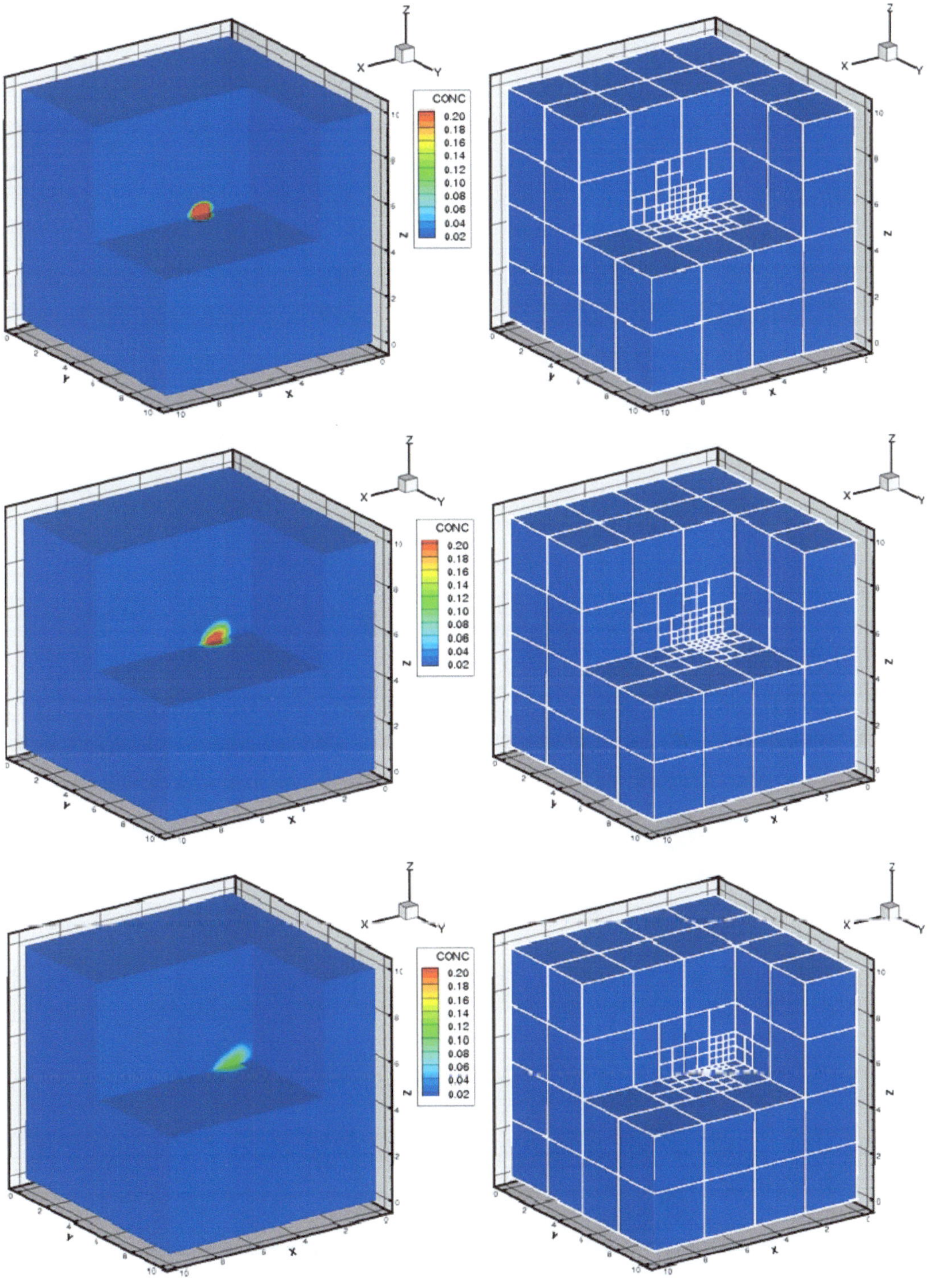

Figure 6: DG results for the 3D example (left column: the concentration in fluid; right column: the mesh structure; from top row to bottom: t=0.1, 0.5 and 1, respectively).

We remark that we have used an 8×8×8 mesh here to demonstrate that adaptive DG works well with a small number of degrees of freedom. Obviously, it is not possible for DG or any other finite element method to resolve the transport phenomena here in a fixed 8×8×8 uniform mesh, because the scale of the concentration plume is much smaller than the size of a single element. Of course, if we use a finer mesh, the adaptive DG provides more accurate concentration profiles in both the long-term and short-term simulations (not shown). The successful application of DG to this example illustrates the useful feature of DG in treating nonconforming meshes. In particular, as local element refinement is independent of neighborhood elements, unnecessary areas do not need to be refined to merely maintain mesh conformity in DG.

While continuous Galerkin methods have to use a tensor-product polynomial space $\mathbf{Q}_r(\mathrm{E})$ for a hexahedron element *E*, DG can employ a complete polynomial function space $\mathbf{P}_r(\mathrm{E})$ to obtain the same order of convergence. The replacement of $\mathbf{Q}_r(\mathrm{E})$ by $\mathbf{P}_r(\mathrm{E})$ leads to a substantial reduction of computational cost, as the latter space has fewer degrees of freedom than the former space. For example, the space $\mathbf{P}_2(\mathrm{E})$ we are currently using in this example has only 10 degrees of freedom, while $\mathbf{Q}_2(\mathrm{E})$ has 27 degrees of freedom. In general, one can show that the number of degrees of freedom for $\mathbf{P}_r(\mathrm{E})$ is $r^3/6+O(r^2)$ while that for $\mathbf{Q}_r(\mathrm{E})$ is $r^3+O(r^2)$. Consequently, even taking into consideration the "repeated" unknowns along element interfaces (there are $O(r^2)$ of them), DG with $\mathbf{P}_r(\mathrm{E})$ has a smaller number of unknowns than continuous Galerkin with $\mathbf{Q}_r(\mathrm{E})$ if the degree *r* is sufficiently high.

Finally, we refer the reader to [59, 64, 73, 74] for more DG simulation results, including dynamically adaptive DG for the ANDRA- Couplex1 case [75] and computation using other types of error estimators.

REFERENCES

[1] T. Arbogast, S. Bryant, C. Dawson, F. Saaf, C. Wang, and M. Wheeler. "Computational methods for multiphase flow and reactive transport problems arising in subsurface contaminant remediation", Journal of Computational and Applied Mathematics., vol. 74, pp. 19–32, 1996.

[2] C. Y. Chiang, C.N.Dawson, and M.F.Wheeler. "Modeling of *in-situ* biorestoration of organic compounds in groundwater", Transport in Porous Media, vol. 6, pp. 667–702, 1991.

[3] P. Engesgaard and K. L. Kipp. "A geochemical transport model for redox-controlled movement of mineral fronts in groundwater flow systems: A case of nitrate removal by oxidation of pyrite", Water Resources Research, vol. 28, pp. 2829– 2843, 1992.

[4] Larry W. Lake, Steven L. Bryant, and Aura N. Araque-Martinez. Geochemistry and fluid flow. Elsevier Science Ltd, New York, 1st ed. edition, 2002.

[5] J. Rubin. "Transport of reacting solutes in porous media: Relation between mathematical nature of problem formulation and chemical nature of reactions", Water Resources Research, vol. 19, pp. 1231–1252, 1983.

[6] J. Rubin and R. V. James. "Dispersion-affected transport of reacting solutes in saturated porous media: Galerkin method applied to equili¬brium-controlled exchange in unidirectional steady water flow", Water Resources Research, vol. 9, no. 5, pp. 1332–1356, 1973.

[7] C. I. Steefel and P. Van Cappellen. "Special issue: Reactive transport modeling of natural systems", Journal of Hydrology, vol. 209, pp. 1–388, 1998.

[8] A. J. Valocchi and M. Malmstead. "Accuracy of operator splitting for advection-dispersionreaction problems", Water Resources Research, vol. 28, no. 5, pp. 1471–1476, 1992.

[9] J. van der Lee and L. De Windt. "Present state and future directions of modeling of geoche¬mistry in hydrogeological systems", Journal of Contaminant Hydrology, vol. 47/2, pp. 265–282, 2000.

[10] M. Th. van Genuchten. "Analytical soultions for chemical transport with simultaneous adsorption, zero-order production and first-order decay", Journal of Hydrology, vol. 49, pp. 213–233, 1981.

[11] G. T. Yeh and V. S. Tripathi. "A critical evaluation of recent developments in hydrogeochemical transport models of reactive multichemical components", Water Resources Research, vol. 25, no. 1, pp. 93– 108, 1989.

[12] G. T. Yeh and V. S. Tripathi. "A model for simulating transport of reactive multispecies components: model development and demon¬stration", Water Resources Research, vol. 27, no. 12, pp. 3075– 3094, 1991.

[13] J. C. Nedelec. "Mixed finite elements in R3 ", Numerische Mathematik., vol. 35, pp. 315–341, 1980.

[14] R. A. Raviart and J. M. Thomas. A mixed finite element method for 2nd order elliptic problems. In Mathematical Aspects of the Finite Element Method, Lecture Notes in Mathematics, vol. 606, pp. 292–315. Springer-Verlag, New York, 1977.

[15] I. Aavatsmark. "An introduction to multipoint flux approximations for quadrilateral grids", Computa¬tional Geosciences, vol. 6, no. (3-4), pp. 405– 432, 2002.

[16] I. Aavatsmark, G. T. Eigestad, R. Klausen, M. F. Wheeler, and I. Yotov. "Convergence of a symmetric mpfa method on quadrilateral grids", Computational Geometry, vol. 11, pp. 333–345, 2007.

[17] M. Shashkov M. F. Wheeler M. Berndt, K. Lipnikov and I. Yotov. "A mortar mimetic finite difference method on non-matching grids", Numerische Mathematik, vol. 102, no. 2, pp. 203–230, 2005.

[18] M. F. Wheeler and I. Yotov. "A multipoint flux finite element method", SIAM Journal on Numerical Analysis, vol. 44, pp. 2082–2106, 2006.

[19] D. N. Arnold. An interior penalty finite element method with discontinuous elements. PhD thesis, The University of Chicage, Chicago, IL, 1979.

[20] D. N. Arnold. "An interior penalty finite element method with discontinuous elements", SIAM Journal on Numerical Analysis, vol. 19, pp. 742–760, 1982.

[21] C. E. Baumann and J. T. Oden. "A discontinuous hp finite element method for convection-diffusion problems", Computer Methods in Applied Mechanics and Engineering., vol. 175, pp. 311–341, 1999.

[22] J. T. Oden, I. Babuska, and C. E. Baumann. "A discontinuous hp finite element method for diffusion problems", Journal of Computational Physics, vol. 146, pp. 491–516, 1998.

[23] B. Rivi`ere. Discontinuous Galerkin finite element methods for solving the miscible displacement problem in porous media. PhD thesis, The University of Texas at Austin, 2000.

[24] B. Riviere, M. F. Wheeler, and V. Girault. "A priori error estimates for finite element methods based on discontinuous approximation spaces for elliptic problems", SIAM Journal on Numerical Analysis, vol. 39, no. 3, pp. 902–931, 2001.

[25] S. Sun and M. F. Wheeler. "Discontinuous Galerkin methods for coupled flow and reactive transport problems", Applied Numerical Mathematics, vol. 52, no. 23, pp. 273–298, 2005.

[26] S. Sun and M. F. Wheeler. "Symmetric and non-symmetric discontinuous Galerkin methods for reactive transport in porous media", SIAM Journal on Numerical Analysis, vol. 43, no. 1, pp.195–219, 2005.

[27] S. Sun and M. F. Wheeler. "A dynamic, adaptive, locally conservative and nonconforming solution strategy for transport phenomena in chemical engineering", Chemical Engineering Communi¬cations, vol. 193, pp. 1527–1545, 2006.

[28] S. Sun and M. F. Wheeler. "Discontinuous Galerkin methods for simulating bioreactive transport of viruses in porous media", Advances in Water Resources, vol. 30, pp. 1696–1710, 2007.

[29] M. F. Wheeler. "An elliptic collocation-finite element method with interior penalties", SIAM Journal on Numerical Analysis, vol. 15, pp. 152–161, 1978.

[30] T. Arbogast, L. C. Cowsar, M. F. Wheeler, and I. Yotov. "Mixed finite element methods on non-matching multiblock grids", SIAM Journal on Numerical Analysis, vol. 37, pp. 1295–1315, 2000.

[31] Q. Lu, M. Peszynska, and M. F. Wheeler. A parallel multi-block black-oil model in multi-model implementation. In 2001 SPE Reservoir Simulation Symposium, Houston, Texas, 2001. SPE 66359.

[32] Q. Lu, M. Peszynska, M. F. Wheeler, and I. Yotov. "Multiphysics and multinumerics couplings for multiphase flow in porous media", In preparation.

[33] M. Balho, M. F. Wheeler, and S. G. Thomas. "Mortar coupling and upscaling of pore-scale models", Comp. Geosc., vol. 12, no. 1, pp. 15–27, 2008. Subject Collection: Mathematics and Statistics.

[34] G. Pencheva, S. G. Thomas, and M. F. Wheeler. Mortar coupling of discontinuous Galerkin and mixed finite element methods. Bergen, Norway, Sep. 8-11 2008. 11th European Conference on the Mathematics of Oil Recovery, ECMOR XI.

[35] Shuyu Sun, Mary F. Wheeler, Ivan Yotov, and Vivette Girault. "Coupling DG with DG and MFE using a mortar space", in preparation.

[36] IPARSv2. Integrated parallel and accurate reservoir simulator. Technical Report TICAM01-25, Center for Subsurface Modeling, ICES, The University of Texas at Austin, Austin, TX, 2000.

[37] S. Sun. Discontinuous Galerkin methods for reactive transport in porous media. PhD thesis, The University of Texas at Austin, 2003.

[38] X. Gai. A coupled geomechanics and reservoir flow model on parallel computers. PhD thesis, The University of

Texas at Austin, 2004.

[39] J. Wheeler, I. Yotov, and M. F. Wheeler. "Enhanced velocity mixed finite element methods for flow in multiblock domains", Computers and Geosciences, vol. 6, no. 3-4, pp. 315–332, 2002.

[40] B. Ganis and I. Yotov. "Implementation of a mortar mixed finite element method using a multi-scale flux basis", Computer Methods in Applied Mechanics and Engineering., submitted.

[41] G. Pencheva, S. G. Thomas, and M. F. Wheeler. Mortar coupling of multiphase flow and reactive transport on non-matching grids. In 5th International Symposium on Finite Volumes for Complex Applications, 1999.

[42] J. C. Parker. "Multiphase flow and transport in porous media", Review of Geophysics, vol. 27, no. 3, pp. 311– 328, 1989.

[43] M. Parashar, J.A. Wheeler, G. Pope, K. Wang, and P. Wang. A new generation eos compo¬sitional reservoir simulator. Part II: Framework and multiprocessing. In Fourteenth SPE Symposium on Reservoir Simulation, Dalas, Texas, pp. 31–38, June 1997.

[44] P. Wang, I. Yotov, M. F. Wheeler, et al. A new generation eos compositional reservoir simulator. part I: Formulation and discretization. In Four¬teenth SPE Symposium on Reservoir Simulation, Dalas, Texas, pp. 55–64. Society of Petroleum Engineers, June 1997.

[45] C. Dawson, S. Sun, and M. F. Wheeler. "Compatible algorithms for coupled flow and transport", Computer Methods in Applied Mechanics and Engineering, vol. 193, pp. 2565– 2580, 2004.

[46] I. Yotov. Mixed finite element methods for flow in porous media. PhD thesis, Houston, TX, 1996.

[47] F. Brezzi and M. Fortin. Mixed and hybrid finite element methods. Springer-Verlag, New York, 1991.

[48] T. Arbogast, M. F. Wheeler, and I. Yotov. "Mixed finite elements for elliptic problems with tensor coefficients as cell-centered finite differences", SIAM Journal on Numerical Analysis, vol. 34, pp. 828–852, 1997.

[49] R. Glowinski and M. F. Wheeler. Domain decomposition and mixed finite element methods for elliptic problems. In R. Glowinski, G. H. Golub, G. A. Meurant, and J. Periaux, editors, First International Symposium on Domain Decom¬position Methods for Partial Differential Equations, pp. 144–172. SIAM, Philadelphia, 1988.

[50] C. N. Dawson, H. Klie, M. F. Wheeler, and C. Woodward. "A parallel, implicit, cell-centered method for two-phase flow with a preconditioned Newton-Krylov solver", Computer and Geosciences, vol. 1, pp. 215– 249, 1997.

[51] S. Lacroix, Yu. Vassilevski, J. A. Wheeler, and M. F. Wheeler. An iterative solution of linear systems in the implicit parallel accurate reservoir simulator (IPARS). In preparation.

[52] C. N. Dawson. "Godunov mixed methods for advection-diffusion equations in multi-dimen¬sions", SIAM Journal on Numerical Analysis, vol. 30, no. 5, pp. 1315–1332, 1993.

[53] R. Eymard, T. Gallou¨et, and R. Herbin. Finite volume methods. In P. G. Ciarlet and J. L. Lions, editors, Handbook of Numerical Analysis, vol. VII, pp. 713–1020. Elsevier, North Holland, Amsterdam, Sep 2000.

[54] S. K. Godunov. "A difference scheme for numerical computation of discontinuous solutions of equations in fluid dynamics", Math Sbornik, vol. 47, pp. 271–306, 1959.

[55] B. Van Leer. "Towards the ultimate conservative difference scheme v.a. second order sequel to godunov's method", Journal of Computational Physics, vol. 32, pp. 101– 136, 1979.

[56] M. Delshad, S. G. Thomas, and M. F. Wheeler. "Modeling CO2 sequestration using a sequen¬tially coupled 'Iterative-IMPEC-Time- Split- Thermal' compositional simulator", In 11th European Conference on the Mathematics of Oil Recovery, ECMOR XI, Bergen, Norway, Sep 2008. EAGE.

[57] B. Lu. Iteratively coupled reservoir simulation for multiphase flow in porous media. PhD thesis, The University of Texas at Austin, 2008.

[58] S. Sun and M. F. Wheeler. "L2(H1) norm aposteriori error estimation for discontinuous Galerkin approximations of reactive transport problems", Journal of Scientific Computing, vol. 22, pp. 501–530, 2005.

[59] S. Sun and M. F. Wheeler. "A posteriori error estimation and dynamic adaptivity for symmetric discontinuous Galerkin approximations of reactive transport problems", Computer Methods in Applied Mechanics and Engineering, vol. 195, pp. 632–652, 2006.

[60] M. Ainsworth and J. T. Oden. A posteriori error estimation in finite element analysis, John Wiley and Sons, Inc., New York, 2000.

[61] R. E. Bank and R. K. Smith. "A posteriori errorestimaters based on hierarchical bases". SIAM Journal on Numerical Analysis, vol. 30, no. 4, pp. 921–935, 1993.

[62] R. E. Bank and A. Weiser. "Some a posteriori error estimators for elliptic partial differential equations", Mathematics of Computation, vol. 44, pp. 283–301, 1985.

[63] L. Demkowicz, W. Rachowicz, and Ph. Devloo. "A fully automatic hp-adaptivity", Journal of Scientific Computing, vol. 17, no. 1-3, pp. 127–155, 2002.

[64] S. Sun and M. F. Wheeler. "Anisotropic and dynamic mesh adaptation for discontinuous Galerkin methods applied to reactive transport", Computer Methods in Applied Mechanics and Engineering., vol. 195, no. 2528, pp. 3382–3405, 2006.

[65] S. Sun and M. F. Wheeler. "Local problem-based a posteriori error estimators for discontinuous galerkin approximations of reactive transport", Computational Geosciences, 2007. accepted.

[66] F. Larsson, P. Hansbo, and K. Runesson. "Strategies for computing goal-oriented a posteriori error measures in non-linear elasticity", Intern¬ational Journal for Numerical Methods in Engineering, vol. 55, no. 8, pp. 879–894, 2002.

[67] S. Prudhomme and J. T. Oden. "On goal- oriented error estimation for elliptic problems: Application to the control of pointwise errors", Computer Methods in Applied Mechanics and Engineering., vol. 176, no. 1-4, pp. 313–331, 1999.

[68] Y. Di, R. Li, T. Tang, and P. Zhang. "Moving mesh finite element methods for the incompressible navier-stokes equations", SIAM Journal on Scientific Computing, vol. 26, pp. 1036–1056, 2005.

[69] R. Li and T. Tang. "Moving mesh discontinuous galerkin method for hyperbolic conservation laws", Journal of Scientific Computing., vol. 27, pp. 347–363, 2006.

[70] T. Tang. "Moving mesh methods for computational fluid dynamics", Contemporary Mathematics, vol. 383, pp. 141–173, 2005.

[71] Q. Du and D. Wang. "Tetrahedral mesh generation and optimization based on centroidal voronoi tessellations", International Journal for Numerical Methods in Engineering, vol. 56, pp. 1355–1373, 2003.

[72] L. Ju, M. Gunzburger, and W. Zhao. "Adaptive finite element methods for elliptic pdes based on conforming centroidal voronoi-delaunay triangu¬lations", SIAM Journal on Scientific Computing, vol. 28, no. 6, pp. 2023–2053, 2006.

[73] S. Sun and M. F. Wheeler. Mesh adaptation strategies for discontinuous Galerkin methods applied to reactive transport problems. In H.-W. Chu, M. Savoie, and B. Sanchez, editors, Proceedings of International Conference on Computing, Communications and Control Technologies (CCCT 2004), vol. I, pp. 223–228, 2004.

[74] S. Sun and M. F. Wheeler. Adaptive discontinuous Galerkin methods for coupled diffusion-and advection-dominated transport phenomena. In H.-W. Chu, M. J. Savoie, and B. Sanchez, editors, Proceedings of The 3rd International Conference on Computing, Communications and Control Technologies (CCCT 2005),vol. I, pp. 130–135, 2005.

[75] A. Bourgeat, M. Kern, S. Schumacher, and J. Talandier. "The COUPLEX test cases: Nuclear waste disposal simulation", Computational Geos¬ciences, vol. 8, no. 2, pp. 83–98, 2004.

CHAPTER 3

TOUGHREACT: A Simulation Program for Subsurface Reactive Chemical Transport under Non-isothermal Multiphase Flow Conditions

T. Xu[1*], E. Sonnenthal[1], N. Spycher[1], G. Zhang[2], L. Zheng[1], and K. Pruess[1]

[1]Earth Sciences Division, Lawrence Berkeley National Laboratory, One Cyclotron Road, Berkeley, CA, USA and [2]Current address: Shell International E&P Inc. Houston, TX, USA

Abstract: TOUGHREACT is a numerical simulation program for chemically reactive non-isothermal flows of multiphase fluids in porous and fractured media. The program was written in Fortran 77 and developed by introducing reactive chemistry into the multiphase fluid and heat flow simulator TOUGH2. A variety of subsurface thermo-physical-chemical-biological processes are considered under a wide range of conditions of pressure, temperature, water saturation, ionic strength, and pH and Eh. Reactions among aqueous species and interactions between mineral assemblages and fluids can occur under local equilibrium or *via* kinetically controlled rates. The gas phase can be chemically active. Precipitation and dissolution reactions can change formation porosity and permeability. Intra-aqueous kinetics, biodegradation and surface complexation have recently been incorporated. The program can be applied to one-, two- or three-dimensional porous and fractured media with physical and chemical heterogeneity. It can accommodate any number of chemical species present in liquid, gas and solid phases. TOUGHREACT can be applied to many geologic systems and environmental problems, including subsurface storage of nuclear waste and CO_2, geothermal systems, diagenetic and weathering processes, acid mine drainage remediation, contaminant transport, and groundwater quality. The methods and approaches used in TOUGHREACT have been extensively published over the last decade. Here we give a general description and summary of the program, including the main features, scope, processes, and solution method. To illustrate its applicability, we present four examples: (1) Denitrification and sulfate reduction, (2) long-term fate of injected CO_2 for geological sequestration, (3) bentonite alteration due to THC processes in a nuclear waste repository, and (4) chemical stimulation of an enhanced geothermal system.

Keywords: TOUGHREACT, CO_2 sequestration, environmental remediation, nuclear waste disposal, geothermal development.

1. INTRODUCTION

Coupled modeling of subsurface multiphase fluid and heat flow, solute transport, and chemical reactions can be applied to many geologic systems and environmental problems, including geothermal systems, diagenetic and weathering processes, subsurface nuclear waste disposal, CO_2 geological sequestration, acid mine drainage remediation, contaminant transport, and groundwater quality. TOUGHREACT is a numerical simulation program for reactive chemical transport that has been developed by introducing reactive chemistry into the existing framework of the non-isothermal multi-component fluid and heat flow simulator TOUGH2 [1]. A wide range of subsurface thermo-physical-chemical processes are considered under various thermo-hydrological and geochemical conditions of pressure, temperature, water saturation, ionic strength, and pH and Eh. TOUGHREACT can be applied to one-, two- or three-dimensional porous and fractured media with physical and chemical heterogeneity. The code can accommodate any number of chemical species present in liquid, gas and solid phases.

TOUGHREACT is written in FORTRAN 77 with some Fortran-90 extensions. It has been tested on various computer platforms, including Microsoft Windows- and Linux-based PCs, Apple Macintosh G4, G5, and Intel-based computers. An effort was made for the TOUGHREACT source code to comply with the

*Address correspondence to T. Xu: Earth Sciences Division, Lawrence Berkeley National Laboratory, One Cyclotron Road, Berkeley, CA, USA; Tel: 1-510-486-7057; Email: Tianfu_Xu@lbl.gov

Fan Zhang, Gour-Tsyh (George) Yeh, Jack C. Parker and Xiaonan Shi (Eds)

ANSIX3.9-1978 (FORTRAN 77) standard. On most machines the code should compile using Fortran 95, Fortran 90, and some Fortran 77 compilers, and run without modification. The computer memory required by TOUGHREACT depends on the problem size such as numbers of grid blocks, aqueous and gaseous species, and minerals. Parameter statements are used in INCLUDE files.

The method and applicability have been extensively published in [2-8]. Here we give a general description of TOUGHREACT including the main features, scope, processes, and solution methods. To illustrate the applicability of TOUGHREACT, we present four examples for practical problems: (1) denitrification and sulfate reduction, (2) long-term fate of injected CO_2 for geological sequestration, (3) bentonite alteration due to THC processes in a nuclear waste repository, and (4) chemical stimulation of an enhanced geothermal system.

2. MAIN SCOPE

TOUGHREACT is applicable to one-, two-, or three-dimensional geologic domains with physical and chemical heterogeneity and can be applied to a wide range of subsurface conditions. The temperature (T) and pressure (P) range is controlled by the applicable range of the chemical thermodynamic database, and the range of the EOS (equation of state) module employed. For example, the range in the commonly-used database of EQ3/6 [9] is 0 to 300°C, 1 bar below 100°C, and water saturation pressure above 100°C. Because the equilibrium constants are generally not as sensitive to pressure as to temperature, the program may be applied to pressures of several hundred bars (corresponding to depths of less than a few km). Water saturation can vary from completely dry to fully water-saturated. Activity coefficients of charged aqueous species are computed using an extended Debye-Huckel equation and parameters derived by [10]. The model can deal with ionic strengths from dilute to moderately saline water (up to 6 molal for a NaCl-dominant solution). TOUGHREACT can cope with any pH and Eh conditions.

Geochemical modeling involving high ionic strength brines is a challenge. A Pitzer ion-interaction model was implemented into a special version of TOUGHREACT for the Yucca Mountain project [7]. This allows the application of this simulator to problems involving much more concentrated aqueous solutions. The Pitzer ion-interaction model, which we refer to as the Pitzer virial approach, and associated ion-interaction parameters have been applied to study non-ideal concentrated aqueous solutions. The formulation and details of the Pitzer model implemented is presented in [7].

3. MAJOR PROCESSES

The major processes for fluid and heat flow are as follows: (1) fluid flow in both liquid and gas phases occurs under pressure, viscous, and gravity forces; (2) interactions between flowing phases are represented by characteristic curves (relative permeability and capillary pressure); (3) heat transfer is by conduction and convection, and (4) diffusion of water vapor and non-condensable gases is included.

Transport of aqueous and gaseous species by advection and molecular diffusion is considered in both liquid (aqueous) and gas phases. Depending on the computer memory and CPU performance, any number of chemical species in the liquid, gas and solid phases can be accommodated. Aqueous complexation, acid-base, redox, gas dissolution/exsolution, and multi-site cation exchange are considered under the local equilibrium assumption. Mineral dissolution and precipitation can proceed either subject to local equilibrium or kinetic conditions. Linear adsorption and decay can be included. Recently, intra-aqueous kinetics and biodegradation [6] and surface complexation using non-electrostatic, constant capacity and double layer models [8], have been incorporated into TOUGHREACT.

Changes in porosity and permeability due to mineral dissolution and precipitation can modify fluid flow. This feedback between flow and chemistry is considered in our model. The formulation for changes in rock properties will be given later.

Hydrodynamic dispersion is an important solute transport mechanism which arises from an interplay between non-uniform advection and molecular diffusion. In geologic media, velocities of fluid parcels are spatially variable due to heterogeneities on multiple scales, all the way from the pore-scale to basin-scale. The process is often represented by a Fickian diffusion analog (convection-dispersion equation), which has fundamental flaws and limitations, as has been demonstrated in numerous studies in the hydrogeology literature of the last 30 years. Although field tracer tests can generally be matched with the convection-dispersion equation, such matching and associated parameters have little predictive power. There is much evidence that when a Fickian dispersion model is calibrated to field tracer data, such success does not indicate that a realistic description of *in situ* solute distribution has been attained. Dispersivities are generally found to increase with space and time scale of observation [11]. Observed dispersivities are only partly due to mixing and dilution *in situ*; they also reflect the mixing that occurs when subsurface flow systems are observed (perturbed) and sampled, as when fluids are extracted from wells. It has been established that Fickian dispersion implies an unrealistically large level of mixing and dilution [12]. Fickian plumes represent a probability distribution, not a distribution of solute; they strongly overestimate dilution in any particular representation of a heterogeneous medium. This can produce erroneous predictions for transport and even more unrealistic consequences for reactions that depend on concentrations in a non-linear manner. Fickian dispersion also gives rise to spurious upstream dispersion opposing the direction of advective flow. For these reasons, we are not using a Fickian model for dispersion. Instead, hydrodynamic dispersion is modeled through appropriate spatial resolution on multiple scales, using multiple continua or multi-region models to describe interactions between fluid regions with different velocities (details are given below).

4. SOLUTION METHODS

4.1. Solution of Flow and Transport Equations

The numerical solution of multi-phase fluid and heat flow proceeds as in TOUGH2. Space discretization is made by means of Integral Finite Differences (IFD) [13]. Because chemical transport equations (derived from mass conservation) have the same structure as fluid and heat flow equations, the transport equations can be solved by the same numerical method. The discretization approach used in the IFD method and the definition of the geometric parameters are illustrated in Fig. **1**. The basic mass- (for water, air, and chemical components) and energy- (for heat) balance equations are written in integral form for an arbitrary domain V_n:

$$V_n \frac{\Delta M_n}{\Delta t} = \sum_m A_{nm} F_{nm} + V_n q_n \tag{1}$$

where subscript n labels a grid block, subscript m labels grid blocks connected to grid block n, Δt is time step size, and M_n is the average mass or energy density in grid block n. Surface integrals are approximated as a discrete sum of averages over surface segments A_{nm}, F_{nm} is the average flux (of mass or energy) over the surface segment A_{nm} between volume elements n and m, and q_n is the average source/sink rate in grid block n per unit volume. Time is discretized fully implicitly as a first-order finite difference to achieve unconditional stability. More detail on the numerical discretization is given in [1]. The IFD method gives a flexible discretization for geologic media that allows the use of irregular unstructured grids, which is well suited for simulation of flow, transport, and fluid-rock interaction in multi-region heterogeneous and fractured rock systems. For systems with regular grids, IFD is equivalent to conventional finite differences.

The time discretization of fluid and heat flow equations results in a set of coupled non-linear algebraic equations for the unknown thermodynamic state variables in all grid blocks. These equations are solved by Newton-Raphson iteration as implemented in the original TOUGH2 simulator [1]. The set of coupled linear equations arising at each iteration step is solved iteratively by means of preconditioned conjugate gradient methods.

TOUGHREACT uses a Sequential Iteration Approach (SIA) similar to [14, 15]. After solution of the flow equations, the fluid velocities and phase saturations are used to simulate chemical transport. Chemical transport equations are solved on a component-by-component basis. The resulting concentrations obtained from solving transport equations are substituted into the chemical reaction model. The system of mixed equilibrium-kinetic chemical reaction equations is solved on a grid block by grid block basis by Newton-Raphson iteration. Optionally, the chemical transport and reactions are solved iteratively until convergence is achieved. An automatic time stepping scheme is implemented in TOUGHREACT, which includes an option to recognize "Quasi-Stationary States" (QSS) [16] and perform a “large” time step towards the end of a QSS.

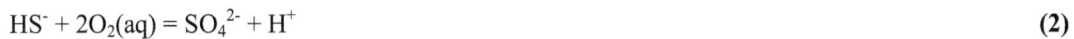

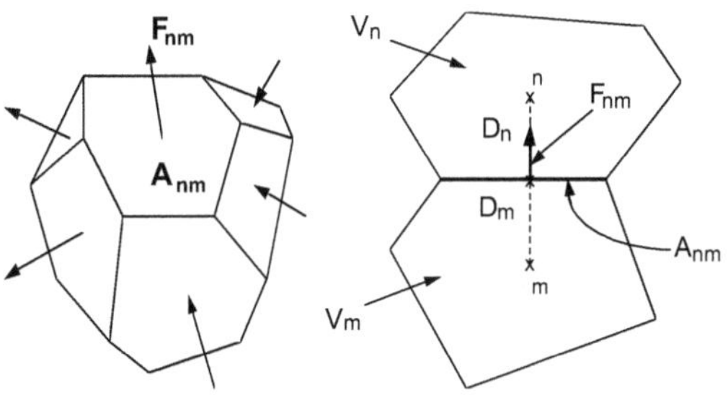

Figure 1: Space discretization and geometric data for the integral finite difference method.

As an alternative to the sequential iterative approach, a Sequential Non-Iterative Approach (SNIA) may be used, in which the sequence of transport and reaction equations is solved only once [17, 18]. The accuracy of SNIA depends mainly on the Courant number, which is defined as $C = v\Delta t/\Delta x$, where v is fluid velocity and Δx is grid spacing. For small Courant numbers, satisfying the stability condition $C \leq 1$, the differences between SNIA and SIA are generally small. The accuracy of SNIA also depends on the type of chemical process. Therefore, the applicability of the decoupling of chemical reactions from transport will depend on time and space discretization parameters, the nature of the chemical reactions and the desired accuracy. When SNIA is used, the Courant number condition $C \leq 1$ can be automatically enforced during the simulation.

4.2. Solution of Chemical System

The primary equations for the chemical system are based on mass balances in terms of primary (basis) species. In contrast to aqueous equilibrium, species involved in kinetic reactions, such as redox couples, are independent and must be considered as primary species [19]. For example, for the reaction:

$$HS^- + 2O_2(aq) = SO_4^{2-} + H^+ \qquad \textbf{(2)}$$

under kinetic conditions, both HS^- and SO_4^{2-} must be placed in the primary species list. Thus, all redox reactions making use of these species must be decoupled in the input of the thermodynamic database.

Details on the formulation for solving the mixed equilibrium-kinetics system of equations are given in [6]. Here we only present the final Jacobian equations. Denoting residuals of mass-balance of each component j as F_j^c (which are zero in the limit of convergence), we have:

$$\begin{aligned} F_j^c = \;& (c_j - c_j^0) && \text{primary species} \\ & + \sum_{k=1}^{N_x} \nu_{kj}(c_k - c_k^0) && \text{equilibrium aqueous complexes} \\ & + \sum_{m=1}^{N_p} \nu_{mj}(c_m - c_m^0) && \text{equilibrium minerals} \\ & - \sum_{n=1}^{N_q} \nu_{nj} r_n \Delta t && \text{kinetic minerals} \\ & - \sum_{l=1}^{N_a} \nu_{lj} r_l \Delta t && \text{kinetics among primary species} \\ = \;& 0 \qquad j = 1 \ldots N_C \end{aligned} \qquad \textbf{(3)}$$

where superscript 0 represents time zero; Δt is the time step; c are concentrations; subscripts j, k, m, and n are the indices of primary species, aqueous complexes, minerals at equilibrium, and minerals under kinetic constraints, respectively; N_c, N_x, N_p, and N_q are the number of corresponding species and minerals; ν_{kj}, ν_{mj}, and ν_{nj} are stoichiometric coefficients of the primary species in the aqueous complexes, equilibrium, and kinetic minerals, respectively; r_n is the kinetic rate of mineral dissolution and precipitation (positive for dissolution and negative for precipitation, units used here are moles of mineral per kilogram of water per time), for which a general multi-mechanism rate law was used (see below Eq. 5); l is the aqueous kinetic reaction (including biodegradation) index, N_a is total number of kinetic reactions among primary species, and r_l is the kinetic rate which is in terms of one mole of product species per unit time. For product species, the stoichometric coefficients v_{lj} are positive, for reactant species they are negative.

According to mass-action equations, concentrations of aqueous complexes c_k can be expressed as functions of concentrations of the primary species c_j. Kinetic rates r_n and r_l are functions of c_j. The expression for r_l and r_n will be presented in the next Section. No explicit expressions relate equilibrium mineral concentrations c_m to c_j. Therefore, N_P additional mass-action equations (one per mineral) are needed. Notice that gas dissolution/ exsolution, cation exchange, and surface complexation are included in TOUGHREACT, but the formulation is not shown here.

5. REACTION KINETICS

5.1. Aqueous Kinetics and Biodegradation

A general rate expression for intra-aqueous kinetic reaction and biodegradation is incorporated into TOUGHREACT. Following the expression of [20] and adding multiple mechanisms (or pathways), a general rate law used is,

$$r_i = \sum_{s=1}^{M} \begin{bmatrix} k_{i,s} & \text{rate constant} \\ \times \prod_{j=1}^{N_l} \left(\gamma_j^{v_{i,j}} C_j^{v_{i,j}} \right) & \text{product terms} \\ \times \prod_{k=1}^{N_m} \dfrac{C_{i,k}}{K_{Mi,k} + C_{i,k}} & \text{Monod terms} \\ \times \prod_{p=1}^{N_P} \dfrac{I_{i,p} + C_{i,p}}{I_{i,p}} & \text{inhibition terms} \end{bmatrix} \quad (4)$$

where r_i is the reaction rate of the i-th reaction, M is the number of mechanisms or pathways and s is the mechanism counter, k is a rate constant, (often denoted v_{max}, maximum specific growth constant for biodegradation), γ_j is the activity coefficient of species j, C_j is the concentration of species j (with biodegradation the product term is usually biomass concentration), $v_{i,j}$ is a stoichiometric coefficient, N_l is the number of reacting species in the forward rate term (called product terms), N_m is the number of Monod factors (Monod terms), $C_{i,k}$ is the concentration of the k-th Monod species, $C_{i,p}$ is the concentration of the p-th inhibiting species, $K_{Mi,k}$ is the k-th Monod half-saturation constant of the i-th species, N_P is the number of inhibition factors (inhibition terms), and $I_{i,p}$ is the p-th inhibition constant. Eq. (4) accounts for multiple mechanisms and multiple products, Monod, and inhibition terms, which can cover many rate expressions.

5.2. Mineral Dissolution and Precipitation

The general rate expression is used for mineral dissolution and precipitation, which is taken from [21]:

$$r_n = \pm k_n A_n \left| 1 - \left(\frac{Q_n}{K_n} \right)^{\theta} \right|^{\eta} \quad (5)$$

where n denotes kinetic mineral index, positive values of r_n indicate dissolution, and negative values precipitation, k_n is the rate constant (moles per unit mineral surface area and unit time) which is temperature dependent, A_n is the specific reactive surface area per kg H_2O, K_n is the equilibrium constant for the mineral-water reaction written for the destruction of one mole of mineral n, and Q_n is the reaction quotient. The parameters θ and η must be determined from experiments; usually, but not always, they are taken equal to one.

For many minerals, the kinetic rate constant k can be summed from three mechanisms [22], or

$$\begin{aligned} k = {} & k_{25}^{nu} \exp\left[\frac{-E_a^{nu}}{R}\left(\frac{1}{T}-\frac{1}{298.15}\right)\right] + \\ & k_{25}^{H} \exp\left[\frac{-E_a^{H}}{R}\left(\frac{1}{T}-\frac{1}{298.15}\right)\right] a_H^{n_H} + \\ & k_{25}^{OH} \exp\left[\frac{-E_a^{OH}}{R}\left(\frac{1}{T}-\frac{1}{298.15}\right)\right] a_{OH}^{n_{OH}} \end{aligned} \quad \textbf{(6)}$$

where superscripts or subscripts nu, H, and OH indicate neutral, acid and base mechanisms, respectively, E_a is the activation energy, k_{25} is the rate constant at 25°C, R is gas constant, T is absolute temperature, a is the activity of the species; and n is an exponent (constant). The rate constant k can be also dependent on other species such as Al^{3+} and Fe^{3+}. Two or more species may be involved in one mechanism. A general form of species dependent rate constants (extension of Eq. 6) is implemented in TOUGHREACT as:

$$k = k_{25}^{nu} \exp\left[\frac{-E_a^{nu}}{R}\left(\frac{1}{T}-\frac{1}{298.15}\right)\right] + \sum_i k_{25}^{i} \exp\left[\frac{-E_a^{i}}{R}\left(\frac{1}{T}-\frac{1}{298.15}\right)\right] \prod_j a_{ij}^{n_{ij}} \quad \textbf{(7)}$$

where superscripts or subscripts i is the additional mechanism index, and j is species index involved in one mechanism that can be primary or secondary species. TOUGHREACT currently considers up to five additional mechanisms and up to five species involved in each mechanism.

6. CHANGES IN POROSITY AND PERMEABILITY

Temporal changes in porosity and permeability due to mineral dissolution and precipitation can modify fluid flow path characteristics. This feedback between flow and chemistry is considered. Changes in porosity are calculated from changes in mineral volume fractions. Four different porosity-permeability relationships were implemented in TOUGHREACT. One is a commonly used cubic Kozeny-Carman grain model. Laboratory experiments have shown that modest decreases in porosity due to mineral precipitation can cause large reductions in permeability. This is explained by the convergent-divergent nature of natural pore channels, where pore throats can become clogged by precipitates while disconnected void spaces remain in the pore bodies. A relationship proposed by [23], with a more sensitive coupling of permeability to porosity than the Kozeny-Carman relationship was found to better capture injectivity losses in a geothermal example [24]:

$$\frac{k}{k_0} = \left(\frac{\varphi - \varphi_c}{\varphi_0 - \varphi_c}\right)^n \quad \textbf{(8)}$$

where ϕ_c is the value of "critical" porosity at which permeability goes to zero, and n is a power law exponent. Eq. (8) is derived from a pore-body-and-throat model in which permeability can be reduced to zero with a finite ("critical") porosity remaining.

Permeability and porosity changes will likely result in modifications to the unsaturated flow properties of the rock, which is treated by modification of the capillary pressure function using the Leverett scaling relation.

7. TREATMENT OF FRACTURED ROCK AND HETEROGENEOUS MEDIA

7.1. Multiple Interacting Continua

For chemical transport in variably saturated fractured rocks, global fluid flow and transport of aqueous and gaseous species occurs primarily through a network of interconnected fractures, while chemical species may penetrate into tight matrix blocks primarily through relatively slow diffusive transport in gas and liquid phases. Methods developed for fluid flow in fractured rock can be applied to geochemical transport.

The method of "Multiple Interacting Continua" (MINC) is used to resolve "global" flow and transport of chemicals in the fractured rock and its interaction with "local" exchange between fractures and matrix rock. This method was developed by [25, 26] for fluid and heat flow in fractured porous media. The extension of the MINC method to reactive geochemical transport is described in detail by [2]. It is well-known that in the case of reactive chemistry, diffusive fluxes may be controlled by reactions occurring near (within millimeters of) the fracture walls. The resolution of concentration gradients in matrix blocks is achieved by appropriate subgridding. The MINC concept is based on the notion that changes in fluid pressures and chemical concentrations propagate rapidly through the fracture system, while invading the tight matrix blocks only slowly. Therefore, changes in matrix conditions will be (locally) controlled by the distance from the fractures and can then be modeled by means of one-dimensional strings of nested grid blocks (Fig. **2**).

7.2 Multi-Region Model

Similar to "Multiple Interacting Continua" (MINC) for fractured rock, multi-region concepts can be employed in TOUGHREACT for modeling biogeo-chemical transport in porous heterogeneous media [6, 27]. The multi-region model was proposed to account for pore structures, the resultant widely distributed pore water velocities, and the effect of local-scale and field-scale heterogeneities on mass transport. Pore regions can either be physically identified as discrete features, or be experimentally determined by separation of water retention curves according to pore classification schemes. Note that the concept of "region" here is not a "regional model" that generally implies a scale measured in square kilometers. The concept being utilized is a sub REV concept. An application of the multi-region model can be found in the application example of denitrification and sulfate reduction (Fig. **3**).

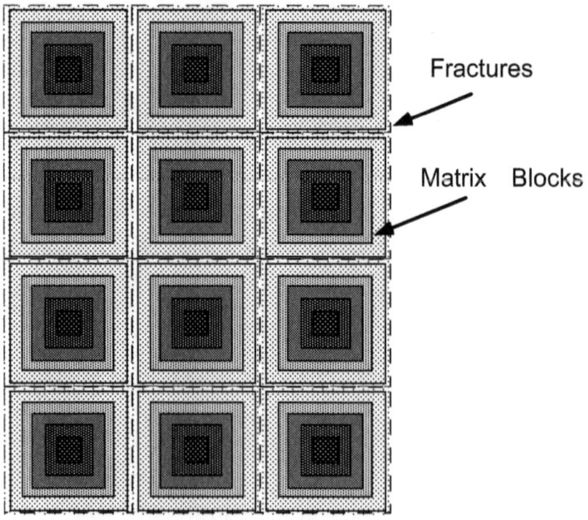

Figure 2: Subgridding in the method of "Multiple Interacting Continua" (MINC).

The basic idea of using multiple interacting continua or multi-region approaches to model dispersion is to explicitly resolve domains with different advective velocities through appropriate spatial discretization (gridding). This approach is applicable for heterogeneous media in which regions of higher permeability form spatially-extensive correlated structures. Hydrodynamic dispersion then arises from an interplay between different advective velocities, and primarily diffusive exchange between subdomains.

8. APPLICATIONS

The TOUGHREACT code was extensively verified against analytical solutions and other numerical simulators. The program has been applied to a wide variety of geological and environmental problems. Here we give four illustrative examples.

8.1. Denitrification and Sulfate Reduction

To test the applicability of TOUGHREACT to reactive transport of denitrification and sulfate reduction, the column experiments of [28] were modeled. Their experiments were designed to simulate infiltration of an organically polluted river into an aquifer. Thus, synthetic river water, including an organic substrate (lactate) and electron acceptors of oxygen, nitrate and sulfate, were injected into columns filled with river sediments.

Three major microbially-mediated reactions are involved in the experiments. Three electron acceptors are reduced, while Dissolved Organic Matter (DOC) represented by lactate ($C_3H_5O_3^-$) in the experiment, is oxidized as follows:

$$3O_2(aq) + C_3H_5O_3^- \rightarrow 3HCO_3^- + 2H^+$$

$$12NO_3^- + 5C_3H_5O_3^- + 2H^+ \rightarrow 15HCO_3^- + 6N_2(aq) + 6H_2O$$

$$3SO_4^{2-} + 4C_3H_5O_3^- \rightarrow 4C_2H_3O_2^- + 4HCO_3^- + 3HS^- + H^+ \qquad \textbf{(9)}$$

The bacterial reaction rates due to three different electron acceptors are given in Eqs. (10)-(12). Denitrification is inhibited by oxygen, and sulfate reduction is inhibited by both oxygen and nitrate.

$$r_b^{O_2} = k_b^{O_2} X_b \left(\frac{C_{DOC}}{K_{DOC}^{O_2} + C_{DOC}} \right) \left(\frac{C_{O_2}}{K_{O_2} + C_{O_2}} \right) \qquad \textbf{(10)}$$

$$r_b^{NO_3} = k_b^{NO_3} X_b \left(\frac{C_{DOC}}{K_{DOC}^{NO_3} + C_{DOC}} \right) \left(\frac{C_{NO_3}}{K_{NO_3} + C_{NO_3}} \right) \left(\frac{I_{O_2 \rightarrow NO_3}}{I_{O_2 \rightarrow NO_3} + C_{O_2}} \right) \qquad \textbf{(11)}$$

$$r_b^{SO_4} = k_b^{SO_4} X_b \left(\frac{C_{DOC}}{K_{DOC}^{SO_4} + C_{DOC}} \right) \left(\frac{C_{SO_4}}{K_{SO_4} + C_{SO_4}} \right) \left(\frac{I_{O_2 \rightarrow SO_4}}{I_{O_2 \rightarrow SO_4} + C_{O_2}} \right) \left(\frac{I_{NO_3 \rightarrow SO_4}}{I_{NO_3 \rightarrow SO_4} + C_{NO_3}} \right) \qquad \textbf{(12)}$$

The overall biotic reaction rate is expressed as:

$$r_b = r_b^{O_2} + r_b^{NO_3} + r_b^{SO_4} - bX_b \qquad \textbf{(13)}$$

where X_b is biomass concentration (mg/l), b is decay constant. In this example, biomass is assumed not subject to transport. Most of the bacteria are fixed on the solid phase within geologic media. The rate parameters for Eqs. (10) through (12) were taken from Table **6** of [6]. A general multi-region model for hydrological transport interacting with microbiological and geochemical processes was used (Fig. **3**). The model consists of: (1) a mobile region (a fraction of the porosity), (2) an immobile region (another fraction of the porosity), and (3) a solid particle region where mineral dissolution/ precipitation and surface reactions may occur. Here the immobile region contains stagnant water and biomass. Currently, TOUGHREACT does not consider the dynamic changes in the volume of immobile region. The dynamic changes in porosity and volume will be implemented in the future.

The applicability of this enhanced multi-region model for reactive transport of denitrification and sulfate reduction was evaluated by comparison with column experiments (Figs. **4** and **5**). The matches with measured nitrate and sulfate concentrations were achieved by adjusting the interfacial area between mobile and immobile regions. The values of 38 m^2 per m^3 bulk medium for the initial period and 75 m^2 for the late period were calibrated. The match and parameter calibration suggest that TOUGHREACT can not only be a useful interpretative tool for biogeochemical experiments, but also can produce insight into the processes and parameters of microscopic diffusion and their interplay with biogeochemical reactions. More on the problem setup and results can be found in [6].

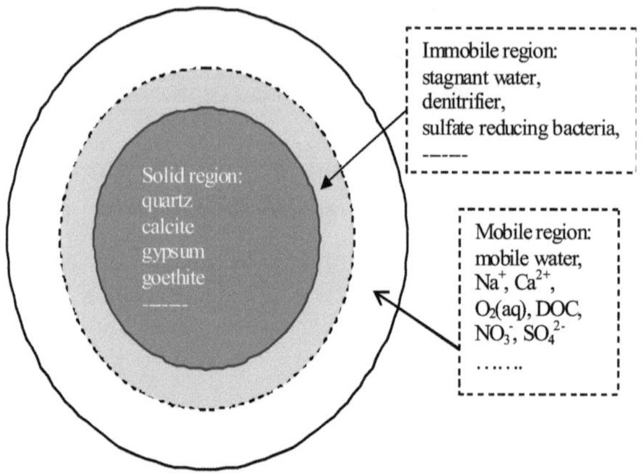

Figure 3: Schematic representation of a multi-region model for resolving local diffusive transport.

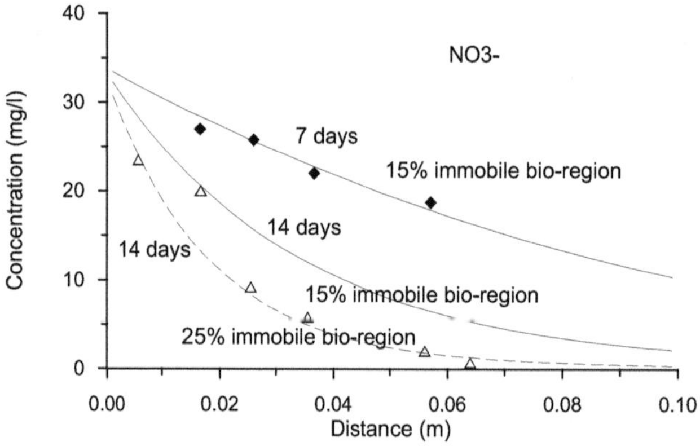

Figure 4: Nitrate concentrations obtained with the multi-region model after 7 and 14 days, together with measured data of [28].

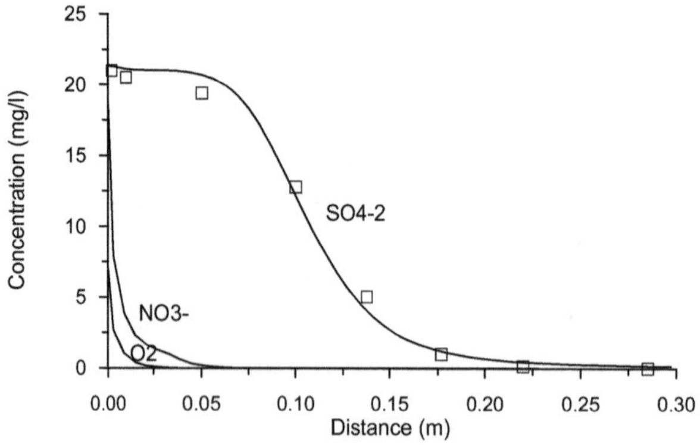

Figure 5: The simulated concentration profiles (lines) of sulfate, nitrate, and oxygen at steady-state (35 days), together with measured data of [28].

8.2. Long-Term Fate of Injected CO_2

8.2.1. Problem Setup

CO_2 injected into deep saline formations will tend to migrate upwards towards the cap-rock because the density of the supercritical CO_2 phase is lower than that of aqueous phase. In the upper portions of the reservoir, CO_2 dissolution into brine decreases pH and induces mineral dissolution and complexing with dissolved ions such as Na^+, Ca^{2+}, Mg^{2+}, and Fe^{2+} to form $NaHCO_3$, $CaHCO_3^+$, $MgHCO_3^+$, and $FeHCO_3^+$. Over time these dissolution and complexing processes will increase CO_2 solubility, enhance solubility trapping, and will increase the density of the aqueous phase. These processes together with changes in rock properties induced by CO_2 injection into a US Gulf Coast sandstone formation have been studied using a generic 2-D radial well flow model (Fig. **6**).

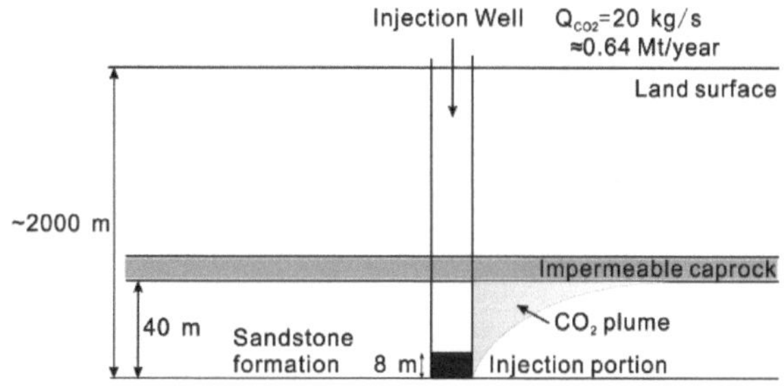

Figure 6: Schematic representation of the 2-D radial flow model for supercritical CO_2 injection into a sandstone formation.

Table 1: List of kinetic rate parameters for minerals considered in the simulation (n = 0 for neutral mechanism, n > 0 for acid mechanism).

Mineral	k_{25} (moles $m^{-2}s^{-1}$)	E_a (KJ/mol)	n	Surface area (cm^2/g)
calcite	Equilibrium			
quartz	$1.26x10^{-14}$	87.50	0	9.8
K-feldspar	$1.00x10^{-12}$	57.78	0	9.8
	$3.55x10^{-10}$	51.83	0.4	9.8
kaolinite	$1.00x10^{-13}$	62.76	0	151.6
	$4.37x10^{-12}$	62.76	0.17	151.6
magnesite	$4.47x10^{-10}$	62.76	0	9.8
	$4.37x10^{-5}$	18.98	1.0	9.8
siderite	$1.26x10^{-9}$	62.76	0	9.8
	$1.02x10^{-3}$	20.90	0.9	9.8
dolomite	$1.26x10^{-9}$	62.76	0	9.8
	$1.02x10^{-3}$	20.90	0.9	9.8
ankerite	$1.26x10^{-9}$	62.76	0	9.8
	$1.02x10^{-3}$	20.90	0.9	9.8
dawsonite	$1.26x10^{-9}$	62.76	0	9.8
	$1.02x10^{-3}$	20.90	0.9	9.8
oligoclase	$1.00x10^{-12}$	57.78	0	9.8
	$3.55x10^{-10}$	51.83	0.4	9.8
albite-low	$1.00x10^{-12}$	67.83	0	9.8
	$2.04x10^{-10}$	59.77	0.5	9.8
Na-smectite	$1.00x10^{-13}$	62.76	0	151.6
	$4.37x10^{-12}$	62.76	0.17	151.6
Ca-smectite	$1.00x10^{-13}$	62.76	0	151.6
	$4.37x10^{-12}$	62.76	0.17	151.6
illite	$1.00x10^{-13}$	62.76	0	151.6
	$4.37x10^{-12}$	62.76	0.17	151.6
pyrite	$4.00x10^{-11}$	62.76	0	12.9
hematite	$4.00x10^{-11}$	62.76	0	12.9
chlorite	$2.51x10^{-12}$	62.76	0	151.6

For numerical simulation, in the vertical direction a total of 20 model layers were used with a constant spacing of 2 m. In the horizontal direction, a radial distance of 100 km was modeled with a radial spacing that increases gradually away from the injection well. A total of 56 radial grid elements were used. The volume of the outer grid element is specified a large value of 10^{30} m^3, representing an infinite lateral boundary for constant pressure, temperature and concentrations. CO_2 injection was applied at the bottom portion over 8 m thickness with a constant rate of 20 kg/s (corresponding to 0.64 Mt/year) for a period of 10 years. The 2-D radial model of fluid flow and geochemical transport was simulated for a period for 1,000 years, which is a relevant time scale for CO_2 geological sequestration. We considered a system with initially homogeneous porosity and permeability. Changes in porosity and permeability due to mineral alteration could seed fluid dynamics instability and promote convective mixing.

Hydrological and geochemical conditions and parameters are taken from [5]. The kinetic parameters for mineral dissolution and precipitation are presented in Table **1**. Calcite is assumed to be at equilibrium with the solution because its reaction rate is rapid relative to the time frame being modeled. Parameters for the rate laws were taken from [22], who compiled and fitted experimental data reported by many investigators. Mineral reactive-surface areas are based on the work of [4], and was calculated assuming a cubic array of truncated spheres constituting the rock framework. The larger surface areas for clay minerals are due to smaller grain sizes.

If the aqueous phase supersaturates with respect to a potential secondary mineral, a small volume fraction such as 1×10^{-6} is used to calculate the seed surface area for the new phase to grow. The precipitation of secondary minerals is represented using the same kinetic expression as that for dissolution. However, because precipitation rate data for most minerals are unavailable, parameters for neutral pH rates only were employed to describe precipitation. Multiple kinetic precipitation mechanisms can be specified in an input file of the TOUGHREACT program, should such information become available.

8.2.2. Results and Discussion

Simulation results indicate that the supercritical CO_2 fluid injected at near the bottom of the storage formation migrates upward rapidly by buoyancy forces because the density of supercritical CO_2 phase is lower than that of aqueous phase (Fig. **7**). A small fraction of CO_2 gas is trapped in the porous rock as residual gas after injection. The residual gas trapping keeps CO_2 dissolving into brine and precipitating carbonate minerals, and gradually disappears at the bottom of the reservoir. With time most of the free CO_2 gas accumulates below the caprock, and then spreads laterally.

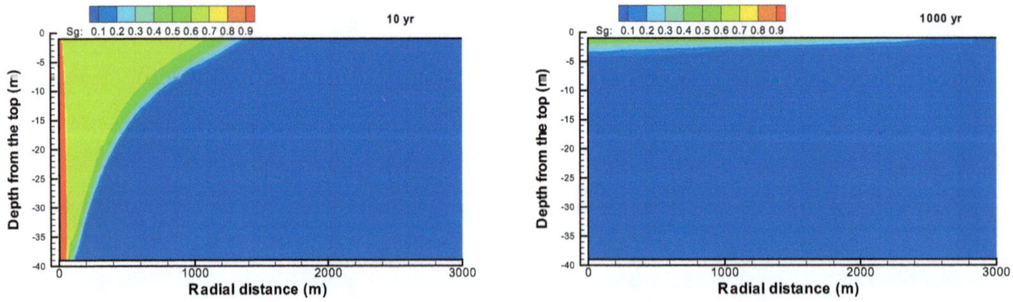

Figure 7: Distribution of supercritical CO_2 phase saturation at 10 and 1000 yr for the 2-D radial injection model.

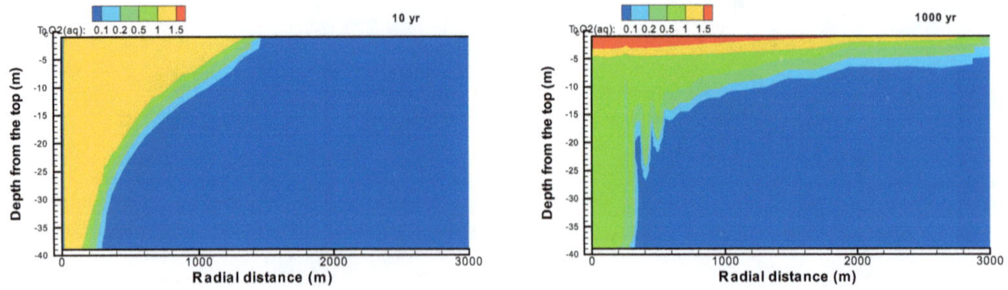

Figure 8: Distribution of total dissolved CO_2 (mol/kg H_2O) at 10 and 1000 yr.

With the migration of CO_2 gas, the concentration of dissolved CO_2 rapidly increases to larger than 1 mol/kg H_2O in the two-phase region (Fig. **8**). The injected CO_2 dissolves in the surrounding formation water, forming H_2CO_3, HCO_3^-, and CO_3^{2-}, and decreasing pH. Then, the increased acidity induces dissolution of many of the primary host rock minerals (discussed later). The mineral dissolution increases concentrations of cations such as Na^+, Ca^{2+}, Mg^{2+}, and Fe^{2+}, which in turn form aqueous complexes with the bicarbonate ion such as $NaHCO_3$, $CaHCO_3^+$, $MgHCO_3^+$, and $FeHCO_3^+$. Over time they tend to increase dissolved CO_2 (solubility) and enhance solubility trapping.

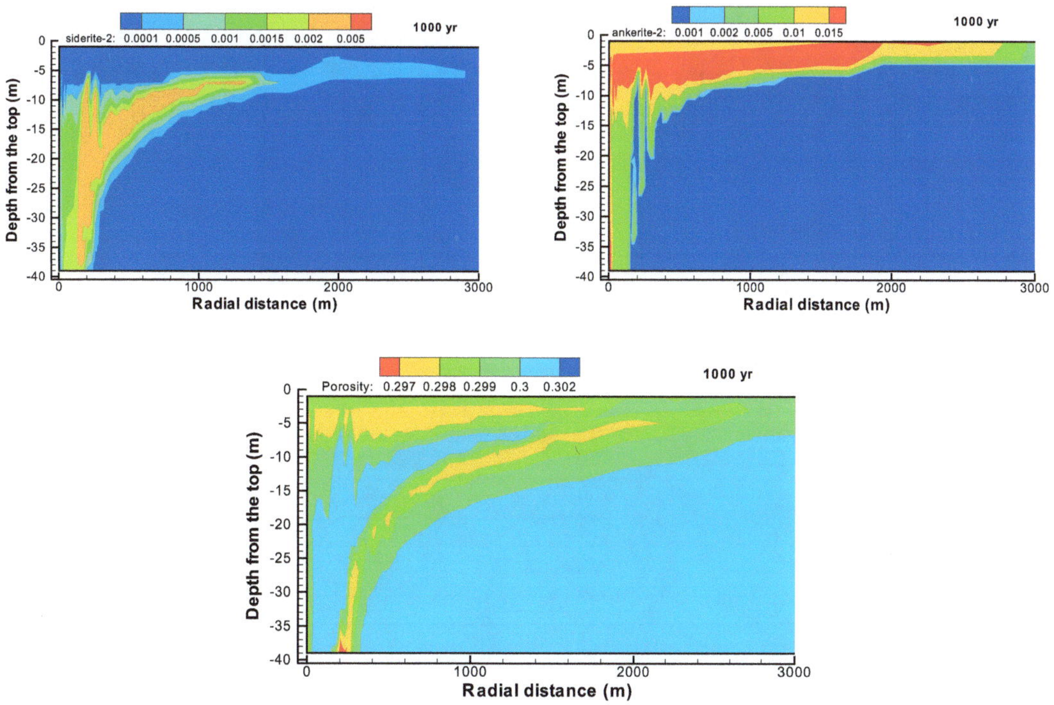

Figure 9: Distribution of siderite and ankerite precipitation, and porosity change at 1000 yr.

Minerals such as oligoclase and chlorite dissolve in the two-phase region and near the front of the single aqueous-phase zone, supplying reactants for carbonate mineral precipitation. Some amount of siderite and dawsonite, and significant amount of ankerite precipitate, sequestering injected CO_2 (mineral trapping). At the same time, clay minerals such as smectite-Na precipitate. Changes in porosity are calculated from variations in mineral volume fractions (the bottom of Fig. **9**). Figs. **8** and **9** show "finger" flow patterns near the bottom of the CO_2 plume. These are because of advection in the aqueous phase that is triggered by an increase in density due to CO_2 dissolution.

The mineral trapping starts at a late stage (about 100 years) and then increases linearly with time. Amounts of CO_2 trapped in different phases and variations of different storage modes with time are estimated. After 1,000 years, 29% of the injected CO_2 could be trapped in the solid (mineral) phase, 28% in the aqueous phase, and 43% in the gas phase.

Mineral alteration and CO_2 trapping capability depends on the primary mineral composition. Precipitation of siderite and ankerite requires Fe^{2+}, which can be supplied by the dissolution of iron-bearing minerals, such as chlorite, or by reduction of Fe^{3+} in small amounts of hematite. Variation in Ca content in plagioclase significantly affects carbonate mineral precipitation, and thus CO_2 mineral trapping. The time required for mineral alteration and CO_2 sequestration depends on the rates of mineral dissolution and precipitation, which are products of the kinetic rate constant and reactive surface area. The current simulated mineral alteration pattern is generally consistent with available mineralogy observed at natural high-pressure CO_2 gas reservoirs. Details are discussed in [5].

8.3. Bentonite Alteration Due to THC Processes During the Early Thermal Period in a Nuclear Waste Repository

After closure of an underground nuclear waste repository, the decay of radionuclides will raise temperatures, and the bentonite buffer will resaturate by water inflow from the host rock. The perturbations from these thermal and resaturation processes are expected to dissipate within a few hundred years. Simulating coupled Thermal-Hydrological-Geochemical (THC) processes in the backfill material and near-field environment of a heat-generating nuclear waste repository requires site-specific and detailed information to assess the coupled processes and their impact at any particular site, and to develop engineering designs. Before moving into site-specific investigations, we explore general features and issues representing characteristics that are common and essential for many such systems. The present study is not related to any particular site. However, the geometric configuration and the hydraulic parameters and mineralogical composition of the clayey formation are abstracted from a nuclear waste repository concept considered in Switzerland [29]. This work was presented in [30]. Here we only give a brief description of the setup and result.

8.3.1. Geometric Configuration

The reference design for canisters to be stored in a repository for spent fuel and high-level waste (SF/HLW) involves a cast steel body with about 20 cm wall thickness. The canisters are about 1 m in diameter and are surrounded by a 0.75-m thick bentonite buffer in emplacement tunnels which are 2.5 m in diameter (Fig. **10**).

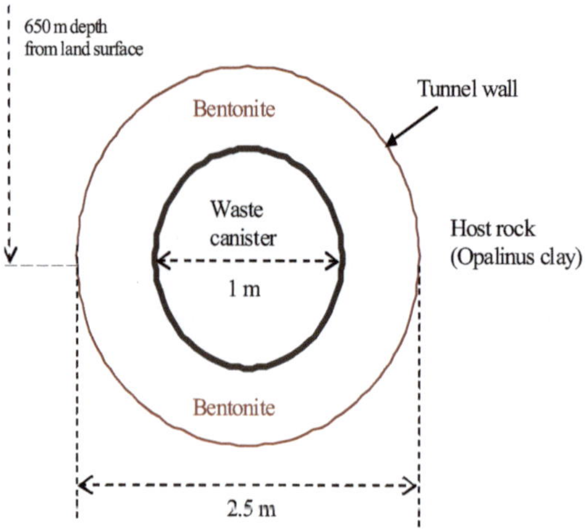

Figure 10: Schematic representation of a nuclear waste repository with a single waste canister, bentonite backfill, and Opalinus Clay host rock [29].

The repository tunnel is assumed to be in the water-saturated zone at a depth of 650 m below the land surface; the host rock is referred to as Opalinus clay. The model is radially symmetric, *i.e.*, it ignores the lateral no flow boundary and gravity effects, which may represent general features and issues related to a nuclear waste repository.

8.3.2. Thermal and Hydrological Conditions

The Opalinus Clay host rock is initially fully water-saturated with a background pressure of 65 bar. The outer boundary at a radial distance of 75 m was held at a constant pressure of 65 bar.

The thermo-physical properties for the bentonite buffer and the Opalinus Clay host rock are summarized in Table **2**. These parameters were taken from [31], who reported simulation results for a 3-D model of the thermo-hydrologic conditions in the vicinity of a backfilled emplacement tunnel for spent fuel. The purpose was to quantify the coupled thermo-hydrologic evolution of temperature, saturation, and pressure through time to determine potential non-uniform resaturation of the bentonite buffer, or potential localized accumulation of pore water in contact with the waste canister.

Table 2: Thermo-physical parameters used for bentonite and Opalinus Clay in the THC model.

	Bentonite	**Opalinus Clay**
Porosity [-]	0.475	0.14
Permeability [m²]	1×10^{-19}	1×10^{-20}
Pore compressibility [Pa⁻¹]	3.58×10^{-9}	1.83×10^{-9}
Rock grain density (kg/m³)	2700	2670
Rock specific heat (J/kg/°C)	964	946.5
Thermal conductivity (W/- m/°C)	1.35	2.5
Two-Phase Parameter Model	van Genuchten	
Residual liquid saturation	0.3	0.5
Residual gas saturation	0.02	0.02
Van Genuchten parameter *n*	1.82	1.67
Gas entry pressure [Pa]	1.8×10^{7}	1.8×10^{7}

Time-dependent heat generation from the waste package was specified. A large density and an initial temperature of 150°C were specified for the grid block representing the waste package, which produced the temperature evolution given in Fig. **11**, providing the inner boundary condition. This heat boundary specification was based on Fig. 4.20a of the report of [31], in which temperature reduced to about 55°C after 1,000 years. An initial temperature of 38°C was used for the remaining grid blocks of the model domain.

8.3.3. Geochemical Conditions

The initial mineral composition of bentonite and Opalinus Clay is given in Table **3**. The MX-80 type of bentonite is used as the buffer material. It contains 75% montmorillonite. Montmorillonite-Na is the dominant clay mineral, which is assigned to have a volume fraction of 75%. The mineral content of Opalinus clay was assigned based on [32]. Smectite was substituted with montmorillonite in the present modeling for the portion of Opalinus Clay. Amorphous silica and anhydrite are considered as secondary minerals with initial volume fractions of zero, and they could be formed during the simulation time.

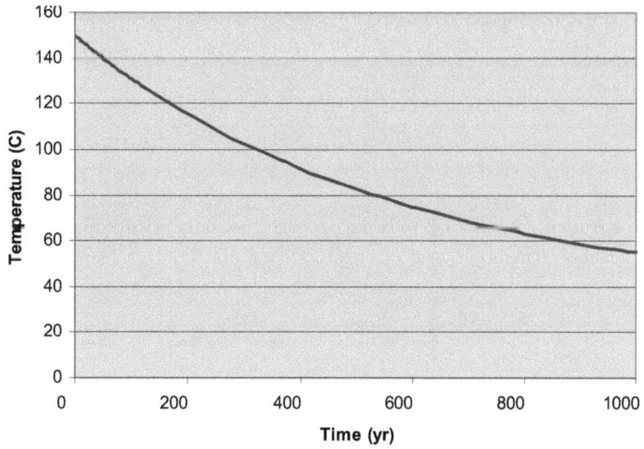

Figure 11: Temperature evolution at canister surface.

Two initial water chemical compositions were used: a dilute water for the bentonite, and the composition measured for BWS-A6 water [32] for the host rock. Prior to simulating reactive transport, batch geochemical modeling of water-rock interaction for the two materials was conducted, equilibrating the initial water with the primary minerals listed in Table **3** at a temperature of 38°C. A reasonably short simulation time (10 years) is needed to obtain nearly steady-state aqueous solution compositions, which were then used as initial chemical conditions for reactive transport (THC) simulations.

Table 3: Initial mineral volume fractions and possible secondary mineral phases used in the THC simulations.

Mineral	Volume percent in term of solid	
	Bentonite	Opalinus clay
Calcite	1.4	15.0
Quartz	15.1	18.0
Kaolinite	1.0	10.0
Illite		20.0
K-feldspar	6.5	3.0
Montmorillonite-Na	75	10.0
Montmorillonite-Ca		10.0
Chlorite		10.0
Dolomite		1.0
Siderite		3.0
Ankerite		1.0
Annite	1.0	
Anhydrite	0.0	0.0
Amorphous silica	0.0	0.0

Calcite and anhydrite are assumed to react at equilibrium because their reaction rates are rapid relative to the time frame being modeled. Parameters for the rate law and reactive-surface areas used are similar to the previous CO_2 sequestration example. Parameters for montmorillonite were set to those of smectite.

8.3.4. Results

After closure of an underground nuclear waste repository, the decay of radionuclides raises temperatures, and the bentonite buffer resaturates through water flow from the surrounding host rock. The perturbations from these thermal and resaturation processes induce mineral dissolution and precipitation. Consequently, the porosity of the bentonite buffer is changed. The simulated porosity distribution is presented in Fig. **12**. Porosity decreases from the initial value of 0.475, indicating precipitation is dominant. Changes in porosity are larger close to the interface between the bentonite buffer and Opalinus Clay host rock because resaturation processes carry chemical constituents from the host rock. Decreases in porosity are smaller close to the canister surface.

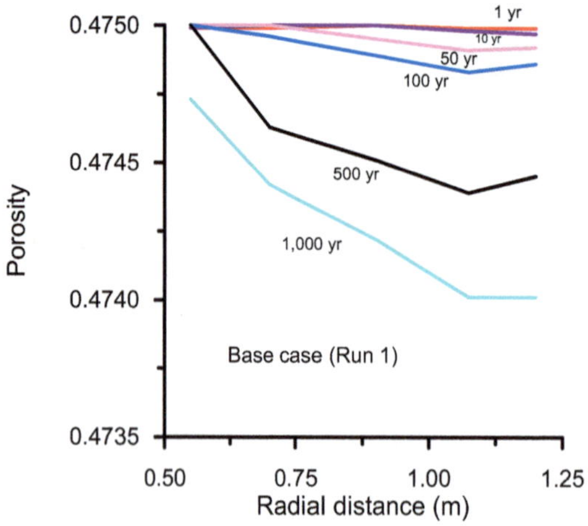

Figure 12: Distribution of porosity in bentonite buffer obtained for the base case.

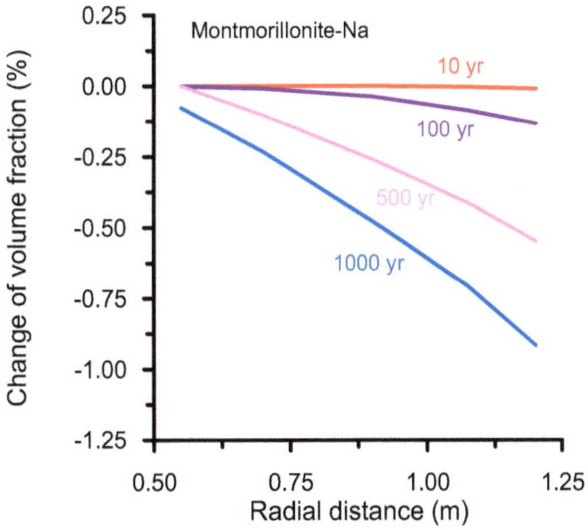

Figure 13: Change in volume fraction of Montmorillonite-Na in bentonite buffer obtained for the base case.

Montmorillonite-Na, the dominant mineral (with an initial volume fraction of 75%), dissolves (Fig. **13**). The pattern of annite dissolution is similar to that of montmorillonite-Na. Calcite precipitates in most parts of the bentonite because its solubility decreases with temperature, but it dissolves close to the interface with the host rock supplying reactants for its precipitation close to the hot end of the canister surface (Fig. **14**). K-feldspar precipitation occurs over the entire radial distance, and more strongly close to the interface with the host rock (Fig. **15**). The pattern of kaolinite and chlorite precipitation is the same as for K-feldspar.

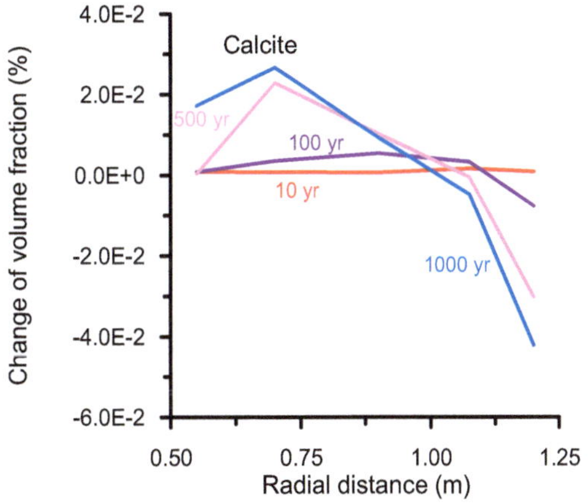

Figure 14: Change in volume fraction of calcite in bentonite buffer obtained for the base case.

Using this model, mineral alteration and changes in porosity for the early thermal and resaturation processes in a nuclear waste repository were examined for more simulation cases with decreasing and increasing reaction rates, and considering cation exchange. More results on the sensitivity simulations can be found in [30]. Overall, mineral alteration and changes in porosity during the 1,000 year period of thermal and resaturation processes are not significant, and do not significantly affect flow and transport properties.

8.4. Chemical Stimulation for an Enhanced Geothermal System

Dissolution of silica, silicate, and calcite minerals in the presence of a chelating agent (NTA) at a high pH has been successfully performed in the laboratory using a high-temperature flow reactor [33]. The mineral dissolution and porosity enhancement in the laboratory experiment has been reproduced by reactive transport simulations using TOUGHREACT. Chemical stimulation of dissolution has been applied to a field geothermal injection well system, to investigate its effectiveness [34]. Parameters from the quartz monzodiorite unit at the Enhanced Geothermal System (EGS) site at Desert Peak (Nevada) were used. Here we only give a brief description of the problem.

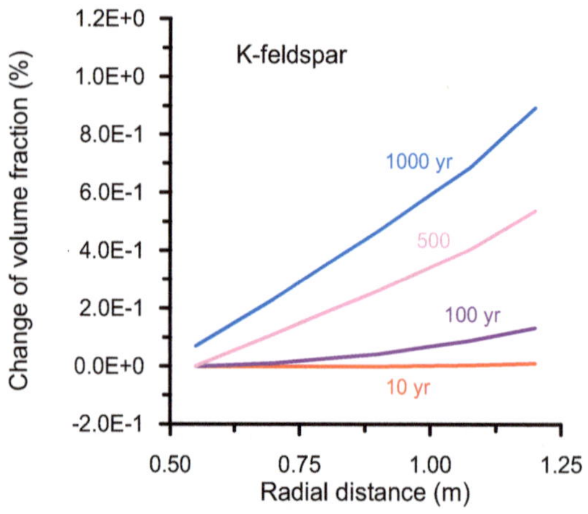

Figure 15: Change in volume fraction of K-feldspar in bentonite buffer obtained for the base case.

8.4.1. Calibration of Kinetic Dissolution Model

Laboratory experiments of calcite and silica dissolution using a high pH solution with NTA were performed by Rose *et al* (2008, unpublished data). Experiments were conducted for a range of temperatures from 150 to 300°C. The injection water was prepared by adding Na-NTA reagent and NaOH to distilled water (representing steam condensate), the resulting injection water had a NTA^{3-} concentration of 0.1 mol/kgw (w denotes H_2O), and a high pH. In the modeling a pH of 11.5 was used because the maximum value of pH decreases with temperature, the highest pH possible at 150°C being 11.63, while at 300°C it is 11.3. Total amounts of silica and calcite dissolved (in percent) after each experiment with six-hour duration are measured. Here we only present data for silica in Fig. **16**. Each data point represents one experiment at a constant temperature.

A 1-D model using TOUGHREACT was developed for the dissolution experiments described above. Experimental data were then compared to model outputs and model parameters for silica and calcite were adjusted as necessary to match the data (Fig. **16**).

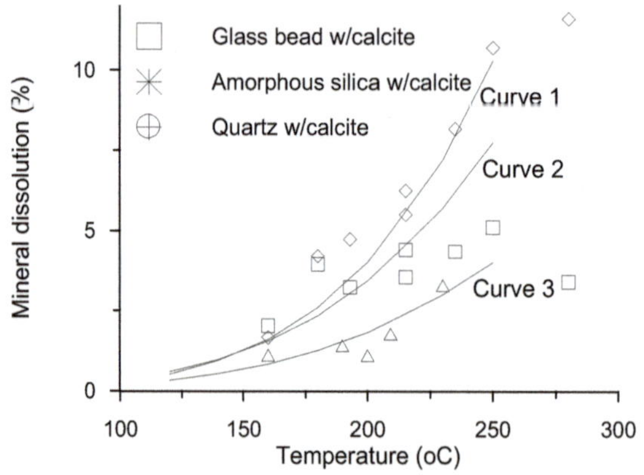

Figure 16: Measured total silica dissolution together with fitted models (curves). Curve 1: $k_{25} = 1.05\times10^{-8}$ mol m^{-2} s^{-1}, $E_a = 39$ kJ mol^{-1}; Curve 2: $k_{25} = 1.85\times10^{-8}$ mol m^{-2} s^{-1}, $E_a = 33.8$ kJ mol^{-1}; Curve 3: $k_{25} = 1.14\times10^{-8}$ mol m^{-2} s^{-1}, $E_a = 32.8$ kJ mol^{-1}.

8.4.2. Geologic Setting and Mineralogy

Initial mineralogical composition used in the modeling is summarized in Table **4**. The compositions specified were based on the original crystalline rock mineralogy considering altered fracture vein mineralogy. Plagioclase was modeled using 50% low-albite and 50% anorthite. Other minerals including epidote, pyrite, and biotite were not considered in the model, because their reactions with injection solution are slow and not important for the chemical stimulation purpose.

8.4.3. Reaction Kinetics

For quartz, specific reactive surface area and kinetic parameters k_{25} and E_a were taken from the calibration of the lab experiment (Curve 3 of Fig. **16**). Specific reactive surface areas for low-albite, anorthite, and K-feldspar are set the same as quartz. Surface area for chlorite and illite was from [4]. Kinetic parameters k_{25} and E_a for low-albite, anorthite, K-feldspar kaolinite, chlorite, and illite were taken from the previous CO_2 sequestration example. Calcite dissolution is controlled by the kinetics of the chelating process and the calibrated parameters mentioned above were used.

Table 4: Initial mineralogical compositions used in the numerical modeling.

Mineral	***Quartz monzodiorite (% in terms of solid).***
Quartz	*9*
Calcite	*12*
Low-albite	*21.5*
Anorthite	*21.5*
K-Feldspar	*13*
Chlorite	*8*
Illite	*7*
Others	*8*

8.4.4. Flow Conditions

A 120 m thick reservoir formation with an injection well was modeled. A simple one-dimensional radial flow model was used, consisting of 50 radial blocks with logarithmically increasing radii. The 50 blocks cover a distance of 1000 m from the wall of the drilled open hole (Fig. **17**). Only the fracture network is considered in the model, with the assumption that the fluid exchange with the surrounding low permeability matrix is insignificant for the short period of chemical stimulation. An initial fracture permeability of 5.2×10^{-12} m^2 was assumed. A fracture porosity of 1% (ratio of fracture volume to the total formation volume) was assumed. A 1% volume of wall rock was included in the fracture domain, to allow minerals on the fracture wall interacting chemically with injection water. Therefore, initial porosity of the modeled fracture domain is 0.5, and permeability is 2.6×10^{-12} m^2. The uncertainty on the permeability specification does not affect modeling results of reactive transport and porosity enhancement, as long as excessive pressure buildup at the wellbore is avoid, because a constant injection rate was specified in the present study.

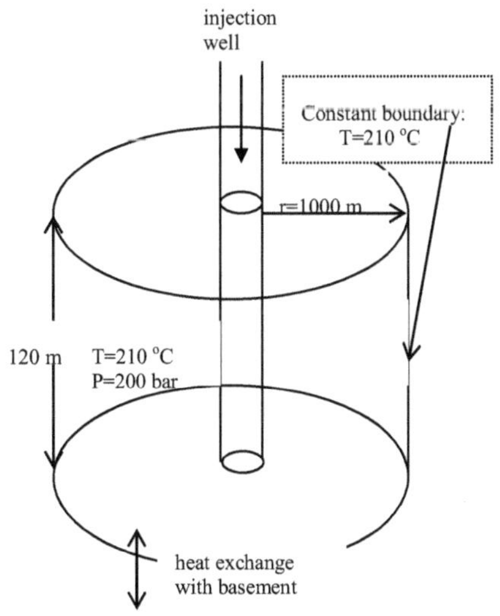

Figure 17: Simplified conceptual model for 1-D radial flow around a geothermal injection well.

Initial reservoir temperature is 210°C. An initial hydrostatic pressure of 20 MPa was assumed for about 2000 m depth. An injection temperature of 160°C was used. Injection water chemistry was the same as in the modeling of the lab experiment, which was prepared by adding NTA agent and NaOH solution to steam condensate, and had a NTA^{3-} concentration of 0.1 mol/kgw, a Na^+ concentration of 1.5 mol/kg_W, and a pH of 11.5. The initial water chemistry is in equilibrium with the initial mineralogy at a reservoir temperature of 210°C. An injection rate of 10 kg/s was applied for a period of half a day. Reactive transport simulations were performed for up to one day, including a no-flow period after the 12-hour injection.

8.4.5. Results and Discussion

Injection of the high pH solution with chelating agent (NTA) results in increases in porosity along the flow path close to the injection well. Overall enhancement of porosity at different times obtained from the simulation is presented in Fig. **18**. Increases in porosity are mainly caused by dissolution of calcite, low-albite and anorthite. The porosity increases to about 60% from an initial value of 50% close to the injection point. The enhancement of porosity extends to a radial distance of about 4 m.

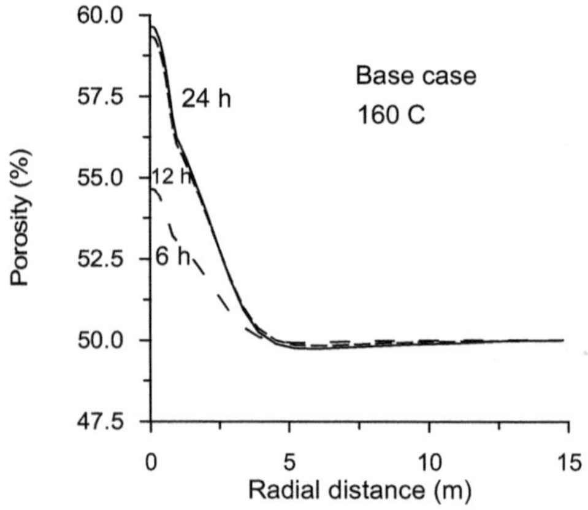

Figure 18: Distribution of porosity enhancements at different times obtained from the simulation (Initial porosity of the fracture domain is 50%).

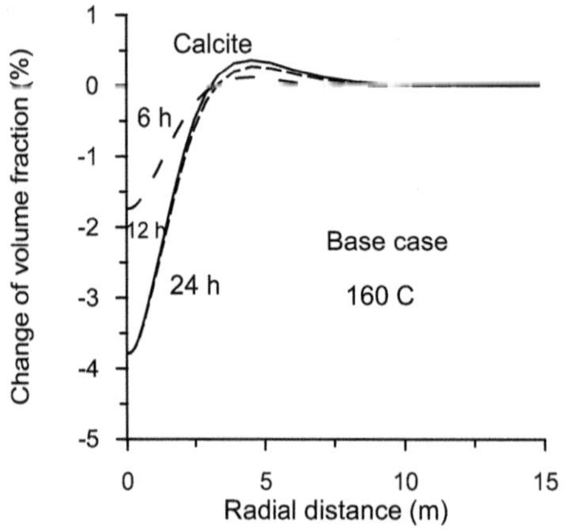

Figure 19: Changes of calcite abundance (in percentage of volume fraction, negative values indicate dissolution, positive precipitation) at different times.

Maximum calcite dissolution of 4% occurs close to the injection point (Fig. **19**). A small amount of precipitation occurs at the moving front because increase in temperature causes decrease in calcite solubility. Amounts of quartz dissolution are very small because of low reaction rate. Generally low-albite dissolution occurs close to the injection point due to high pH, but later its precipitation occurs along the flow path because of lowered pH and high injected Na^+ concentration (Fig. **20**). More results on sensitivity simulations can be found in [33].

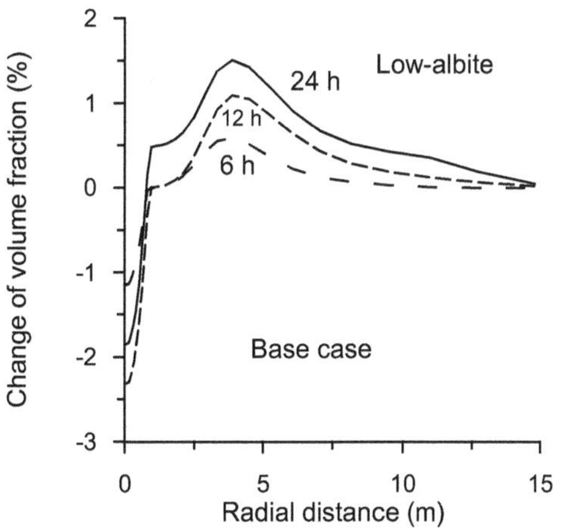

Figure 20: Changes of low-albite abundance (in percentage of volume fraction, negative values indicate dissolution, positive precipitation) at different times.

CONCLUSIONS

A non-isothermal reactive transport program, TOUGHREACT, has been developed, which allows comprehensive modeling of chemical interactions between liquid, gaseous and solid phases that are coupled to solute transport and subsurface multiphase fluid and heat flow. The program is applicable to porous media as well as to fractured rocks. An Integral Finite Difference (IFD) technique is employed for space discretization. The IFD methodology can deal with irregular grids, does not require reference to a global system of coordinates, and includes classical dual-continua, multiple interacting continua, and multi-region models for heterogeneous and fractured rocks as special cases. Non-isothermal effects are considered, including water-vapor phase change and air partitioning between the liquid and gas phases, temperature-dependence of thermophysical properties such as phase density and viscosity, and chemical properties such as thermo-dynamic and kinetic parameters. Chemical reactions considered under the local equilibrium assumption include aqueous complexation, acid-base, redox, gas dissolution/exsolution, and cation exchange. Intra-aqueous kinetics and biodegradation are incorporated. Mineral dissolution/precipitation can proceed either subject to local equilibrium or kinetic conditions.

TOUGHREACT is applicable to one-, two-, or three-dimensional geologic domains with physical and chemical heterogeneity, and can be applied to a wide range of subsurface conditions of pressure, temperature, water saturation, ionic strength, and pH and Eh. The program can be applied to a variety of problems, including contaminant transport and environmental remediation, natural ground-water quality evolution under ambient conditions, assessment of nuclear waste disposal sites, sedimentary diagenesis and CO_2 sequestration in deep formations, mineral deposition such as supergene copper enrichment, and mineral alteration and silica scaling in hydrothermal systems under natural and production conditions. Many other types of geologic, experimental, and engineered systems may be analyzed using the program.

ACKNOWLEDGEMENTS

The development of TOUGHREACT was supported by various DOE program offices, and documentation of this paper was supported by the Zero Emission Research and Technology project (ZERT), of the U.S. Department of Energy under Contract No. DE-AC02-05CH11231 with Lawrence Berkeley National Laboratory. The third application example (bentonite alteration) was supported by the National Co-operative for the Disposal of Radioactive Waste (NAGRA) of Switzerland.

REFERENCES

[1] K. Pruess, "The TOUGH codes: A family of simulation tools for multiphase flow and transport processes in permeable media", *Vadose Zone Journal*, vol. 3, pp. 738-746, 2004.

[2] T. Xu, and K. Pruess, "Modeling multiphase non-isothermal fluid flow and reactive geochemical transport in variably saturated fractured rocks: 1. Methodology", *American Journal of Science*, vol. 301, pp. 16-33, 2001.

[3] N. F. Spycher, E. L. Sonnenthal, and J. A. Apps, "Fluid flow and reactive transport around potential nuclear waste emplacement tunnels at Yucca Mountain, Nevada", *Journal of Contaminant Hydrology*, vol. 62-63, pp. 653-673, 2003.

[4] E. Sonnenthal, A. Ito, N. Spycher, M. Yui, J. Apps, Y. Sugita, M. Conrad, S. Kawakami. "Approaches to modeling coupled thermal, hydrological, and chemical processes in the Drift Scale Heater Test at Yucca Mountain", *International Journal of Rock Mechanics and Mining Sciences*, vol. 42, pp. 6987-719, 2005.

[5] T. Xu, E.L. Sonnenthal, N. Spycher, and K. Pruess, "TOUGHREACT - A simulation program for non-isothermal multiphase reactive geochemical transport in variably saturated geologic media: Applications to geothermal injectivity and CO_2 geological sequestration", *Computers & Geosciences*, vol 32/2, pp. 145-165, 2006.

[6] T. Xu, "Incorporation of aqueous reaction kinetics and biodegradation into TOUGHREACT: Application of a multi-region model to hydrobiogeochemical transport of denitrification and sulfate reduction", *Vadose Zone Journal*, vol. 2008-7, pp. 305-315, 2008.

[7] G. Zhang, N. Spycher, E. Sonnenthal, C. Steefel, and T. Xu, "Modeling reactive multiphase flow and transport of concentrated aqueous solutions", *Nuclear Technology*, vol. 164, pp. 180-195, 2008.

[8] L. Zheng, J. A. Apps, Y. Zhang, T. Xu, J. Birkholzer, On mobilization of lead and arsenic in groundwater in response to CO2 leakage from deep geological storage', *Chemical Geology*, vol. 268, pp. 281-297, 2009.

[9] T. J. Wolery, "EQ3/6: Software package for geochemical modeling of aqueous systems: Package overview and installation guide (version 8.0)", Lawrence Livermore National Laboratory Report UCRL-MA-110662 PT I, Livermore, California, 1992.

[10] H. C. Helgeson, D. H. Kirkham, and D. C. Flowers, "Theoretical prediction of the thermodynamic behavior of aqueous electrolytes at high pressures and temperatures: IV. Calculation of activity coefficients, osmotic coefficients, and apparent molal and standard and relative partial molal properties to 600 C and 5 kb", *American Journal of Science*, vol. 281, pp. 1249–1516, 1981.

[11] L. W. Gelhar, C. Welty, and K. R. Remfeldt, "A critical review of data on field-scale dispersion in aquifers", *Water Resource Research*, vol. 28, pp. 1955-1974, 1992.

[12] V. Kapoor, L. W. Gelhar, and F. Miralles-Willem, "Bimolecular second-order reactions in spatially varying flows: Segregation induced scale-dependent transformation rates", *Water Resource Research*, vol. 33, pp. 527-536, 1997.

[13] T. N. Narasimhan, and P. A. Witherspoon, "An integrated finite difference method for analyzing fluid flow in porous media", *Water Resources Research*, vol. 12, pp. 57–64, 1976.

[14] G. T. Yeh, and V. S. Tripathi, "A model for simulating transport of reactive multispecies components: model development and demonstration", *Water Resource Research*, vol. 27, pp. 3075-3094, 1991.

[15] J. Simunek, and D. L. Suares, "Two-dimensional transport model for variably saturated porous media with major ion chemistry", *Water Resource Research*, vol. 30, pp. 1115-1133, 1994.

[16] P. C. Lichtner, "The quasi-stationary state approximation to coupled mass transport and fluid-rock interaction in a porous medium", *Geochimica et Cosmochimica Acta*, vol. 52, pp. 143-165, 1988.

[17] T. Xu, J. Samper, C. Ayora, M. Manzano, and E. Custodio, Modeling of non-isothermal multi-component reactive transport in field-scale porous media flow system", *Journal of Hydrology*, vol. 214, pp. 144-164, 1999.

[18] J. Samper, T. Xu, and C. Yang, "A sequential partly iterative approach for multicomponent reactive transport with CORE2D", *Computational Geosciences*, vol. 13, pp. 301-316, 2009.

[19] C. I. Steefel, and K. T. B. MacQuarrie, "Approaches to modeling of reactive transport in porous media", In Lichtner, P. C., Steefel, C. I., and Oelkers, E. H. (eds.), *Reactive transport in porous media: Reviews in Mineralogy, Mineral Society of America*, vol. 34, pp. 83-129, 1996.

[20] G.P. Curtis, "Comparison of approaches for simulating reactive solute transport involving organic degradation reactions by multiple terminal electron acceptors", *Computers & Geosciences*, vol. 29, pp. 319-329, 2003.

[21] A. C. Lasaga, J. M. Soler, J. Ganor, T. E. Burch, and K. L. Nagy, "Chemical weathering rate laws and global geochemical cycles", *Geochimica et Cosmochimica Acta*, vol. 58, pp. 2361-2386, 1994.

[22] J. Palandri, and Y.K. Kharaka, "A compilation of rate parameters of water-mineral interaction kinetics for application to geochemical modeling", US Geol. Surv. Open File Report 2004-1068, 64 pp, 2004.

[23] A. Verma, and K Pruess, "Thermohydrological conditions and silica redistribution near high-level nuclear wastes emplaced in saturated geological formations", *Journal of Geophysical Research*, vol. 93, pp. 1159-1173, 1988.

[24] T. Xu, Y. Ontoy, P. Molling, N. Spycher, M. Parini, and K. Pruess, "Reactive transport modeling of injection well scaling and acidizing at Tiwi Field, Philippines", *Geothermics*, vol. 33(4), pp. 477-491, 2004.

[25] K. Pruess, K. Karasaki, "Proximity functions for modeling fluid and heat flow in reservoirs with stochastic fracture distributions", in Proceedings, Eighth workshop on geothermal reservoir engineering: Stanford University, Stanford, California, 1982; p. 219-224.

[26] K. Pruess, T. N. Narasimhan, "A practical method for modeling fluid and heat flow in fractured porous media", *Society of Petroleum Engineers Journal*, vol. 25, pp. 14-26, 1985.

[27] J. P. Gwo, P. M. Jardine, G. V. Wilson, and G. T. Yeh, "Using a multiregion model to study the effects of advective and diffusive mass transfer on local physical nonequilibrium and solute mobility in a structured soil", *Water Resource Research*, vol. 32, pp. 561-570, 1996.

[28] U. von Gunten, and J. Zobrist, "Biogeochemical changes in groundwater infiltration systems: column studies", *Geochim. Cosmochim. Acta*, vol. 57, pp. 3895-3906, 1993.

[29] NAGRA, "Project Opalinus Clay: Safety Report: Demonstration of Disposal Feasibility (Entsorgungs-nachweis) for Spent Fuel, Vitrified High-Level Waste and Long-Lived Intermediate-Level Waste", NAGRA Technical Report NTB 02-05, NAGRA, Wettingen, Switzerland, 2002.

[30] T. Xu, R. Senger, and S. Finsterle, "Bentonite alteration due to thc processes during the early thermal period in an nuclear waste repository", In Proc. of TOUGH Symposium, Lawrence Berkeley National Laboratory, Berkeley, California, Sept. 14-16, 2009.

[31] R. Senger, and J. Ewing, "Evolution of temperature and water content in the bentonite buffer: Detailed modeling of two-phase flow processes associated with the early closure period –complementary simulations", Report NAB 08-53 for NAGRA, INTERA Inc., Austin, Texas 78758, 2008.

[32] R. Fernández, J. Cuevas, L. Sánchez, R. V. de la Villa, and S. Leguey, "Reactivity of the cement-bentonite interface with alkaline solutions using transport cells", *Applied Geochemistry*, vol. 21, pp. 977–992, 2006.

[33] M. Mella, K. Kovac, T. Xu, P. Rose, J. Mcculloch, and K. Pruess, "Calcite dissolution in geothermal reservoirs using chelants", In: Proceedings of Geothermal Resources Council (ed. Hamblin D.M.), pp. 151–57. Davis, California, USA, 2006.

[34] T. Xu, P. Rose, S. Fayer, and K. Pruess, "On modeling of chemical stimulation of an enhanced geothermal system using a high pH solution with chelating agent", *Geofluid*, vol. 9, pp. 167-177, 2009c.

CHAPTER 4

RT3D: Reactive Transport in 3-Dimensions

T. P. Clement[1*] and C. D. Johnson[2]

[1]Department of Civil Engineering, Auburn University, Auburn, Alabama, USA and [2]Environmental Sustainability Division, Pacific Northwest National Laboratory, Richland, Washington, USA

Abstract: RT3D is a Fortran-based software for simulating three-dimensional, multi-species, reactive transport of chemical compounds (solutes) in groundwater. RT3D is a MODFLOW-based solute transport code derived from MT3DMS, but with greatly expanded reaction capabilities. Although RT3D is often discussed in the context of accelerated *in situ* bioremediation (ISB) and natural/enhanced attenuation scenarios, RT3D is a general-purpose reactive transport code suitable for simulating a multitude of scenarios. Potential capabilities include simulation of inorganic reactions, geochemistry reactions, NAPL dissolution, mobile/immobile dual porosity, colloid transport, virus transport, heat transport, and risk analysis. With some degree of effort, RT3D can be linked to other codes to include time-varying porosity, interaction with the unsaturated zone, or full geochemistry. Commercial third-party graphical user interface software is typically used to define RT3D simulation model configurations and to visualize contours or isosurfaces of results. Results consist of whole-grid data sets at points in time and location-specific time series data sets. Multiple examples of RT3D application in the published literature are discussed, with a more in depth look at case studies for a monitored natural attenuation application and the design of an active remediation system.

Keywords: User-defined reactions, natural attenuation, bioremediation, reactive transport groundwater modeling, solute transport, three-dimensional, multi-species, finite difference, operator split, MODFLOW-based.

INTRODUCTION

Reactive transport modeling plays an important role in understanding the fate and transport of pollutants in groundwater aquifers [1-5]. A member of the MT3D [6] family of codes, RT3D (Reactive Transport in 3-Dimensions) [7-11] is a Fortran-based software for numerical simulation of three-dimensional, coupled, multi-species, reactive transport in groundwater where the reactions are mediated by complex processes such as microbial degradation and geochemical reactions. The original impetus for developing RT3D was for use in accelerated *in situ* bioremediation (ISB) and natural attenuation scenarios, hence RT3D comes with a suite of reaction modules targeted at Monitored Natural Attenuation (MNA) [12] and Enhanced Attenuation (EA) [13] for organic contaminants. However, the software structure provides the flexibility to easily adopt any type of user-specified reaction kinetics, allowing RT3D to be used to simulate a multitude of scenarios. This chapter describes the RT3D software, its capabilities, and how the software can be applied, including descriptions of two case studies involving bioremediation.

DESCRIPTION OF RT3D

RT3D is a software code derived from MT3DMS [14] for simulating reactive solute transport in the saturated subsurface (*i.e.*, groundwater aquifers). Similar to MT3DMS, RT3D does not solve groundwater flow equations and relies on an externally supplied head/velocity solution. The groundwater flow solution, which may be either steady state or time varying, is obtained from the U.S. Geological Survey's MODFLOW code [15, 16]. The MODFLOW LMT6 Package [17] (operated in "standard" mode) provides the linkage between flow and transport. Because MODFLOW is formulated for a finite difference domain, MODFLOW-based transport codes such as RT3D and MT3DMS are naturally also finite difference codes.

*Address correspondence to T.P. Clement: Department of Civil Engineering, Auburn University, Auburn, Alabama, USA; Tel: 1-334-844-6268; Email: clement@auburn.edu

Fan Zhang, Gour-Tsyh (George) Yeh, Jack C. Parker and Xiaonan Shi (Eds)

RT3D solves the coupled partial differential equations that describe reactive transport of multiple mobile and/or immobile chemical species in three-dimensional saturated groundwater systems.

The general reactive transport equations solved by RT3D can be written as shown in Equations (1) and (2) [7, 14, 18, 19].

rate of change = dispersion - advection + sources/sinks = internal source/sink + reaction- sorption

$$\theta\frac{\partial C_k}{\partial t}=\frac{\partial}{\partial x_i}\left(\theta D_{ij}\frac{\partial C_k}{\partial x_j}\right)-\frac{\partial\left(\theta v_i C_k\right)}{\partial x_i}+q_s C_{s,k}-q_s' C_k+r_{c,k}-r_{x,k}\ \forall\ \begin{matrix}k=1,2,\ldots,m\\ i,j=x,y,z\end{matrix} \qquad \textbf{(1)}$$

$$\rho_b\frac{\partial \breve{C}_p}{\partial t}=\breve{r}_{c,p}-r_{x,p}\ \forall\ p=m+1,m+2,\ldots,n \qquad \textbf{(2)}$$

In Equations (1) and (2), θ is the (effective) porosity of the aquifer (L^3L^{-3}), C_k is the aqueous-phase concentration of the kth chemical species (ML^{-3}), t is time (T), x_i is the distance along the respective axis (x, y, or z) of the Cartesian coordinate system (L), D_{ij} is the hydrodynamic dispersion coefficient tensor (L^2T^{-1}), v_i is the linear pore water velocity (LT^{-1}), q_s is the volumetric flow rate of sources (positive) or sinks (negative) per unit volume of aquifer (T^{-1}), $C_{s,\,k}$ is the concentration of the kth species in the sources or sinks (ML^{-3}), q'_s is the rate of change in transient groundwater storage (T^{-1}), $r_{c,\,k}$ is the aqueous reaction rate (consumption or production) per unit volume for the kth chemical species ($ML^{-3}T^{-1}$), $r_{x,\,k}$ ($r_{x,\,p}$) is the rate of mass transfer for kinetic (non-equilibrium) exchange of the kth (pth) chemical species between aqueous and solid phases ($ML^{-3}T^{-1}$), ρ_b is the dry bulk density of the subsurface sediments (M/L^3), $\breve{C}_p$ is the solid-phase concentration of the pth species on the solid phase (MM^{-1}), $\breve{r}_{c,\,p}$ is the solid-phase reaction rate (consumption or production) per unit mass for the pth chemical species ($MM^{-1}T^{-1}$), n is the total number of chemical species, and m is the number of mobile chemical species (*i.e.*, dissolved in the aqueous phase). Units are specified for these variables generically as M = mass, L = length, and T = time. Note that the linear pore water velocity times the porosity is equal to the specific discharge (Darcy flux). The reaction terms $r_{c,\,k}$ and $\breve{r}_{c,\,p}$ may be comprised of multiple terms to account for specific reactions that occur. Under the local equilibrium assumption (*i.e.*, that sorption is a much faster process than solute transport), the $r_{x,\,k}$ term is reconfigured into a retardation factor on the aqueous phase solute transport (with implicit tracking of solid-phase species mass).

The multi-species reactive transport equations for a set of chemical species are solved by RT3D though an Operator-Split (OS) numerical strategy [8]. The operator-split approach allows the sequential solution of the advection, dispersion, source/sink, and reaction/kinetic terms, in that order. RT3D employs the solution routines of MT3DMS [14] to solve the basic transport problem (advection, dispersion, and sources/sinks). This provides the user with multiple choices of advection solvers, including finite difference (solved either implicitly or explicitly), particle tracking, and total variation diminishing solvers. Dispersion and source/sink terms are always solved with finite difference equations (implicit or explicit).

The kinetic reaction terms are assembled as a set of ordinary differential equations (ODEs) that are always solved implicitly, using one of two types of solvers. One type is based on Runge-Kutta methods [20], which is a good choice for "well behaved" reaction kinetics. However, for stiff kinetics, which are oftentimes a result of widely varying time scales in the rate of change in concentrations, the automatically switching stiff/non-stiff "Gear" solver of Hindmarsh [21] (*i.e.*, LSODA) is the better choice.

Model Discretization and Boundary Conditions

A RT3D simulation model consists of a three-dimensional finite difference grid, a temporal scheme, initial/boundary conditions, and associated property/parameter values. Being a finite difference formulation, the model domain should be defined by a rectilinear grid (Fig. **1**). Row, column, and layer spacing is allowed to vary, but should avoid excessively rapid changes. The finite difference formulation assumes a uniform thickness for each layer, but it is possible to use a deformed grid, as discussed in the MODFLOW documentation [16]. The intent with a deformed grid is to represent a volume with uniform property values that correlates to a geological unit, thus layer thickness varies with lateral position (Fig. **2**). In an orthogonal grid, layer grid cells may have property values (*e.g.*, hydraulic conductivity) that represent an average value for multiple geological units.

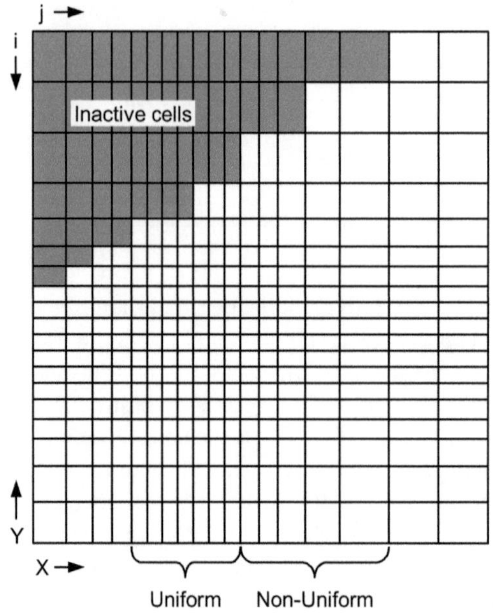

Figure 1: Plan view of a model grid showing uniform and non-uniform grid discretization approaches. Inactive grid cells can be specified to represent an irregular boundary. Increasing coordinate (X, Y) and grid cell index (j, i) directions are also noted.

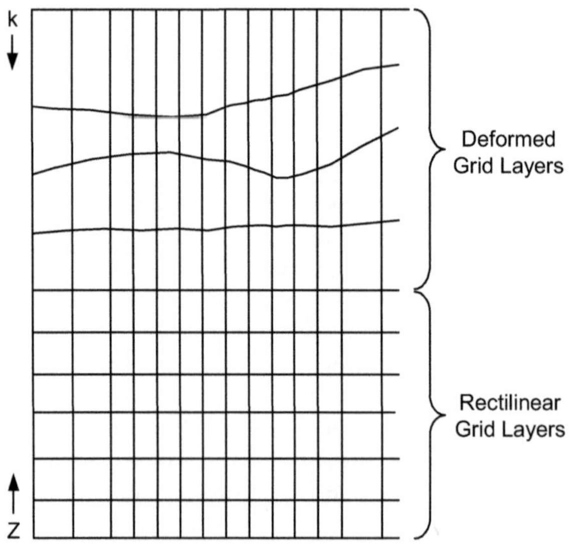

Figure 2: Cross section view of a model grid showing deformed and rectilinear grid layers. Increasing coordinate (Z) and grid cell index (k) directions are also noted.

A solute transport simulation requires that the user specify information about the time course of flow and concentration conditions. Each MODFLOW and RT3D simulation consists of one or more stress periods, which are subdivided into flow time steps and transport time steps (Fig. **3**).

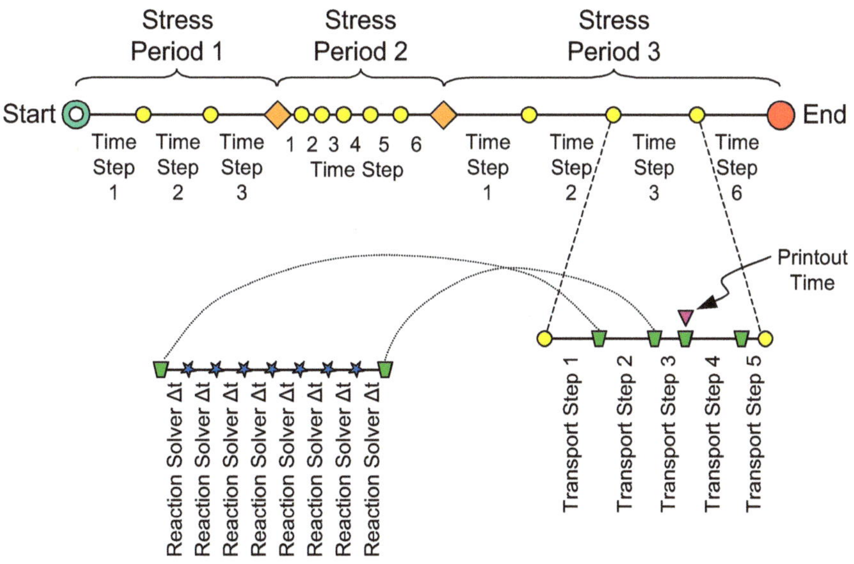

Figure 3: Depiction of the relationship between stress periods and the different types of time steps (in order of decreasing size – flow, transport, and reaction).

A stress period is a block of time that represents a particular set of flow and/or concentration conditions (*e.g.*, the on/off status of an extraction well). A head solution is calculated for each flow time step within a stress period. RT3D must include the same number of stress periods and flow time steps as were defined for the MODFLOW simulation except when the MODFLOW simulation uses a single stress period/flow time step to represent steady-state flow conditions. Multiple RT3D stress periods may be defined for a steady-state flow scenario to represent non-flow-related sources/boundary conditions. RT3D uses subdivisions of the flow time steps, termed transport time steps, in the solution of the contaminant transport problem. Transport time steps are the smallest user-configurable unit for temporal discretization in RT3D and have multiple options for explicit specification by the user (subject to calculated stability constraints) or automated calculation (with bounding guidance from the user). The differential equation solver uses subdivisions of each transport time step when solving the kinetic reaction equations.

RT3D requires that starting concentrations (for all species in all active grid cells of the model domain) and boundary condition concentrations (for mobile species) be specified. RT3D understands the MODFLOW areal recharge, evapotranspiration, well, drain, river, general head-dependent boundary, and constant-head boundary (constant or time-varying) types of source/sink/boundary conditions. Output flows (*e.g.*, an extraction well) remove mass based on the concentration in the model grid cell from which the flow is removed. In addition to flow-based sources/sinks/boundary conditions, RT3D provides a set of non-flow-related boundary condition/source types. These source types include constant concentration (for the duration of a simulation), specified concentration (stepwise transient), and exponentially decaying source.

Assumptions

As a numerical model, RT3D incorporates a number of assumptions about the processes represented by the code and the solution methods used. The assumptions applicable to MODFLOW [15] and MT3DMS [14] codes are relevant to RT3D simulations, since RT3D relies on the MODFLOW for the groundwater flow solution and on MT3DMS routines for most non-reaction components of solute transport. Assumptions specific to built-in RT3D reaction modules are discussed in the module documentation [7, 11].

Capabilities of RT3D

While RT3D is an excellent tool for a wide variety of groundwater reactive transport problems, there are some scenarios beyond RT3D's capabilities. Table **1** lists capabilities of RT3D that are available or can be achieved and features beyond the scope of the current RT3D framework.

Table 1: RT3D capabilities and limitations.

RT3D Capabilities	Achievable with RT3D	Beyond the Scope of RT3D
• Multi-species, 3-D reactive transport • Biodegradation kinetics • Inorganic reaction kinetics • Geochemical reactions • NAPL dissolution • Dual porosity (mobile/immobile) • Colloid transport	• Virus transport • Heat transport • Risk analysis • Variable porosity over time • Interaction with vadose zone or surface water • Full geochemistry with speciation	• Variable density flow • Dual permeability (mobile/mobile) • Surface water flow and transport • Multiphase transport • Unsaturated transport

RT3D was originally developed based on an interest in modeling bioremediation reactions for accelerated *in situ* bioremediation and monitored natural attenuation scenarios. Building off of the advection, dispersion, and source/sink framework of MT3DMS [14],

RT3D development provided multi-species and advanced general reaction capabilities. Thus, RT3D is a sister code to MT3DMS; the two codes share much of their core functionality, but they differ significantly in their reaction capabilities and each is a standalone software. The reaction capabilities of RT3D include built-in bioremediation/MNA reaction modules, as well as the capability to represent other processes (such as inorganic reactions [22], geochemistry reactions [23], NAPL dissolution [24], mobile/immobile dual porosity [25], and colloid transport [26]) *via* the user-defined reaction module.

In addition to the designed attributes, a number of other capabilities can be achieved with varying amounts of effort. Virus transport [27], heat transport [28, 29], and risk analysis [30, 31] take relatively little effort to implement. With a moderate effort, the capabilities of RT3D can be extended to include features such as time-varying porosity [23, 32], interaction with vadose zone or surface waters [33-35], and full geochemistry [36-[38]. Bringing about these types of extended capabilities generally entails a manual, loose, or tightly integrated coupling of an external code to RT3D. Significant effort would be needed to implement features such as variable density [39] or dual permeability flow/transport [40-43]. Note that in examples cited above, implementations in MT3DMS are used interchangeably as examples of extended capabilities for its sister code, RT3D.

Reaction Modules

The reaction terms used in Equations (1) and (2) depend on the reaction kinetics relevant to the problem of interest. With the operator-split numerical strategy, any type of reaction terms can be assembled in a set of differential equations of the form shown in Equation (3), where the local equilibrium assumption has been applied and R_k is the retardation factor (unitless). The expression for $r_{c,k}$ may be comprised of a number of terms, depending on the processes involved and assumptions made. Example reaction terms include first-order decay (often used for natural attenuation or as a simplification of complex processes), Monod growth (*i.e.*, to represent bacterial growth), and inhibition (*e.g.*, lack of oxygen would inhibit an aerobic reaction).

$$R_k \frac{dC_k}{dt} = r_{c,k} \quad \forall \quad k = 1,2,\ldots,n \tag{3}$$

A suite of pre-programmed reaction modules targeted at MNA and EA for organic contaminants are included with RT3D. A number of other pre-programmed reaction modules (no reaction, two-species instantaneous reactions, rate-limited sorption, double Monod model, and sequential first-order decay) provide general tools that can be applied for bioremediation or other scenarios. The pre-programmed reaction modules provide a ready-to-use reaction scenarios based on specific assumptions about reaction mechanisms and pathways. The user must, however, define appropriate reaction parameter values (*e.g.*, stoichiometry, reaction rates), since any "default" values have little bearing on a specific actual site. Details of the pre-programmed reaction modules and their data input requirements can be found in the RT3D documentation [7, 11].

For scenarios with contaminants or reaction kinetics other than those available in the pre-programmed reaction modules, the RT3D software provides a "user-defined" reaction module. The RT3D user can define their own reaction kinetics and, with relative ease, modularly plug those kinetics into RT3D. This approach is essentially as simple as writing the relevant set of differential equations in the same manner as one might for a computational algebra program. To use the reaction equations, the user must take the additional step of compiling the designated subroutine (not the entire RT3D code) with a Fortran compiler. The RT3D documentation [7, 9] discusses how to go about writing and implementing a user-defined reaction module and a template Fortran subroutine is provided to the user.

VALIDATION

The validity of the RT3D code is derived from several sources, including the code pedigree, comparison to analytical results, comparison to results from other numerical codes, and widespread application. The code verification efforts for MT3DMS [14] (itself a well-established code) generally apply to RT3D, since RT3D is a sister code sharing the same basic transport routines. Clement *et al.* [8] validated RT3D code performance by comparing RT3D results for several test problems with existing analytical solutions as well as results from other numerical codes. RT3D solutions for non-reactive test cases were found to produce the same results as MT3D, confirming the equivalence of the basic transport routines. RT3D results closely resembled the analytical solution from the literature for a problem with one-dimensional transport, adsorption, and sequential first-order degradation. Solutions for hydrocarbon and oxygen plumes in a uniform, two-dimensional aerobic aquifer with instantaneous reactions were obtained using the method of characteristics advection solver using RT3D and the numerical code BIOPLUME II [44]; again, the results compared favorably. Clement [45] and Quezada *et al.* [46] have compared RT3D results against a set of analytical model results. Clement [45] compared his analytical solution for one-dimensional transport of non-adsorbing species undergoing serial, parallel, and reversible reactions with first-order reaction kinetics to the RT3D results and found that they were nearly identical. Quezada *et al.* [46] also found nearly identical results between RT3D and an analytical solution for test problems of one-dimensional transport of adsorbing species (distinct retardation factors) with reversible reactions under zero and non-zero initial conditions and for a three-dimensional problem involving serial-parallel reactions. The widespread successful application of RT3D (and MT3DMS) lends credence to the usefulness of the code.

In the published literature, RT3D has been used to solve multiple types of field- and laboratory-scale reactive transport problems. Truex *et al.* [47] provide the details of a modeling effort for predicting the fate and transport of a chlorinated solvent plume at a complex Superfund site location in Baton Rouge, Louisiana. Tartakovsky *et al.* [22] discuss the use of RT3D to design a bioremediation system to treat a nitrate-contaminated site located within a Canadian airport. As a part of this field study, extensive groundwater data was collected at a nitrate-contaminated site within the airport. Ethanol was used as a carbon source to simulate the growth and activity of indigenous denitrifying bacterial population. The numerical-model-based predictions were used to evaluate the nitrate distribution patterns during this bioremediation effort. Researchers at Michigan State University used a numerical model to design a bioremediation system, which employed denitrifying bacteria to remediate a carbon tetrachloride plume [48]. Phanikumar and McGuire [36] reported the development and application of a reactive transport model that coupled RT3D with PHREEQC-2. The combined code, BGTK, was used to model test problems involving microbial transport in a laboratory column and redox zonation in a contaminated aquifer. Researchers at Cornell University developed a comprehensive reaction description for modeling an active biotreatment system for remediating chlorinated solvent plumes. The reactions were integrated within the RT3D code and the resulting tool was used to explore various optimal design strategies for managing solvent plumes [49]. Huang *et al.* [50] used a laboratory model to study electron-acceptor and electron-donor mixing patterns under instantaneous reaction conditions; they used a numerical model to recreate their experimental results. Lu *et al.* [51] used the RT3D code to model a petroleum plume at the Hill Air Force Base. Clement *et al.* [52] used RT3D to simulate the natural attenuation processes occurring at a chlorinated solvent site located at Dover Air Force Base.

Clement *et al.* [24] present a numerical formulation to model a complex transport system that described multiple, coupled, rate-limited processes including DNAPL dissolution, rate-limited sorption, and biological reaction kinetics. Lee *et al.* [53] used the RT3D code to model biotransformation and transport occurring within a remediation barrier used for treating nitrogenous waste products. Gomez *et al.* [54] used RT3D to evaluate the effects of the common fuel additive ethanol on benzene fate and transport in fuel-contaminated groundwater systems. Lim *et al.* [55] applied RT3D for predicting microbial reduction and transport of arsenic in laboratory-scale column experiments. Rolle *et al.* [56] used RT3D to model redox processes occurring under a landfill site in Piedmont, Italy.

USING RT3D

Requirements to set up RT3D are generally minimal. The standalone distribution package requires about 8 MB of hard disk space for the executable, source files, documentation, and examples. A version of MODFLOW with the LMT6 package [17] is required to provide the groundwater flow solution. A text editor is required for manual creation of model input data files or a graphical user interface software can be used for fielded entry of data and creation of model input data files. A Fortran compiler is required to employ a user-defined reaction module.

Performance for RT3D simulations is highly dependent on the modeling scenario (model grid size, solver choices, nature of boundary conditions/source/sinks, and the reaction kinetics) and the available computer hardware (processor speed and available memory being the key issues). Recognizing this variability, it is still illustrative to present example performance information. Table **2** gives several examples of performance and resource usage on an Intel® Pentium® 4, 3.2 GHz, single-processor machine with 3.25 GB of physical RAM and using Microsoft® Windows® XP for the operating system. Increased performance is expected on more modern machines.

Table 2: Examples of RT3D computational performance.

Scenario	Run Time	Output File Size (per species)
• 6324 grid cells • BTEX degradation, sequential electron acceptors – 6 chemical species • simulation time of 2 years in 5-day steps • 9 output times	4 minutes	0.5 MB
• 2.18×10^5 grid cells • sequential first-order reactions – 4 chemical species • simulation time of 20 years in 73 day steps • 11 output times	15 minutes	18.3 MB
• 7.3×10^5 grid cells • first-order anaerobic/aerobic reactions – 9 chemical species • simulation time of 50 years in 10 day steps • 11 output times	19 hours	61 MB (< 10 MB compressed)

RT3D reactive solute transport simulation models are specified in a manner similar to that used for MODFLOW models. Input files defining a site model can be defined manually using a text editor or can be assembled using a Graphical User Interface (GUI). GUI software helps organize the process of defining the required RT3D input information, provides data management, and supplies visualization capabilities for model configuration and the RT3D output. Several commercial GUI software products that support interfaces to RT3D are available, including the Groundwater Modeling System (GMS; www.aquaveo.com), Visual Modflow (www. swstechnology.com/groundwater-software), Processing Modflow for Windows

(PMWIN; www.pmwin.net), and Groundwater Vistas (www.groundwatermodels.com). All of these GUI software products provide suitable interfaces for defining a RT3D simulation and the selection of which one to use will generally depend on user needs and the economics. However, GMS currently has an advantage of a built-in flexible interface for applying any user-defined reaction module.

The parameters required for a RT3D simulation fall into two categories: aquifer-related and reaction-related. Aquifer-related parameters include porosity of the geological media and diffusion/dispersion parameters. The porosity parameter represents the effective porosity through which flow occurs (*i.e.*, does not include pore space with stagnant water). Diffusion is the movement of molecules as a result of concentration gradients in the porous medium and is often negligible relative to structurally induced dispersion, but may be important at low groundwater velocities. Dispersion represents the divergence of mass transport (flow pathways) due to the irregular structure and heterogeneity of the porous medium. RT3D allows the specification of longitudinal (along the primary flow direction), transverse (perpendicular to the primary flow direction), and vertical dispersivity coefficients (α_L, α_T, α_Z, respectively), which are combined with groundwater velocity vector components to arrive at the dispersion coefficients used in solving Equation (1). The number and type of reaction-related parameters are highly dependent on the contaminant species, reaction pathways, and chemical transformation processes that comprise the reaction scenario of interest. Reaction parameters may include bulk density, adsorption coefficients, stoichiometric/mass yields, reaction rate coefficients, half-saturation coefficients, inhibition coefficients, mass transfer coefficients, *etc.* If local equilibrium is assumed, then coefficients for linear, Freundlich, or Langmuir equilibrium partitioning to soil are required. Parameters needed for built-in RT3D reaction modules are discussed in the RT3D documentation [7, 11].

Spatial variation in solute and geochemical conditions may warrant a non-uniform spatial distribution of reaction activity, which can be implemented using zones of activity. Johnson *et al.* [57] successfully applied this reaction zone approach in the assessment of MNA at a chlorinated solvent site. For both spatial and temporal variation in reaction activity, reaction activity could be linked to the concentration of an indicator chemical species such that changes in the indicator species concentration influence the rates of contaminant transformation. All built-in RT3D reaction module reaction parameters and the sorption parameters have the capability to be spatially variable on a cell-by-cell basis.

RT3D can provide several types of output. The primary output from RT3D is a set of whole-grid data sets (one for each chemical species) describing solute concentrations for all grid cells in the model at times specified by the user. Whole-grid data sets may be used to examine the spatial distribution of time-varying solute concentrations in the form of concentration contours or isosurfaces (in third-party visualization software). RT3D can also generate concentration data at specified observation point locations over time, which can be used to generate time series plots for fixed points in space. Several ancillary output files are generated with each simulation run to echo back the model configuration input, report calculated quantities, note solver progress, and describe the model mass balance.

NATURAL ATTENUATION CASE STUDY – PPI BROOKLAWN SITE

The predicted fate and transport of groundwater plumes can be used as a line of evidence in assessing MNA and can provide a basis of comparison for long-term monitoring to verify the forecasted attenuation. These modeling functions can provide an acceptable technical basis for regulatory approval of MNA as site remedy. The Petro-Processors, Inc. (PPI) Brooklawn site [47, 58, 59] provides a case study where RT3D simulations indicated that natural attenuation would be an appropriate remedial strategy.

Following U.S. EPA guidelines [2], lines of evidence were assessed to determine the suitability of MNA as a site remedy at the PPI Brooklawn site near Baton Rouge, Louisiana. One line of evidence was reactive transport modeling to assess migration of the chlorinated ethenes (tetrachloroethene and daughter products) and chlorinated ethanes (1, 1, 2, 2 - tetrachloroethane and daughter products).

The Brooklawn site is located about a mile from the Mississippi River at the interface between inland Pleistocene sedimentary clay deposits and near-river alluvial sediments. Complexities at the site included highly variable geology, the transient flow conditions, and a large number of waste components. Hydraulic gradients at the site are controlled by seasonal Mississippi River stage fluctuations and can vary significantly over time (including reversal in the gradient direction). An estimated 160,000 tons (145 million kg) of multiple-constituent NAPL had been disposed of in the Pleistocene clay at the Brooklawn site. Subsequent NAPL migration resulted in a dissolved groundwater contamination plume containing chlorinated ethenes and chlorinated ethanes. The initial remedial action at the site was hydraulic containment and NAPL source recovery. MNA was investigated as an alternative remedial option for the dissolved plume.

Degradation pathways and biodegradation reaction rates were assessed in unamended laboratory microcosms with site sediment and groundwater. Chlorinated ethane and chlorinated ethene contaminants were found to be completely dechlorinated rapidly. Reaction rates and degradation pathways from the microcosm tests were incorporated into a user-defined RT3D reaction module.

Due to the transient nature of the hydraulic gradient and the changes imparted by the hydraulic control actions, direct field measurements could not be used to estimate natural attenuation processes. Instead, field and laboratory data were used in conjunction with reactive transport modeling to evaluate MNA within the framework of the EPA protocol [2]. A numerical flow and transport model was developed to describe the transient flow conditions, the complex hydrogeology, and the complex contaminant degradation pathways. The transport model was calibrated by matching simulation results to measured constituent data along transects downgradient of the contaminant source area and at plume boundaries.

Simulations were conducted to predict the fate of the plume in multiple scenarios, including discontinuation of current hydraulic containment activities. As shown in Fig. **4**, the predictive simulations indicated that the plume would reach a steady-state condition and cease migration prior to reaching the receptors of concern. These results, in combination with other lines of evidence, led to regulatory acceptance of MNA as the remedial action for the dissolved contamination plume at the site. The long term monitoring plan is now being implemented.

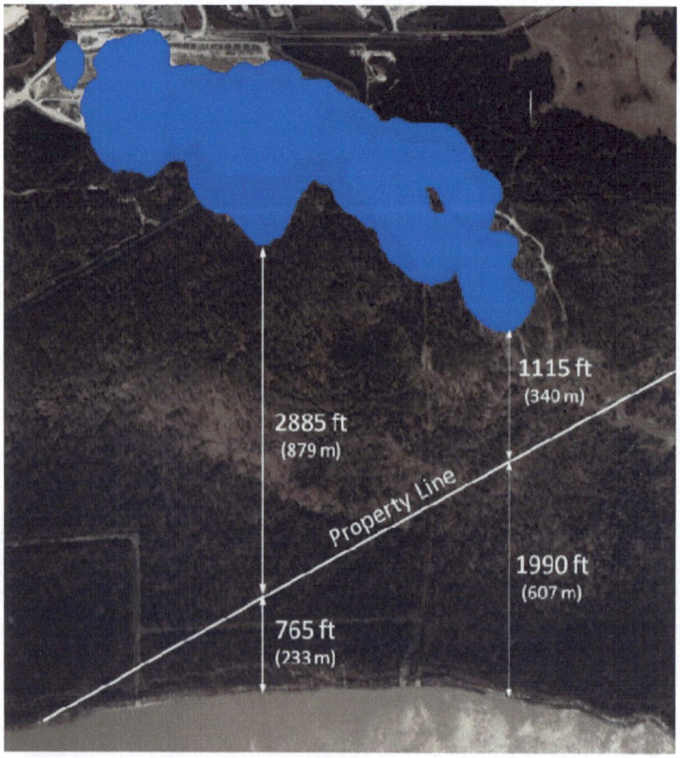

Figure 4: Maximum extent of VC plume relative to the property line and the Mississippi river.

ACTIVE REMEDIATION CASE STUDY – POINT MUGU SITE 24

RT3D can be used for simulation of active remediation (*e.g.*, accelerated bioremediation, chemical oxidation, *etc.*). Either an existing reaction module or a user-defined reaction module is required that can represent the relevant processes identified in the conceptual model. The accelerated *in situ* bioremediation system implemented at Point Mugu by Johnson *et al.* [60] is an example where RT3D was used to design an effective treatment system.

Installation Restoration Program (IRP) Site 24 at Naval Base Ventura County's Point Mugu facility in California had groundwater contaminated with Trichloroethene (TCE) as a result of releases from an underground oil/water separator. The chlorinated ethene plume was located in a sandy zone underneath a confining clay layer. Although highly permeable, the groundwater gradient was very low. A natural attenuation evaluation [61] concluded that active remediation was needed for this site. A pilot-scale demonstration of anaerobic ISB began in December 1998 using a 2-well recirculation cell to distribute lactic acid volumetrically [62, 63]. TCE and DCE were rapidly dechlorinated to VC. Monitoring through April 2002 showed that the rate of decline in VC concentrations was at least two orders of magnitude less than that found in lactate-fed laboratory microcosms with site sediments and groundwater. *In situ* aerobic cometabolic biodegradation was selected as a follow-on treatment to accelerate destruction of VC.

Aerobic cometabolism of chlorinated ethenes is mediated by non-specific monooxygenase (MMO) enzymes of methanotrophic bacteria (and other microbial species) in an epoxidation process that inserts an oxygen atom into the carbon-carbon double of the ethene. The resulting epoxide is not stable and breaks apart into non-hazardous constituents. In the epoxidation process the MMO enzyme is inactivated, implying a limited capacity for transformation of chlorinated ethenes unless the bacteria are stimulated to produce more MMO enzyme. Yet, the growth substrate (methane) inhibits chlorinated ethene destruction because the MMO enzyme has a higher affinity for the methane than for chlorinated ethenes. A user-defined RT3D reaction module was developed for this work to address aerobic cometabolism of VC, considering both the transformation capacity and chlorinated ethene degradation inhibition processes.

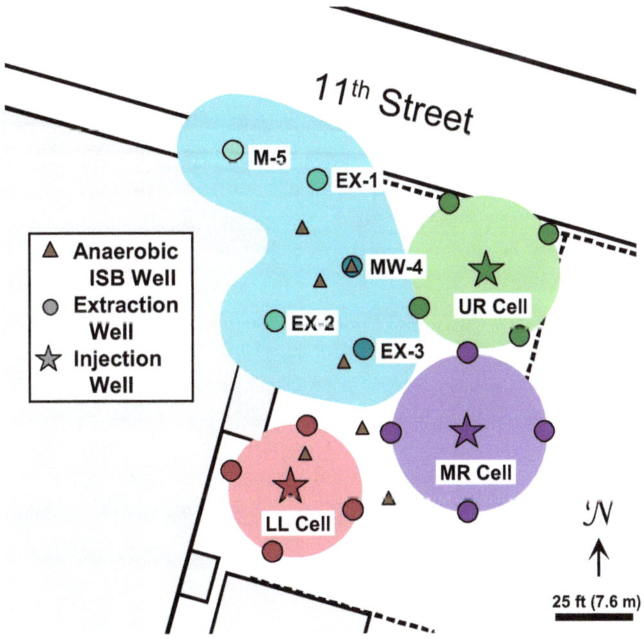

Figure 5: *In situ* biofilter design for Pt. Mugu Site 24 showing the locations of five-spot recirculation cells and the extended groundwater extraction zone.

The ISB design for aerobic cometabolism of VC was based on the concept of an *in situ* biofilter [64]. This concept involves creation of an *in situ* bioactive zone around an injection well. Injected contaminants are destroyed within the bioactive zone. RT3D was used to derive a pulsed-flow design [64] using repeated cycles of high-flow rate injection (12 hours) and rest (about 3 days) was used to deliver nutrients (methane, oxygen, nitrate) for stimulating the *in situ* biofilter (MMO enzyme production) and VC destruction. A pilot-scale (single recirculation cell) system was fielded in April 2002 and demonstrated that the pulsed injection operating strategy was effective in treating VC. Fig. **5** shows the location of the single recirculation cell pilot test (the LL cell) and Fig. **6** shows the contours of VC after the pilot-test, based on observed concentration data.

Figure 6: Estimated concentration contours after application of the single recirculation cell pilot-test (based on field data) for the Pt. Mugu Site 24 site.

The single recirculation cell pilot-scale system was expanded to a demonstration of treatment with multiple recirculation cells. RT3D was used to assess design options for placement of treatment cells and the groundwater withdrawal strategy, with the objective of achieving maximum VC capture while maintaining high hydraulic control (to minimize plume spreading). The multi-cell system design was conducted with several technical, logistical, and budgetary constraints. Placement of bioactive zones was restricted to locations outside the former anaerobic treatment zone to avoid the need for re-oxidation of the heavily reduced anaerobic zone sediment. Logistic and budgetary constraints limited operations to no more than 3 recirculation cells, operated in series over a time frame of less than 16 weeks.

The multi-cell design process determined the placement of wells and the operational parameters to best treat the VC contamination within the specified constraints. Fig. **5** shows the locations of the three recirculation cells for the selected multi-cell system design. The operational strategy consisted of two phases: an initial phase of biofilter stimulation followed by a phase of VC destruction. In the first phase, groundwater was extracted using the wells of each treatment cell to produce zones of low VC concentration and high biological activity (Fig. **7**).

The second phase supplemented the injected groundwater by extracting (with a fifth extraction well) from high VC concentration areas (in the former anaerobic treatment zone). Predictions estimated a significant reduction in the VC plume by 8 weeks (Fig. **8**) and that the plume would consist of a strip of nominally 50 μg/L VC underneath the difficult-to-access building footprint after 12 weeks. Note that the high concentration contours to the south belong to a previously unidentified chloroethene plume, which was not addressed in these demonstrations.

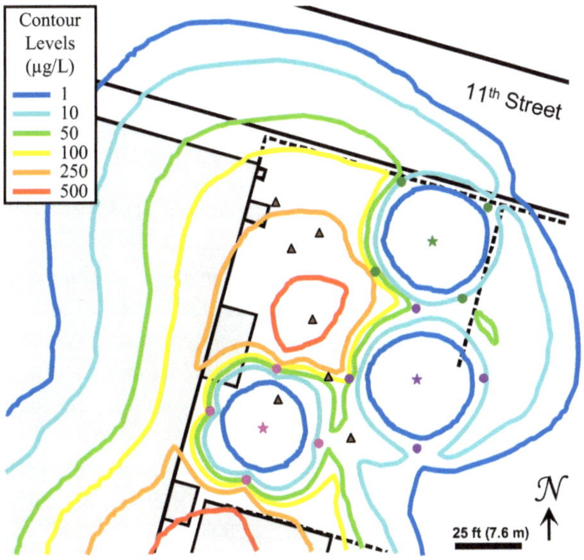

Figure 7: Simulation predictions of the VC concentration distribution for the selected multi-cell system design after three weeks of within-cell recirculation.

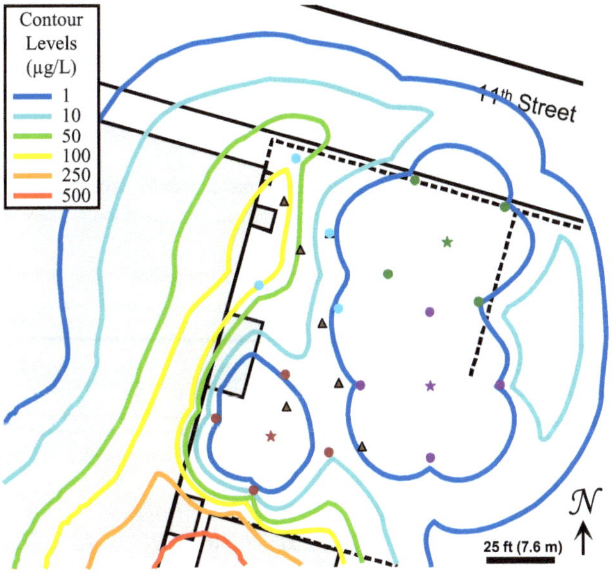

Figure 8: Simulation predictions of VC concentration distribution for the selected multi-cell recirculation design after 8 weeks of operation with groundwater extraction from a fifth well in the former anaerobic treatment zone.

Project constraints limited the actual time available for field operations to about 9 weeks, but this was sufficient to demonstrate process effectiveness. Process monitoring indicated that VC was successfully being destroyed in the *in situ* biofilter zones. Results for several one-cycle-long periods, representing the single-pass biofilter effectiveness, showed an average reduction in VC concentration of 81%, with concentrations decreasing from a range of 18 to 88 µg/L to a range of 3 to 14 µg/L. Expectations were that continued operation would have treated the remainder of the plume, although verification of concentrations under the building would have been challenging.

ACKNOWLEDGEMENT

This work was, in part, funded by the Office of Science (BER), U.S. Department of Energy Grant No. DE-FGO2-06ER64213 at Auburn University.

Microsoft and Windows are either registered trademarks or trademarks of Microsoft Corporation in the United States and/or other countries. Intel and Pentium are trademarks of Intel Corporation in the U.S. and other countries.

REFERENCES

[1] ASTM, "Standard guide for subsurface flow and transport modeling", D 5880-95 (2000), *ASTM International*, West Conshohocken, Pennsylvania. 1995a.

[2] U.S. EPA, *Technical protocol for evaluating natural attenuation of chlorinated solvents in ground water.* EPA/600/R-98/128, U.S. Environ-mental Protection Agency, Office of Research and Development, Washington, D.C. 1998.

[3] M. J. Truex, C. J. Newell, B. B. Looney, and K. M. Vangelas, *Scenarios evaluation tool for chlorinated solvent monitored natural attenuation.* WSRC-STI-2006-00096, Savannah River National Laboratory, Washington Savannah River Company, Aiken, South Carolina, 2006.

[4] AFCEE, *Principles and practices of enhanced anaerobic bioremediation of chlorinated solvents.* Air Force Center for Environmental Excellence, Brooks City-Base, Texas. Available online at: http://www.afcee.af.mil/shared/media/document/AFD-071130-020.pdf (accessed July 9, 2009), 2004.

[5] V. Batu, *Applied flow and solute transport modeling in aquifers.* Taylor and Francis Group, CRC Press, Boca Raton, Florida, 2006.

[6] C. Zheng, *MT3D: A Modular Three-dimensional transport model for simulation of advection, dispersion and chemical reactions of contaminants in groundwater systems.* S.S. Papadopulos & Associates, Inc., Rockville, Maryland. Available online at: http://hydro.geo.ua.edu/mt3d/Mt3d_1990.pdf (accessed on July 9, 20 09), 1990.

[7] T. P. Clement, *RT3D-A modular computer code for simulating reactive multi-species transport in 3-dimensional groundwater aquifers.* PNNL-11720, Pacific Northwest National Laboratory, Richland, Washington, 1997.

[8] T. P. Clement, Y. Sun, B. S. Hooker, and J. N. Petersen, "Modeling multispecies reactive transport in ground water", *Ground Water Monitoring Remediation.*, vol. 18, pp. 79-92, 1998.

[9] T.P. Clement and C. D. Johnson, *RT3D v2.5 Update document.* Pacific Northwest National Laboratory, Richland, Washington. Available online (in the documentation section of the downloads page) at: http://bioprocess.pnl.gov/-rt3d.htm (accessed July 9, 2009), 2002.

[10] C. D. Johnson, M. J. Truex, and T. P. Clement, *Natural and enhanced attenuation of chlorinated solvents using RT3D.* PNNL-15937, Pacific Northwest National Laboratory, Richland, Washington, 2006.

[11] C. D. Johnson and M. J. Truex, *RT3D: Reaction modules for natural and enhanced attenuation of Chloroethanes, Chloroethenes, Chloromethanes, and Daughter products.* PNNL-15938, Pacific Northwest National Laboratory, Richland, Washington, 2006.

[12] U.S. EPA, *Use of monitored natural attenuation at superfund, RCRA corrective action, and underground storage tank sites.* OSWER Directive number 9200.4-17P, United States Environmental Protection Agency, Office of Solid Waste and Emergency Response, Washington, D.C. 1999a.

[13] T. Early, B. Borden, M. Heitkamp, *et al.*, *Enhanced attenuation: A reference guide on approaches to increase the natural treatment capacity of a system.* WSRC-TR 2005-00198, Rev.0, Westinghouse Savannah River Company, Aiken, South Carolina, 2006.

[14] C. Zheng and P. P. Wang, *MT3DMS: A modular three-dimensional multispecies transport model for simulation of advection, dispersion, and chemical reactions of contaminants in groundwater systems; documentation and user's guide.* SERDP-99-1, U.S. Army Corps of Engineers, Engineer Research and Development Center, Vicksburg, Mississippi, 1999.

[15] A. W. Harbaugh, E. R. Banta, M. C. Hill, and M. G. McDonald, *MODFLOW-2000, the U.S. geological survey modular ground-water model - user guide to modularization concepts and the ground-water flow process.* Open file report 00-92, U.S. Geological Survey, Reston, Virginia, 2000.

[16] A. W. Harbaugh, *MODFLOW-2005: the U.S. geological survey modular ground-water model the ground-water flow process.* Techniques and Methods 6-A16, U.S. Geological Survey, Reston, Virginia, 2005.

[17] C. Zheng, M. C. Hill, and P. A. Hsieh, *Modflow-2000: the U.S. geological survey modular ground-water model - user guide to the LMT6 package, the linkage with MT3DMS for multi-species mass transport modeling.* Open file report 01-82, U.S. Geological Survey, Reston, Virginia, 2001.

[18] C. Zheng and G. D. Bennett, *Applied Contami-nant transport modeling.* John Wiley and Sons, New York, 2002.

[19] L. F. Konikow, "Use of numerical models to simulate groundwater flow and transport", in *Environmental Isotopes in the Hydrological Cycle*: Principles and Applications, vol. VI: Modeling, W.G. Mook, ed. Isotope Hydrology Section,

International Atomic Energy Agency, Vienna, Austria. Available online at: http://www-naweb.iaea.org/napc/ih/IHS_resources3_ publication_en.html (accessed July 9, 2009), 2000.

[20] J. C. Butcher and G. Wanner, "Runge-Kutta methods: some historical notes", *Applied Numerical Mathematics*, vol. 22, pp. 113-151, 1996.

[21] A. C. Hindmarsh, "ODEPACK, A systematized collection of ODE solvers", in *Scientific Computing* (vol. 1 of IMACS Transactions on Scientific Computation), R.S. Stepleman (ed.). North-Holland, Amsterdam. pp. 55-64. Available online at: https://computation.llnl.gov/casc/nsde/pubs/u88007.pdf (accessed July 9, 2009), 1983.

[22] B. Tartakovsky, D. Millette, S. Delisle, and S. R. Guiot, "Ethanol-stimulated bioremediation of nitrate-contaminated ground water", *Ground Water Monitoring Remediation.*, vol. 22, pp. 78-87, 2002.

[23] L. Li, C. H. Benson, and E. M. Lawson, "Modeling porosity reductions caused by mineral fouling in continuous-wall permeable reactive barriers", *Journal of Contaminant Hydrology*, vol. 83, pp. 89-121, 2006.

[24] T. P. Clement, T. R. Gautam, K. K. Lee, M. J. Truex, and G. B. Davis, "Modeling of DNAPL-dissolution, rate-limited sorption and biodegra-dation reactions in groundwater systems", *Bioremediation Journal*, vol. 8, pp.47-64, 2004.

[25] Y. Sun, J. N. Petersen, J. Bear, T. P. Clement, and B. S. Hooker, "Modeling microbial transport and biodegradation in a dual-porosity system", *Transport in Porous Media*, vol. 35, pp. 49-65, 1999.

[26] K. M. Coulibaly, C. M. Long, and R. C. Borden, "Transport of edible oil emulsions in clayey sands: one-dimensional column results and model development", *Journal of Hydrologic Engineering*, vol. 11, pp. 230-237, 2006.

[27] G. R. Barth and M. C. Hill, "Parameter and observation importance in modeling virus transport in saturated porous media-investigations in a homogenous system", *Journal of Contaminant Hydrology*, vol. 80, pp. 107-129, 2005.

[28] A. Battermann, J. M. Gablonsky, A. Patrick, *et al.*, "Solution of a groundwater control problem with implicit filtering", *Optimization and Engineering.*, vol. 3, pp. 189-199, 2002.

[29] R. Sethi and A. Di Molfetta, "Heat transport modeling in an aquifer downgradient a municipal solid waste landfill in Italy", *American Journal of Environmental Sciences*, vol. 3, pp. 106-110, 2007.

[30] I. M. Khadam and J. J. Kaluarachchi, "Multi-criteria decision analysis with probabilistic risk assessment for the management of contaminated ground water", *Environmental Impact Assessment Review*, vol. 23, pp. 683-721, 2003.

[31] G. Whelan, K. J. Castleton, and M. A. Pelton, *FRAMES-2.0 Software system: linking to the groundwater modeling system (GMS) RT3D and MT3DMS models*. PNNL-16758, Pacific Northwest National Laboratory, Richland, Washington, 2007.

[32] T. Bolisetti and S. Reitsma, "Numerical simulation of chemical grouting in heterogeneous porous media", in *Grouting and Ground Treatment*, L.F. Johnsen, D.A. Bruce, and M.J. Byle, eds. ASCE Publications, Reston, Virginia. pp. 1454-1465. doi:10.1061/40663 (2003)123, 2003.

[33] A. Davis, S. Kamp, G. Fennemore, R. Schmidt, and M. Keating, "A risk-based approach to soil remediation modeling", *Environmental Science & Technology*, vol. 31, pp. 520A-525A, 1997.

[34] J. Weaver, J. Zhang, M. Tonkin, and R. J. Charbeneau. "Modeling the transport of ethanol fuel blends with the combined HSSM and MT3D models", Presented at *21st Annual National Tanks Conference and Expo*, Sacramento, CA, March 30-April 01, 2009. Available online at: http://oaspub.epa.gov/eims/eimscomm.getfile?p_download_id= 488039 (accessed July 9, 2009), 2009.

[35] R. B. Thoms, *Simulating fully coupled overland and variably saturated subsurface flow using MODFLOW*. Master's Thesis, Oregon Health and Science University, Department of Environmental Science and Engineering, Portland, Oregon, 2003.

[36] M. S. Phanikumar and J. T. McGuire. "A 3D partial-equilibrium model to simulate coupled hydrogeo-logical, microbiological, and geochemical processes in subsurface systems", *Geophysical Research Letters*, 31(11): L11503. doi:10. 1029/2004GL 019468, 2004.

[37] H. Prommer, D. A. Barry, and G. B. Davis. "Numerical modeling for design and evaluation of groundwater remediation schemes", *Ecological Modeling*, vol. 128, pp. 181-195, 2000.

[38] H. Prommer, D. A. Barry, and C. Zheng, "MODFLOW/MT3DMS-based reactive multi-component transport modeling", *Ground Water*, vol. 41, pp. 247-257, 2003.

[39] W. Guo and C. D. Langevin, *User's guide to SEAWAT: A computer program for simulation of three-dimensional variable-density ground-water flow*. Techniques of Water-Resources Investigations 6-A7, U.S. Geological Survey, Tallahassee, Florida, 2002.

[40] G. Teutsch and M. Sauter, "Groundwater modeling in Karst Terranes: scale effects, data acquisition, and field validation", in: *Proc. Third Conference on Hydrogeology, Ecology, Monitoring, and Management of Ground Water* in Karst Terranes, Nashville, Tennessee. U.S. Environmental Protection Agency and National Ground Water Association, 1991.

[41] M. Beyer and U. Mohrlok, "A double continuum approach for determining contaminant transport in fractured porous media", in *Groundwater in Fractured Rocks*, J. Krásný and J.M. Sharp, eds. Taylor & Francis/Balkema, Leiden, The Netherlands, 2007.

[42] S. L. Painter, R. T. Green, and A. Y. Sun, "Dual Conductivity Module (DCM), A MODFLOW package for modeling flow in Karst aquifers", in *U.S. Geological Survey Karst Interest Group Proceedings*, Rapid City, South Dakota, September 12-15, 2005, E.L. Kuniansky, ed. SIR 2005-5160, U.S. Geological Survey, Reston, Virginia, 2005.

[43] W. B. Shoemaker, E. L. Kuniansky, S. Birk, S. Bauer, and E. D. Swain. *Documentation of a conduit flow process (CFP) for MODFLOW-2005*. Techniques and Methods 6-A24, U.S. Geological Survey, Reston, Virginia, 2008.

[44] H. S. Rifai, P. B. Bedient, R. C. Borden, and J. F. Haasbeek, *BIOPLUME II: Computer model of two-dimensional contaminant transport under the influence of oxygen limited biodegradation in ground water*. U.S. Environmental Protection Agency, Office of Research and Development, Ada, Oklahoma, 1987.

[45] T. P. Clement, "Generalized solution to multi-species transport equations coupled with a first-order reaction network", *Water Resources Research*, vol. 37, pp. 157-163, 2001.

[46] C. R. Quezada, T. P. Clement, and K. K. Lee, "Generalized solution to multi-dimensional multispecies transport equations coupled with a first-order reaction network involving distinct retardation factors", *Advances in Water Resources*, vol. 27, pp. 507-520, 2004.

[47] M. J. Truex, C. D. Johnson, J. R. Spencer, and T. P. Clement, "Evaluating natural attenuation of chlorinated solvents at a complex site", in *Proceedings of the Third International Conference on Remediation of Chlorinated and Recalcitrant Compounds*, A.R. Gavaskar and A.S.C. Chen (eds.). Battelle Press, Columbus, Ohio. Paper 2D-06, 2002.

[48] M. S. Phanikumar, D. W. Hyndman, and C. S. Criddle, "Biocurtain design using reactive transport models", *Ground Water Monitoring Remediation*, vol. 22, pp. 113-123, 2002.

[49] M. B. Willis and C. A. Shoemaker. "Engineered PCE dechlorination incorporating competitive biokinetics: optimization and transport modeling", in *Bioremediation and Phytoremediation of Chlorinated and Recalcitrant Compounds*, G.B. Wickramanayake, A.R. Gavaskar, B.C. Alleman and V.S. Magar, eds. Battelle Press, Columbus, Ohio. pp. 311-318, 2000.

[50] W. E. Huang, S. E. Oswald, D. N. Lerner, C. C. Smith, and C. Zheng, "Dissolved oxygen imaging in a porous medium to investigate biodegradation in a plume with limited electron acceptor supply", *Environmental Science & Technology*, vol. 37, pp. 1905-1911, 2003.

[51] G. Lu, T. P. Clement, C. Zheng, and T. H. Wiedemeier, "Natural attenuation of BTEX compounds: model development and field-scale application", *Ground Water*, vol. 37, pp. 707-717, 1999.

[52] T. P. Clement, C. D. Johnson, Y. Sun, G. M. Klecka, and C. Bartlett, "Natural attenuation of chlorinated solvent compounds: model development and field-scale application at the Dover site", *Journal of Contaminant. Hydrology*, vol. 42, pp. 113-140, 2000.

[53] M-S. Lee, K-K. Lee, Y. Hyun, T. P. Clement, and D. Hamilton. "Nitrogen transformation and transport modeling in groundwater aquifers", *Ecological Modeling*, vol. 192, pp. 143-159, 2006.

[54] D. E. Gomez, P. C. de Blanc, W. G. Rixey, P. B. Bedient, and P. J. J. Alvarez, "Modeling benzene plume elongation mechanisms exerted by ethanol using RT3D with a general substrate interaction module", *Water Resources Research*, vol. 44, W05405, doi:10.1029/2007WR00 6184, 2008.

[55] M-S. Lim, I. W. Yeo, T. P. clement, Y. Roh, and K-K. Lee, "Mathematical model for predicting microbial reduction and transport of Arsenic in groundwater systems", *Water Research*, vol. 41, pp. 2079-2088, 2007.

[56] M. Rolle, T. P. Clement, R. Sethi, and A. D. Molfetta, "A kinetic approach for simulating redox-controlled fringe and core biodegradation processes in groundwater: model development and application to a landfill site in Piedmont, Italy", *Hydrological Processes Journal*, vol. 22, pp. 4905-4921, 2008.

[57] C. D. Johnson, M. J. Truex, and J. R. Spencer, "Use of redox zones in modeling natural attenuation of a chlorinated solvent plume", in: *Proceedings of the Seventh in Situ and on-Site Bioremediation Symposium*, V.S. Magar and M.E. Kelley, eds. Battelle Press, Columbus, Ohio. Paper H-20, 2003.

[58] J. R. Spencer, C. D. Johnson, M. J. Truex, and T. P. Clement, "Modeling biological transformation of chlorinated Ethanes and Ethenes in support of natural attenuation", in: *Proceedings of the Third International Conference on Remediation of Chlorinated and Recalcitrant Compounds*, A.R. Gavaskar and A.S.C. Chen (eds.). Battelle Press, Columbus, Ohio. Paper 2D-09, 2002.

[59] M. J. Truex, C. D. Johnson, J. R. Spencer, T. P. Clement, and B. B. Looney, "A deterministic approach to evaluate and implement monitored natural attenuation for chlorinated solvents", *Remediation Journal*, vol. 17, pp. 23-40, 2007.

[60] C. D. Johnson, M. J. Truex, D. P. Leigh, and S. Granade, "Successful implementation of aerobic cometabolism of Vinyl Chloride *via* an *in situ* biofilter", in *Proceedings of Remediation of Chlorinated and Recalcitrant Compounds*: The Fifth International Conference, B.M. Sass, ed. Battelle Press, Columbus, Ohio. Paper A-43, 2003.

[61] C. D. Johnson, R. S. Skeen, D. P. Leigh, T. P. Clement, and Y. Sun, "Modeling natural attenuation of chlorinated Ethenes using the RT3D code", in *Proceedings of the Water Environment Federation 71st Annual Conference and Exposition*, WEFTEC '98, vol. 3. Water Environment Federation, Alexandria, Virginia. pp. 225-247, 1998.

[62] C. D. Johnson, R. S. Skeen, M. G. Butcher, *et al.* "Accelerated *in situ* bioremediation of chlorinated Ethenes in groundwater with high Sulfate concentrations", in *Engineered Approaches for In Situ Bioremediation of Chlorinated Solvent Contamination*, A. Leeson and B.C. Alleman, eds. Battelle Press, Columbus, Ohio. pp. 165-170, 1999.

[63] D. P. Leigh, C. D. Johnson, R. S. Skeen, *et al.*, "Enhanced anaerobic *in situ* bioremediation of chloroethenes at NAS point Mugu", in: *Bioremediation and Phytoremediation of Chlorinated and Recalcitrant Compounds*, G.B. Wickramanayake, A.R. Gavaskar, B.C. Alleman, and V.S. Magar (eds.). Battelle Press, Columbus, Ohio. pp. 229-235, 2000.

[64] M. J. Truex, C. D. Johnson, D. P. Leigh, and S. Granade, "Pulsed injection flow strategy for aerobic co-metabolism of Vinyl Chloride", in: *Proceedings of the Third International Conference on Remediation of Chlorinated and Recalcitrant Compounds*, A.R. Gavaskar and A.S.C. Chen (eds.). Battelle Press, Columbus, Ohio. Paper 2B-33, 2002.

CHAPTER 5

STOMP-ECKEChem: An Engineering Perspective on Reactive Transport in Geologic Media

M. D. White[*] and Y. Fang

Hydrology Group, Pacific Northwest National Laboratory, Richland, Washington, USA

Abstract: ECKEChem (Equilibrium, Conservation, Kinetic Equation Chemistry) is a reactive transport module for the STOMP suite of multifluid subsurface flow and transport simulators that was developed using an engineering perspective. STOMP comprises a suite of operational modes with capabilities for a variety of subsurface applications (*e.g.*, environmental remediation and stewardship, geologic sequestration of greenhouse gases, gas hydrate production, and oil shale production). The ECKEChem module was designed to provide integrated reactive transport capabilities across the suite of STOMP simulator operational modes. The initial application for the ECKEChem module was for the simulation of mineralization reactions that were predicted to occur with the injection of supercritical carbon dioxide into deep Columbia River basalt formations, using STOMP-CO2 which solves sequestration flow and transport problems for deep saline formations. The STOMP-ECKEChem solution approach to modeling reactive transport in multifluid geologic media is founded on an engineering perspective: 1) geochemistry can be expressed, input and solved as a system of coupled nonlinear equilibrium, conservation and kinetic equations, 2) the number of kinetic equation forms used in geochemical practice are limited, 3) sequential non-iterative coupling between the flow and reactive transport is sufficient, 4) reactive transport can be modeled by operator splitting with local geochemistry and global transport. This chapter describes the conceptual approach to converting a geochemical reaction network into a series of equilibrium, conservation and kinetic equations, the implementation of ECKEChem in STOMP, the numerical solution approach, and a demonstration of the simulator on a complex application involving desorption of uranium from contaminated sediments.

Keywords: Equilibrium, conservation, kinetic equations, operator splitting, reaction network translation, numerical Newton-Raphson iteration.

1. STOMP

Developed at the Pacific Northwest National Laboratory (PNNL), Subsurface Transport Over Multiple Phases (STOMP) [1, 2], is a suite of numerical simulators for multifluid subsurface flow and transport processes in geologic media. Individual simulators within the suite are categorized into operational modes, according to the solved coupled conservation equations. For example, STOMP-WAE solves the conservation equations for water mass, air mass, and thermal energy for nonisothermal, variably saturated aqueous geologic media, with mobile aqueous and gas phases. The STOMP-WCSME operational mode, also referred to as STOMP-CO2e, solves flow and transport problems for nonisothermal carbon dioxide (CO_2) sequestration systems in aqueous saline formations.

The current suite of STOMP simulators includes capabilities for modeling nonisothermal environments, compositional gases, compositional Nonaqueous Phase Liquids (NAPLs), supercritical and subcritical CO_2, mixed CO_2-CH_4 gas hydrates, immobile ice, salt precipitation. The most recent operational mode under development is STOMPOS, one specifically designed for studying the in-situ production of oil shales.

Several STOMP operational modes have been parallelized using FORTRAN 90 combined with a preprocessor library called FP [3] to process directives in the FORTRAN 90 code to generate parallel code using MPI [4]. One of the drawbacks of developing the scalable implementations using FP is that each

[*]**Address correspondence to M.D. White:** Hydrology Group, Pacific Northwest National Laboratory, Richland, Washington, USA; Tel: 1-509-372-6070; Email: mark.white@pnl.gov

Fan Zhang, Gour-Tsyh (George) Yeh, Jack C. Parker and Xiaonan Shi (Eds)

operational mode had to be rewritten to take advantage of the parallel-programming structures built into the FORTRAN 90 language. A new parallel STOMP version currently is being developed using the Global Arrays Toolkit (GA) [5] to present a global shared memory view of the data to the programmer while providing the underlying mapping and communication of distributed data. The implementation is based on full-featured STOMP serial code without rewriting the code in FORTRAN 90.

The official development of STOMP started with the publication of a simulator design document in 1992 [6]. Immediately prior to this time, PNNL had been developing and applying the short-lived Multiphase Subsurface Transport Simulator (MSTS) to nuclear waste repository problems [7, 8]. For the next fourteen years, STOMP development continued without a fully integrated reactive transport capability. During this period a derivative code, Subsurface Transport Over Reactive Multiphases (STORM) [9] was developed from the STOMP-WAE simulator for variably saturated nonisothermal systems. Otherwise reactive transport capabilities have been generally separated from the coupled flow and transport components.

In 1999 the Reactive Flow and Transport of Groundwater Contaminants (RAFT) simulator [10] was partially integrated into the first scalable version of STOMP (STOMP-SC). The unique features of RAFT included a suite of solvers for transport, reactions, and regression that allowed a combination of numerical methods to be used during a single simulation. The simulator also offered user specified coupled equilibrium and kinetic reaction systems; where, arbitrary reaction systems were integrated as subroutines into the simulator coding using automated code generation *via* an integrated symbolic computational language utility, MAPLE.

In 2002 the RAFT simulator capabilities were replaced in STOMP-SC with CRUNCH [11, 12]. The CRUNCH simulator used an automated thermodynamic and kinetic data-base reader and was designed for reactive transport problems of arbitrary complexity and size (*i.e.*, no a priori restriction on the number of species or reactions considered). Designed from a geochemical perspective, the simulator's features addressed specific geochemical processes:

- Multicomponent aqueous complexation.
- Kinetically-controlled mineral precipitation and dissolution.
- Multicomponent ion exchange on multiple sites.
- Multicomponent surface complexation on multiple sites including an electro-static correction based on the double layer model [13].
- Biologically-mediated reactions based on Monod-type formulations.
- Radioactive decay chains.
- Multicomponent diffusion with an electrochemical migration term to correct for electroneutrality where diffusion coefficients of charged species differ.
- Automatic read of a reformatted version of the EQ3/EQ6 database augmented with a kinetic database.

In 2005 development of fully integrated reactive transport capabilities were initiated for the STOMP-CO2 simulator; the outcome of which would become the ECKEChem module [14, 15]. Two seemingly contradictive principals guided the development of ECKEChem: 1) minimize implementation time and effort, and 2) maximize the software applicability. Ultimately these objectives were realized through an engineering perspective.

Mathematical Model

Expressed in integral form, the mass conservation equations solved by STOMP equate the rate of change of mass within a volume with the net flux of mass into the volume over the volume surface plus the rate of mass generation within the volume:

$$\frac{d}{dt}\int_V \sum_{\gamma=l,n,g,h,i,p} \left(\phi\omega_\gamma^i \rho_\gamma s_\gamma\right) dV = \int_V m^i dV + \int_\Gamma \sum_{\gamma=l,n,g} \omega_\gamma^i \rho_\gamma \mathbf{V}_\gamma \cdot \mathbf{n} d\Gamma + \int_\Gamma \sum_{\gamma=l,n,g} \mathbf{J}_\gamma^i \cdot \mathbf{n} d\Gamma \quad \textbf{(1)}$$

where, variables are defined in the List of Abbreviations at the end of the chapter. The volumetric flux of a mobile phase is computed according to Darcy's law:

$$\mathbf{V}_\gamma = -\frac{k_{r\gamma}\mathbf{k}}{\mu_\gamma}\left(\nabla P_\gamma + \rho_\gamma g \mathbf{z}\right) \quad \textbf{(2)}$$

The diffusive/dispersive component mass fluxes through the mobile phases are computed from gradients in molar concentration:

$$\mathbf{J}_\gamma^i = -\phi\rho_\gamma s_\gamma \frac{M^i}{M_\gamma}\left(\tau_\gamma D_\gamma^i + \mathbf{D}_{h\gamma}\right)\nabla\chi_\gamma^i \quad \textbf{(3)}$$

As with the mass conservation equations, the conservation of energy equation solved by STOMP equates the rate of change of energy within a volume with the net flux of heat into the volume over the volume surface, plus the rate of heat generated within the volume:

$$\begin{aligned} &\frac{d}{dt}\int_V \left\{ \sum_{\gamma=l,n,g,h,i,p} \left(\phi u_\gamma \rho_\gamma s_\gamma\right) + (1-\phi) u_s \rho_s \right\} dV = \\ &\int_\Gamma \sum_{\gamma=l,n,g} h_\gamma \rho_\gamma \mathbf{V}_\gamma \cdot \mathbf{n} d\Gamma + \int_\Gamma \sum_{i=w,a,o} h_g^i \mathbf{J}_g^i \cdot \mathbf{n} d\Gamma + \int_\Gamma \mathbf{F}\cdot\mathbf{n} d\Gamma + \int_V q dV \end{aligned} \quad \textbf{(4)}$$

The diffusive heat flux is computed from a linear combination of phase conductivity and gradients in temperature:

$$\mathrm{F} = -\left\{ \sum_{\gamma=l,n,g,h,i,p} \left(\phi k_\gamma s_\gamma\right) + (1-\phi) k_s \right\} \nabla T \quad \textbf{(5)}$$

Except for operational modes that consider kinetic phase partitioning of components (*e.g.*, STOMP-WAOk (solving water, air, nonaqueous-oil and aqueous-oil conservation) and STOMP-HYDk (solving energy, water, non-hydrate CH_4, non-hydrate CO_2, hydrate CH_4, and hydrate CO_2 conservation), a fundamental assumption taken in developing STOMP is that the components are in thermodynamic equilibrium between phases.

Gibb's phase rule states the number of degrees of freedom is equal to the number of components minus the number of phases in thermodynamic equilibrium plus two:

$$N_F = N_C - N_P + 2 \quad \textbf{(6)}$$

A porous media, however, has N_P distinct phase pressures, compared with a single pressure in a nonporous media, resulting in $(N_P\text{-}1)$ additional variables [16]:

$$N_F = N_C + 1 \quad \textbf{(7)}$$

This results in the phase rule for porous media being independent of the number of phases. The porous media phase rule indicates that the number of unknowns needed to specify the state of the system is one more than the number of components. For isothermal systems the temperature is fixed, which reduces the number of unknowns to the number of components. Therefore STOMP-WO, a two component isothermal simulator, has two primary variables that need to be solved for the two mass conservation equations for water and oil. STOMP-HYD, a four component nonisothermal simulator, has five primary variables that need to be solved for the four mass conservation equations and one energy conservation equation.

The integral conservation equations are solved in STOMP for a set of primary variables using integral volume differencing on structured orthogonal grids for spatial discretization and a fully implicit formulation for temporal discretization. Secondary variables are those variables in the conservation equations that are not solved directly, but are solved from the primary variables through the constitutive equations, which generally are nonlinear functions. Nonlinearities in the discretized governing equations and associate constitutive equations are resolved using Newton-Raphson iteration with continuous property updating. Whereas phase rules indicate the number of primary variables needed to resolve the conservation equations, primary variable sets must provide closure of the system state. In STOMP, system closure is assured *via* a primary variable switching scheme that alters the primary variable set with phase transitions. The implemented scheme assigns the primary variable set in each grid cell according to the current-iterate phase condition in that grid cell (*i.e.*, phase condition changes occur between Newton-Raphson iterations). The number of possible phase conditions varies across the STOMP simulator operational modes, but the overall solution scheme involving local primary variable switching is implemented consistently.

2. ECKECHEM

Reactive transport solution schemes generally are categorized as direct substitution or operator splitting approaches. The direct substitution schemes involve solving the reactive species transport and chemistry equations simultaneously. This approach has the advantage of yielding integrated solutions for reactive transport, but suffers from the computational costs of solving large Jacobian matrices. The operator splitting schemes involve solving the reactive species transport separately from the reactive species chemistry equations. Sequential schemes can be iterative or non-iterative between the transport and chemistry equations. The iterative schemes yield more integrated solutions, approaching those of the fully coupled schemes, but require additional computational effort. The non-iterative schemes suffer from not yielding fully integrated solutions for the reactive transport and coupled flow and transport equations, but have the lowest computational costs.

For the current ECKEChem implementation a non-iterative sequential scheme was chosen. To reduce the number of transported species only mobile component and kinetic species are transported, which requires that transport properties, such as diffusion and dispersion coefficients are species independent.

The following sections describe the mathematical formulation and numerical solution to the governing equations for the multifluid subsurface transport of reactive species. A principal objective in developing this mathematical formulation and code algorithms for solving multifluid subsurface flow and reactive transport problems was computational efficiency. The kernel of the mathematical formulation is two sequential solution operations. The coupled nonisothermal multifluid flow and transport equations are solved sequentially with the reactive transport equations; and the reactive transport equations are solved sequentially as two components: 1) multifluid component and kinetic species transport and 2) batch chemistry. In this mathematical formulation reactive species are either components of the coupled flow and transport equations (*e.g.*, water, air, oil, CO_2, CH_4) or dilute solutes; where, the principal assumption associated with dilute solutes is that phase properties are independent of solute concentrations. Reactive species that are components of the flow and transport equations are linked to the components *via* source/sink terms. Although the reactive transport sequential scheme starts with the transport component, the description of the mathematical formulation will start with the batch chemistry component, as the concepts and species descriptions introduced in the chemistry component are required to describe the transport component.

Batch Geochemistry

Chemical reactions can be classified as being either sufficiently fast enough to be reversible or in equilibrium or insufficiently fast enough for equilibrium conditions to apply, requiring a kinetic description. Biochemical and geochemical reactive systems that occur in subsurface environments generally require both equilibrium and kinetic reaction types to describe the system. The principal objective in developing the batch chemistry component of the reactive transport solver for the STOMP simulator was to create a generic multifluid chemistry module that could describe biochemical and geochemical reactive processes in subsurface environments.

Following the approaches of Fang *et al.* [17], a reaction-based model for the chemical system is used; where, reactions are assumed to be either fast/equilibrium or slow/kinetic. Non-user specified equilibrium reactions are modeled with an infinite reaction rate *via* mass-action equilibrium equations, and kinetic reactions are modeled using finite reaction rates; where, the available rate equation forms are dictated by need. Incorporating a new kinetic rate form requires roughly a day of coding and verification. Incorporation of user-specified equilibrium reactions will require coding for the user.

More importantly, the batch chemistry formulation used in the STOMP simulator depends on the techniques developed by Fang *et al.* [17] for translating a biochemical/geochemical reaction network, following a formal decomposition approach, into a system of equilibrium, conservation, and kinetic equations. The decomposition approach of Fang *et al.* [17] has been coded into a preprocessor that allows rapid translation of a complex biochemical/geochemical reaction network into a system of mixed differential/algebraic equations.

Equilibrium, Conservation, and Kinetic Equations

The details of translating a biochemical/geochemical reaction network into a system of equilibrium, conservation, and kinetic equations and the implementation of the associated preprocessor can be found in the manuscript by Fang *et al.* [17]. The objective here is to describe the form of the resulting equilibrium, conservation, and kinetic equations. Because biochemical/geochemical reactions in STOMP are modeled as a set of equilibrium, conservation, and kinetic equations, the reactive transport module was named ECKEChem, an acronym for Equilibrium, Conservation, and Kinetic Equation Chemistry. Equilibrium equations, often referred to as mass action equilibrium equations, relate species activities through an equilibrium constant:

$$(C_j) = \{K_{eq}\}_j \prod_{i=1}^{\{N_S^{eq}\}_j} (C_i)^{e_i} \, (i \neq j); for \quad j = 1, N_{eqn}^{eq} \tag{8}$$

where, the exponents are stochiometric coefficients in the reaction network and the equilibrium constant can be temperature dependent. Conservation equations define the component species, which essentially are a set of species, whose collective stochiometrically weighted summed concentration is invariant with time:

$$\frac{d \sum_{i=1}^{\{N_S^{tc}\}_j} (a_i C_i)}{dt} = 0; \quad for \quad j = 1, N_{eqn}^{cn} \tag{9}$$

Kinetic equations define kinetic components and are similar in form to conservation equations, except that stochiometrically weighted sum of species concentrations vary in time according to a stoichiometrically weighted sum of kinetic rates:

$$\frac{d \sum_{i=1}^{\{N_S^{tk}\}_j} (b_i C_i)}{dt} = \sum_{i=1}^{\{N_r^{tk}\}_j} \left(c_i \{R_k\}_i\right); for \; j = 1, N_{eqn}^{kn} \tag{10}$$

Kinetic Reaction Rates

The current set of kinetic reaction rate models available with the STOMP/ECKEChem simulator are:

- Atkins Forward-Backward

- Steefel-Lasaga Dissolution-Precipitation (Transition-State-Theory Dissolution-Precipitation)
- Valocchi Monod
- Valocchi Sorption
- Valocchi Biomass
- Liu Log-Normal
- Liu Dual-Domain
- Langmuir Sorption
- Borden Emulsion Sorption

The Atkins Forward-Backward reaction rate model [18] is an elementary reaction rate model that includes forward and backward rate constants, weighted by species concentration factors:

$$R_k = \left[k_f \prod_{i=1}^{N_{reactants}} C_i^{(e_i)} - k_b \prod_{j=1}^{N_{products}} C_j^{(e_j)} \right] \tag{11}$$

The Steefel-Lasagna Dissolution-Precipitation reaction rate model [19-21] was implemented to test the simulator against a benchmark problem involving mineral trapping of supercritical CO_2 being injected into a glauconitic sandstone aquifer [22]. As expressed below positive rate values indicate dissolution of the species and negative values indicate mineralization and precipitation:

$$R_k = A_m k \left[1 - \left(\frac{Q}{K_{eq}} \right) \right]; \; k = k_{ref} \exp \left[\frac{-E_a}{R} \left(\frac{1}{T} - \frac{1}{T_{ref}} \right) \right] \tag{12}$$

The Valocchi Monod reaction rate model [23, 24] is a multiplicative Monod rate expression that has been included to test the simulator against a benchmark problem involving kinetic biodegradation, cell growth, and sorption:

$$R_k = -q_m X_m \left(\frac{C_d}{K_d + C_d} \right) \left(\frac{C_a}{K_a + C_a} \right) \tag{13}$$

The Valocchi Sorption reaction rate model [23, 24] is a generic kinetic sorption model that includes an equilibrium sorption coefficient:

$$R_k = -k_m \left(C \quad \frac{S}{K_d} \right) \tag{14}$$

The Valocchi Biomass reaction rate model [23, 24] is dependent on the rate of substrate utilization and first-order decay for the biomass:

$$R_k = -Y \left[-q_m X_m \left(\frac{C_d}{K_d + C_d} \right) \left(\frac{C_a}{K_a + C_a} \right) \right] - b X_m \tag{15}$$

The Liu Log-Normal reaction rate model [25] is for multiple-site sorption-desorption and assumes that the kinetic rate at a particular site *m* is proportional to its deviation from local equilibrium:

$$R_k^m = \alpha^m \left[S_i^{m*} - S_i^m \right] \tag{16}$$

The rate constants are assumed to follow a lognormal probability distribution; where the probability of a site having a rate constant is defined by:

$$p(\alpha) = \frac{1}{\alpha\sigma\sqrt{2\pi}} \exp\left[-\frac{1}{2\sigma^2} \left(\ln(\alpha) - \mu \right)^2 \right] \tag{17}$$

The rate constants for a particular site are a function of the integrated probability function and site density:

$$f^m = S_T \int_{\alpha^m - \frac{\Delta\alpha}{2}}^{\alpha^m + \frac{\Delta\alpha}{2}} p(\alpha) d\alpha \tag{18}$$

The Liu Dual-Domain reaction rate model considers mass transfer between mobile and immobile domains within the geologic media. The Langmuir Sorption reaction rate model [26] is a classical kinetic sorption form:

$$R_k = k_S C\left(S^* - S\right) - k_d S \tag{19}$$

The Borden Emulsion Sorption reaction rate model [27] incorporates a Langmuirian blocking function to account for saturation of the oil emulsion attachment sites with previously retained oil droplets. This kinetic reaction rate model was specifically incorporated into ECKEChem to provide the oil operational modes of STOMP with kinetic sorption modeling capabilities:

$$R_k = \left(\frac{3}{2}\right) V_l \eta \left(\frac{1-\phi}{d_c}\right) \alpha \left(\frac{S^*_{oil} - S_{oil}}{S^*_{oil}}\right) \left(\frac{\phi}{\rho_b}\right) C_{oil} \tag{20}$$

Example Reaction Network Translation

A critical component to the engineering perspective for solving reactive transport in geologic media embodied in ECKEChem is the translation of reaction networks into a system of equilibrium, conservation and kinetic equations. The following example shows the translation of a geochemical system for the dissolution of calcite with dissolved CO_2. This example geochemical system involves 13 species (*i.e.*, N = 13) and 9 reactions (*i.e.*, R = 9):

$$\begin{aligned}
&(R_1)\ H_2O \rightleftharpoons H^+ + OH^- \\
&(R_2)\ CaCO_3(aq) \rightleftharpoons Ca^{2+} + CO_3^{2-} \\
&(R_3)\ CaHCO_3^+ \rightleftharpoons Ca^{2+} + CO_3^{2-} + H^+ \\
&(R_4)\ CaOH^+ \rightleftharpoons Ca^{2+} + OH^- \\
&(R_5)\ HCO_3^- \rightleftharpoons H^+ + CO_3^{2-} \\
&(R_6)\ H_2CO_3 \rightleftharpoons 2\,H^+ + CO_3^{2-} \\
&(R_7)\ Ca(OH)_2 \rightleftharpoons Ca^{2+} + 2\,OH^- \\
&(R_8)\ CO_2(g) + H_2O \rightleftharpoons H_2CO_3 \\
&(R_9)\ CaCO_3(s) \rightleftharpoons Ca^{2+} + CO_3^{2-}
\end{aligned} \tag{21a-i}$$

where, reactions (R_1) through (R_8) are equilibrium reactions and reaction (R_9) is kinetic. Excluding H_2O, which has an activity of 1, the corresponding 12 species mass balance equations can be written as:

$$\begin{aligned}
&\frac{d\,H^+}{dt} = R_1 + R_3 + R_5 + 2\,R_6 \\
&\frac{d\,OH^-}{dt} = R_1 + R_4 + 2\,R_7 \\
&\frac{d\,Ca^{2+}}{dt} = R_2 + R_3 + R_4 + R_7 + R_9 \\
&\frac{d\,CO_3^{\ 2-}}{dt} = R_2 + R_3 + R_5 + R_6 + R_9 \\
&\frac{d\,CaCO_3(aq)}{dt} = -R_2;\ \frac{d\,CaHCO_3^{\ +}}{dt} = -R_3 \\
&\frac{d\,CaOH^+}{dt} = -R_4;\ \frac{d\,HCO_3^{\ -}}{dt} = -R_5 \\
&\frac{d\,H_2CO_3}{dt} = -R_6;\ \frac{d\,Ca(OH)_2}{dt} = -R_7 \\
&\frac{d\,CO_2(g)}{dt} = -R_8;\ \frac{d\,CaCO_3(s)}{dt} = -R_9
\end{aligned} \qquad \textbf{(22a-h)}$$

Analytical solutions to species mass balance equations generally are not possible, making numerical solutions necessary. A common approach to numerically solving the species mass balance equations, involving mixed kinetic and equilibrium reaction rates, is to formulate the equations into mixed Differential and Algebraic Equations (DAEs). For the calcium carbonate dissolution geochemical network, a DAE formulation can be created involving eight algebraic mass-action equations, one differential kinetic equation, and three algebraic conservation equations. For larger geochemical reaction systems, with numerous kinetic and equilibrium reactions, developing a DAE formulation is not straightforward.

Although the DAE formulations can overcome problems that arise with direct integration of the ordinary differential equations, there are inherent difficulties with developing a DAE formulation. First, defining the addition or subtraction of infinite reaction rates is not possible; and second, redundant equilibrium and irrelevant kinetic reactions must be excluded from the geochemical system. These difficulties inherent in the DAE approach can be eliminated if all reactions are written in a basic form. The systematic approach developed by Fang *et al.* [17], using Gauss-Jordan matrix decomposition, overcomes the inherent difficulties with the DAE approach. For the calcite dissolution reaction network shown in equations above the resulting equilibrium, conservation, and kinetic equations, expressed in the "kinetic," "conservation," "equilibrium" forms are:

$$\begin{aligned}
&\left(C_{OH^-}\right) = 10^{-14}\left(C_{H^+}\right)^{-1} \\
&\left(C_{CaCO_3(aq)}\right) = 10^{-3}\left(C_{Ca^{2+}}\right)^1\left(C_{CO_3^{\ 2-}}\right)^1 \\
&\left(C_{CaHCO_3^{\ +}}\right) = 10^{-11.6}\left(C_{Ca^{2+}}\right)^1\left(C_{CO_3^{\ 2-}}\right)^1\left(C_{H^+}\right)^1 \\
&\left(C_{CaOH^+}\right) = 10^{-1.8}\left(C_{Ca^{2+}}\right)^1\left(C_{H^+}\right)^{-1} \\
&\left(C_{HCO_3^{\ -}}\right) = 10^{-10.2}\left(C_{CO_3^{\ 2-}}\right)^1\left(C_{H^+}\right)^1 \\
&\left(C_{H_2CO_3}\right) = 10^{-16.5}\left(C_{CO_3^{\ 2-}}\right)^1\left(C_{H^+}\right)^2 \\
&\left(C_{Ca(OH)_2}\right) = 10^{-6.1}\left(C_{Ca^{2+}}\right)^1\left(C_{H^+}\right)^{-2} \\
&\left(CO_2(g)\right) = K_{eq}\left(T; P_g\right)\left[C_{H_2CO_3}\right]
\end{aligned} \qquad \textbf{(23a-h)}$$

$$\frac{d\begin{pmatrix} C_{Ca^{2+}} + C_{CaCO_3(aq)} + C_{CaHCO_3^+} \\ + C_{CaOH^+} + C_{Ca(OH)_2} + C_{CaCO_3(s)} \end{pmatrix}}{dt} = 0$$

$$\frac{d\begin{pmatrix} C_{CO_3^{2-}} + C_{CaCO_3(aq)} + C_{CaHCO_3^+} \\ + C_{HCO_3^-} + C_{H_2CO_3} + C_{CaCO_3(s)} + C_{CO_2(g)} \end{pmatrix}}{dt} = 0 \qquad \textbf{(23i-k)}$$

$$\frac{d\begin{pmatrix} C_{H^+} - C_{OH^-} + C_{CaHCO_3^+} - C_{CaOH^+} \\ + C_{HCO_3^-} + 2C_{H_2CO_3} - 2C_{Ca(OH)_2} + 2C_{CO_2(g)} \end{pmatrix}}{dt} = 0$$

$$\frac{d\left(C_{CaCO_3(s)}\right)}{dt} = 10^{3.3}\left[C_{Ca^{2+}}\right]\left[C_{CO_3^{2-}}\right] - 10^{-5.0}\left[C_{CaCO_3(s)}\right]$$

The ECKE system shown in equations above are nonlinear, requiring an iterative solution. Newton-Raphson iteration is used in the ECKEChem module of the STOMP simulator to solve the batch reaction system. A critical component of the nonlinear solution scheme are the initial guesses of the species concentrations. The Newton-Raphson solution scheme and associated algorithms developed to create initial guesses of species concentrations are described in the Numerical Solution Section below.

Total Species Transport

For increased computational efficiency, only the mobile fractions of the total-component and total-kinetic species are transported, which requires the restriction that physical transport parameters, such as diffusion and dispersion are species independent. The theoretical basis for transporting the mobile total-component and total-kinetic species are thoroughly described by Xu *et al.* [28] and Steefel *et al.* [29]. Following this approach the governing equation for the transport of total-component species, written in partial differential form, equates the time rate of change in moles of mobile and immobile fractions of the total-component species within a control volume, minus the flux moles of mobile constituent of the total-component species into the control volume:

$$\frac{dC_j^{tc}}{dt} + \sum_{\gamma=g,l,n} \left(\nabla\left[\{C_j^{tc}\}_\gamma \mathbf{V}_\gamma\right]\right) - \sum_{\gamma=g,l,n} \left(\nabla\left[\tau_\gamma s_\gamma \phi D_\gamma + s_\gamma \phi \{D_h\}_\gamma\right]\nabla\{C_j^{tc}\}_\gamma\right) - \sum_{i=1}^{\{N_S^{tc}\}_j} q_i = \sum_{k=1}^{\{N_\gamma^{tc}\}_j} \{R_k^{tc}\}_j \qquad \textbf{(24)}$$

The governing equation for the transport of total-kinetic species is similar to that for total-component species, but needs to include the produced moles from kinetic reactions. Written in partial differential form the governing equation appears as:

$$\frac{dC_j^{tk}}{dt} + \sum_{\gamma=g,l,n} \left(\nabla\left[\{C_j^{tk}\}_\gamma \mathbf{V}_\gamma\right]\right) - \sum_{\gamma=g,l,n} \left(\nabla\left[\tau_\gamma s_\gamma \phi D_\gamma + s_\gamma \phi \{D_h\}_\gamma\right]\nabla\{C_j^{tk}\}_\gamma\right) - \sum_{i=1}^{\{N_S^{tk}\}_j} q_i = \sum_{k=1}^{\{N_\gamma^{tk}\}_j} \{R_k^{tk}\}_j \qquad \textbf{(25)}$$

As an aid to understanding the various fractions of the total-component species, the reaction network described in the previous section involving the dissolution of calcite with the dissolution of CO_2 will provide an example. The total-component species concentration equals the stochiometrically weighted sum of mobile and immobile species concentrations:

$$C^{tc}_{CO_3^{2-}} = \begin{pmatrix} C_{CO_3^{2-}} + C_{CaCO_3(aq)} + C_{CaHCO_3^+} + \\ C_{HCO_3^-} + C_{H_2CO_3} + C_{CaCO_3(s)} + C_{CO_2(g)} \end{pmatrix} \qquad \textbf{(26)}$$

The mobile fraction of the total-component species concentration equals the stochiometrically weighted sum of mobile aqueous and gas species concentrations:

$$\left\{ C^{tc}_{CO_3^{2-}} \right\}_m = \begin{pmatrix} C_{CO_3^{2-}} + C_{CaCO_3(aq)} + C_{CaHCO_3^+} + \\ C_{HCO_3^-} + C_{H_2CO_3} + C_{CO_2(g)} \end{pmatrix} \quad \textbf{(27)}$$

The aqueous-mobile fraction of the total-component species concentration equals the stochiometrically weighted sum of only the aqueous species concentrations, and like-wise for the gas-mobile fraction of the total-component species concentration:

$$\left\{ C^{tc}_{CO_3^{2-}} \right\}_l = \begin{pmatrix} C_{CO_3^{2-}} + C_{CaCO_3(aq)} + \\ C_{CaHCO_3^+} + C_{HCO_3^-} + C_{H_2CO_3} \end{pmatrix} \quad \textbf{(28)}$$

$$\left\{ C^{tc}_{CO_3^{2-}} \right\}_g = C_{CO_2(g)} \quad \textbf{(29)}$$

The immobile fraction of the total-component species concentration equals the stochiometrically weighted sum of the immobile mineral, sorbed, or exchanged species concentrations:

$$\left\{ C^{tc}_{CO_3^{2-}} \right\}_{im} = C_{CaCO_3(s)} \quad \textbf{(30)}$$

For the calcite-dissolved CO_2 reaction network system there are 3 total-component species, each with mobile components, which would result in 3 transport equations. The total-kinetic species involves only $CaCO_3$ (s), which is immobile, eliminating the kinetic transport equation from the set of solved transport equations. Solution of the transport equations yields total-component and total-kinetic species concentrations, which are then used in the batch chemistry solver to obtain local values of all species.

Linked Species Transport

Depending on the STOMP operational mode and the specified reactive species, a particular chemical species may be a component in the coupled flow and transport and reactive transport systems. Under these circumstances, the user can choose to ignore or acknowledge this connection between components and reactive species. If the user chooses to ignore this connection, then the coupled flow and transport system is used only to provide fluxes of phases (*i.e.*, aqueous, gas, nonaqueous phase liquid fluxes) and no feedback occurs between the two systems. (*i.e.*, reactive transport does not impact the coupled flow and transport). However, if the user chooses to acknowledge the species connection, then the coupled flow and transport and reactive transport systems will be linked. The linkage mechanisms will depend on the operational mode and the particular species. As the reactive transport capabilities in STOMP were initially developed for investigating geochemical reactions associated with geologic sequestration of CO_2, the linkage between the CO_2 mass in the coupled flow equations (*i.e.*, STOMP-CO2) and the CO_2 mass in the reactive transport equations (*i.e.*, ECKEChem) will be described as an example of linked species transport.

STOMP-CO2 solves the coupled flow and transport equations for the conservation of water, CO_2, and NaCl mass for subsurface systems involving aqueous, gas, precipitated salt and solid phases; where, the aqueous phase comprises liquid water, dissolved CO_2, and dissolved NaCl, the gas phase comprises water vapor and gaseous CO_2, the precipitated salt phase comprises solid NaCl, and the solid phase comprises the porous media. If the carbonate dissolution system described above were solved with STOMP-CO2, then the CO_2 in the coupled flow and transport system could be linked to the mobile total-component CO_2 in the reactive transport system. These results in the CO_2 species linkages shown below where, these linkages are specified *via* user input:

$$Gas\ Phase\ CO_2 \Leftrightarrow CO_2(g) \quad (31)$$

$$Aqueous\ Phase\ CO_2 \Leftrightarrow H_2CO_3 \quad (32)$$

CO_2 mass conserved in the coupled flow and transport solution is either dissolved in the aqueous phase, a component of the gas phase or the principal component of the liquid-CO_2 phase. The mode that solves flow and transport problems for nonisothermal carbon dioxide sequestration systems where liquid-CO_2 phase is included is called STOMP-CO_2ae. To link CO_2 mass between the coupled flow and transport solution and the reactive transport solution, while conserving mass in both systems of equations, an operator splitting scheme was implemented. Prior to transporting the total-component CO_3^{2-} the mobile CO_2(aq) (*i.e.*, H_2CO_3 in this example) and CO_2(g) concentrations are separated out and the remaining stoichiometrically weighted mass (partial-component CO_3^{2-}) is then transported. Prior to solving the reaction equations the partial-component H_2CO_3 mass is recombined with the H_2CO_3 and CO_2(g) mass from the coupled flow and transport solution and the immobile fraction to reform the total-component CO_3^{2-} concentration. After the reactions are solved, changes in the combined H_2CO_3 and CO_2(g) concentrations are converted into a CO_2 mass source/sink for the next coupled flow and transport time step. The overall scheme is a time-lagged one in which the H_2CO_3 and CO_2(g) fractions of the total-component CO_3^{2-} are obtained from the solution of the coupled flow and transport equations and then changes in H_2CO_3 and CO_2(g) concentrations (*e.g.*, mineralization) are incorporated into the coupled flow and transport solution through CO_2 mass source/sink quantities.

3. NUMERICAL SOLUTION

Two types of approaches are popular for solving reactive transport problems in multi-fluid geologic media: 1) direct substitution and 2) operator splitting [11, 28, 30, 31]. The direct substitution approach involves substituting the reaction equations into the transport equations and solving them simultaneously, using a nonlinear equation solution approach (generally Newton-Raphson). This approach yields a direct solution of the species concentration and is considered to be more robust than the operator splitting approaches. Within the operator splitting approaches, there are two schemes: 1) sequential iteration and 2) non-sequential iteration. The sequential iteration solves the transport and batch chemistry governing equations sequentially, using Picard iteration (*i.e.*, successive substitution) until convergence on the species concentration is reached. Yeh and Tripathi [30] compared the direct substitution approach against the sequential iteration approach and concluded that computationally the sequential iteration approach was preferred for sufficiently large grids. More recently Saaltink [32] compared the two approaches in terms of computational domain and quality of solution for varying grid sizes. They concluded that the direct substitution method was more computationally efficient for smaller grid systems, especially when the chemical reactions were highly nonlinear or included significant retardation, but noted that the computational advantages diminished or reversed as the grid size increased.

The reactive transport capabilities of the ECKEChem module, developed for the STOMP simulator have been aimed at field-scale applications involving large computational domains and complex multifluid nonisothermal coupled flow and transport systems. The computational demands of these types of simulations are significant, even without including the reactive transport component. Because of the simulation objectives and intended applications, only the operator splitting numerical solution method has been implemented for reactive transport in the STOMP simulator.

Species Transport

The operator splitting approach involves sequentially solving the species transport and batch chemistry, with species transport being first in the sequence. For equilibrium or mixed equilibrium and kinetic reaction systems, the number of transported species can be reduced, increasing computational efficiency, by transporting only the mobile total-component and total-kinetic species [11, 28]. The governing equations for species transport of mobile total-component and total-kinetic species (shown in the SPECIES TRANSPORT section) are discretized temporally using a backward-Euler time differencing (fully implicit) and spatially using the integral finite difference method [1, 2] applied to structured orthogonal grid systems with a seven-point stencil.

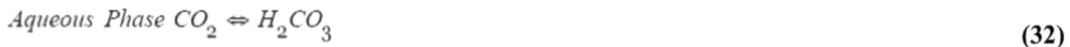

First, the capacitance terms are split into mobile and immobile fractions:

$$\int_{V_p} \frac{dC_j^{tc}}{dt} \cong \left(\frac{V_p \, s_\gamma \, \varphi}{\Delta t} \right) \begin{bmatrix} \left\{C_j^{tc}\right\}_{mo}^{t^+} - \left\{C_j^{tc}\right\}_{mo}^{t^o} \\ + \left\{C_j^{tc}\right\}_{im}^{t^o} - \left\{C_j^{tc}\right\}_{im}^{t^-} \end{bmatrix} \quad \textbf{(33)}$$

where, the mobile fractions are time differenced between the new and current time, and the immobile fractions are time differenced between the current and old time. This time differencing makes the immobile contributions part of the problem vector. The mobile fraction of the total-component species comprises both aqueous and gas species:

$$\left\{C_j^{tc}\right\}_{mo} = \sum_{\gamma=g,l} \left\{C_j^{tc}\right\}_\gamma \quad \textbf{(34)}$$

and, the immobile fraction of the total-component species comprises precipitated, exchanged, and sorbed species:

$$\left\{C_j^{tc}\right\}_{im} = \sum_{k=1}^{N_{mp}} \left\{ \left\{\chi_j^{tc}\right\}_{mp} \right\}_k \left\{C_{mp}\right\}_k + \sum_{k=1}^{N_{ss}} \left\{ \left\{\chi_j^{tc}\right\}_{ss} \right\}_k \left\{C_{ss}\right\}_k \quad \textbf{(35)}$$

Integration and spatial discretization of the advective and diffusive-dispersive flux operators yields summations over the six surfaces surrounding a seven-point stencil grid system:

$$\int_{\Gamma_p} \sum_{\gamma=g,l,n} \left(\nabla \left[\left\{C_j^{tk}\right\}_\gamma \mathrm{V}_\gamma \right] \right) \cdot \mathrm{n} d\Gamma \cong \sum_{\gamma=g,l,n} \left(\sum_{s=1}^{6} A_s \left\{V_s\right\}_\gamma \left\{C_j^{tk}\right\}_\gamma \right) \quad \textbf{(36)}$$

$$\int_{\Gamma_p} \sum_{\gamma=g,l,n} \left(\sum_{\gamma=g,l,n} \left(\nabla \begin{bmatrix} \tau_\gamma \, s_\gamma \, \varphi \, D_\gamma \\ + s_\gamma \, \varphi \, \left\{D_h\right\}_\gamma \end{bmatrix} \nabla \left\{C_j^{tk}\right\}_\gamma \right) \right) \cdot \mathrm{n} d\Gamma \cong \sum_{\gamma=g,l,n} \left(\sum_{s=1}^{6} A_s \begin{bmatrix} \tau_\gamma \, s_\gamma \, \varphi \, D_\gamma \\ + s_\gamma \, \varphi \, \left\{D_h\right\}_\gamma \end{bmatrix}_s \right) \frac{\left(\nabla \left\{C_j^{tk}\right\}_\gamma \right)_s}{\Delta x_s} \quad \textbf{(37)}$$

Combining the advective and diffusive-dispersive flux operators for species transport depends on the local grid Peclet number (*i.e.*, the ratio of advective to diffusive-dispersive flux). As described in the STOMP Theory Guide [1, 2], three combination schemes are available for increasing Peclet numbers: Patankar, Roe-Superbee, and Leonard Total Variation Diminishing (TVD). The species source terms stem from sources specified for the node, boundary conditions specified for a node surface, sources specified for the node from the coupled flow and transport equations *via* linked species:

$$\int_{V_p} \sum_{i=1}^{\left\{N_s^{tc}\right\}_j} q_i \, dV \cong V_p \left[\sum_{i=1}^{\left\{N_s^{tc}\right\}_j} \begin{pmatrix} q_i^{source} \\ + q_i^{boundary} \\ + q_i^{coupled\ flow} \end{pmatrix} \right] \quad \textbf{(38)}$$

Kinetic reactions are computed based on the nodal species concentrations at the current time step, which makes these terms part of the problem vector:

$$\int_{V_p} \sum_{k=1}^{\left\{N_r^{tc}\right\}_j} \left\{R_k^{tc}\right\}_j dV \cong V_p \left[\sum_{k=1}^{\left\{N_s^{tc}\right\}_j} \left\{R_k^{tc}\right\}_j \right] \quad \textbf{(39)}$$

Temporal and spatial discretization of the total-kinetic species transport equations follows that for total-component species.

Geochemistry Solution

The batch chemistry system comprises equilibrium, conservation, and Kinetic Equations (ECKEs); where, the number of equations equals the total number of reactive species. The general form of the batch chemistry equations are described in the BATCH GEOCHEMISTRY section. The number of equilibrium equations equals the number of equilibrium reactions. The number of kinetic equations, which also equals the number of total-kinetic species, but not necessarily equals the number of kinetic reactions because some reactions can be kinetically dependent, is determined from the preprocessor decomposition of the reaction network. The number of conservation equations equals the number of reactive species less the number of equilibrium and kinetic reactions, which also equals the number of total-component species. The transport equations yield concentrations of the mobile total-component and total kinetic species concentrations. The total-component and total-kinetic species concentrations can be determined from their mobile and immobile fractions by adding the mobile fraction concentrations from the transport equations with the immobile fraction concentrations from the previous time step:

$$\left\{C_j^{tc}\right\}^{t^+} = \left\{C_j^{tc}\right\}_{mo}^{t^+} + \left\{C_j^{tc}\right\}_{im}^{t^o} \quad \textbf{(40)}$$

$$\left\{C_j^{tk}\right\}^{t^+} = \left\{C_j^{tk}\right\}_{mo}^{t^+} + \left\{C_j^{tk}\right\}_{im}^{t^o} \quad \textbf{(41)}$$

Therefore, the species transport solution yields the total-component and total-kinetic species concentrations. The solution to the batch chemistry equations are the concentrations for all species. The ECKEs are generally strongly nonlinear and are solved using the Newton-Raphson iterative approach. A critical component of the Newton-Raphson scheme is establishing initial guess of the species concentrations. Poor initial guesses of the species concentrations generally result in additional iterations to achieve convergences or no convergence of the batch chemistry system. Initial guesses for the species concentrations from the total-component and total-kinetic species concentrations are derived from an iterative searching scheme designed to minimize the residuals of the conservation equations, ignoring the kinetic equations.

To start the initial species concentration search scheme, the component and kinetic species concentrations are assigned a value equal to 0.02 times [33] their respective total-component or total-kinetic species concentrations:

$$C_j = 0.02 C_j^{tc} \quad \textbf{(42)}$$

The pH-species (*i.e.*, the reactive species linked to pH, which generally is named H^+) is set to its previous time step value, except for the initial time step; where, the pH-species is assigned *via* the input file or set to neutral pH:

$$C_{H^+} = \left\{C_{H^+}\right\}^{t^o} ; \; C_{H^+} = \frac{10^{-7}}{s_l \, \phi \, \rho_l \, \omega_l^w} \quad \textbf{(43)}$$

With the component and kinetic species concentration approximated, the next step is to use the equilibrium equations to calculate the equilibrium species concentrations from the component and kinetic species concentrations. This step is possible. Because equilibrium reactions should be linearly independent, the equilibrium species concentrations can be defined only in terms of component species and/or kinetic species concentrations:

$$C_i = \frac{K_{eq} \prod_{j=1}^{\{N_s^{eq}\}_i} (C_j)^{e_j}}{\gamma_i} \quad for\ i = 1, N_{eqn}^{eq} \tag{44}$$

The next step involves computing the total residuals of conservation equations:

$$\mathrm{R}^{tc} = \sum_{j=1}^{N_s^{tc}} \mathrm{R}_j^{tc} = \sum_{j=1}^{N_s^{tc}} \left[C_j^{tc} - \sum_{i=1}^{\{N_s^{tc}\}_i} (a_i\, C_i) \right] \tag{45}$$

Once the initial total residuals are computed the algorithm proceeds by decreasing and increasing the component species concentrations by a factor of 1.667. After each decrease or increase in a component species concentration, the equilibrium species and total residual of the conservation equations are recalculated. If an improvement (*i.e.*, decrease) in the total residual of the conservation equations is determined then the component species concentration and total residual are reassigned. This procedure is repeated until no improvements are found in the total residual for the conservation equations with either a decrease or increase of 1.667 of any conservation species is found. The resulting conservation species and associated equilibrium species are used as initial guesses for the Newton-Raphson iteration scheme.

Formulation of the Newton-Raphson scheme requires expressing the ECKEs in residual form. The equilibrium equations are time independent and their residual form appears as:

$$\mathrm{R}_i^{eq} = C_i - \frac{K_{eq} \prod_{j=1}^{\{N_s^{eq}\}_i} (C_j)^{e_j}}{\gamma_i} \quad for\ i = 1, N_{eqn}^{eq} \tag{46}$$

The conservation and kinetic equations are time dependent and using backward-Euler temporal differencing, their residual forms appear as:

$$R_j^{cn} = \frac{\sum (a_j C_j)^{t^+} - (a_j C_j)^{t^o}}{\Delta t} \quad for\ j = 1, N_{eqn}^{cn} \tag{47}$$

$$R_k^{kn} = \frac{\sum (b_k C_k)^{t^+} - (b_k C_k)^{t^o}}{\Delta t} = \sum (c_k R_k)^{t^+} \quad for\ j = 1, N_{eqn}^{kn} \tag{48}$$

The above three residual equations provide ${N_{eqn}}^{eq} + {N_{eqn}}^{cn} + {N_{eqn}}^{kn}$ equations for the set of unknown equilibrium, component, and kinetic species concentrations. The Jacobian matrix, vector of unknowns and solution vector is arranged as follows to avoid having zeros on the diagonal elements:

$$\begin{vmatrix} \frac{\partial R_i^{eq}}{\partial C_i} & \frac{\partial R_i^{eq}}{\partial C_j} & \frac{\partial R_i^{eq}}{\partial C_k} \\ \frac{\partial R_i^{cn}}{\partial C_i} & \frac{\partial R_i^{cn}}{\partial C_j} & \frac{\partial R_i^{cn}}{\partial C_k} \\ \frac{\partial R_i^{kn}}{\partial C_i} & \frac{\partial R_i^{kn}}{\partial C_j} & \frac{\partial R_i^{kn}}{\partial C_k} \end{vmatrix} \begin{vmatrix} \Delta C_i \\ \Delta C_j \\ \Delta C_k \end{vmatrix} = \begin{vmatrix} -R_i^{eq} \\ -R_i^{cn} \\ -R_i^{kn} \end{vmatrix} \begin{matrix} for\ i = 1, N_{eqn}^{eq} \\ for\ j = 1, N_{eqn}^{cn} \\ for\ k = 1, N_{eqn}^{kn} \end{matrix} \tag{49}$$

The solution vector contains corrections to the equilibrium-, conservation-, and kinetic-equation species concentrations and are used to update these concentrations between iterations:

$$\begin{aligned} \{C_i\}^{s+1} &= \{C_i\}^{s} + \Delta C_i \ \ for\ i = 1, N_{eqn}^{eq} \\ \{C_j\}^{s+1} &= \{C_j\}^{s} + \Delta C_j \ \ for\ j = 1, N_{eqn}^{cn} \\ \{C_k\}^{s+1} &= \{C_k\}^{s} + \Delta C_k \ \ for\ k = 1, N_{eqn}^{kn} \end{aligned} \tag{50}$$

where, superscript s indicates the current iterate level and superscript s+1 indicates the next iterate level. The system of equations is considered converged when the maximum relative residual or change in concentration falls below a specified tolerance level. To avoid negative concentrations of the species, the change in species concentration is restricted.

In the Newton-Raphson iteration scheme, the species concentrations are the primary variables. For the Newton-Raphson scheme to execute properly the equation residual must have a dependence on the primary variable (*i.e.*, species concentration) selected for that equation. The primary unknowns for the equilibrium equations are equilibrium species, and for the conservation equations are the component species. For the equilibrium and conservation equation there is a direct correspondence between equations and primary unknowns. For the kinetic equations, however, the primary unknown species must be chosen such that equation residuals are strongly dependent on the species concentration. An equation sequencing algorithm is used to correlate species concentrations with chemistry equations.

Algorithm Flow Path

Reactive transport capabilities of the ECKEChem module have been designed to work with all operational modes of the STOMP simulator, without significantly altering the STOMP coding. This section describes the algorithm structure, the modular coding, and the flow path for solving multifluid subsurface flow and reactive transport problems with the STOMP simulator and ECKEChem module. The core capabilities of the STOMP simulator, prior to developing the ECKEChem module, included coupled multifluid subsurface flow and non-reactive solute transport; where, the two systems are solved sequentially. As the numerical solution approach for the reactive transport system uses an operator splitting scheme, which sequentially computes species transport and batch chemistry, the non-reactive solute transport schemes of the STOMP simulator can be used for transporting reactive species with only minor modifications. The flow path for the coupled STOMP-CO2 and ECKEChem module are shown in Fig. **1**. A time step begins with the solution of the coupled flow and transport system, using Newton Raphson iteration, yielding aqueous and gas phase fluxes, state properties (*e.g.*, temperature, pressure, phase viscosity, phase density, phase saturation), and the concentrations of gaseous and aqueous dissolved CO_2. If transport of nonreactive dilute solutes is specified for the simulation, then solute transport is computed, with the option for sub-time stepping as a function of the maximum Courant number. Inputs to the solute transport solver include aqueous and gas phase fluxes, state properties, and the previous time-step or sub-time-step solute concentrations. Multiple solutes are transported sequentially. Using the operator splitting scheme, the reactive transport solution begins with transport of all the mobile total-component and total-kinetic species, with the exception of the total-CO_2, which has previously been transported *via* the coupled flow and transport solution. The resulting concentrations of the mobile total-component and total-kinetic species, including the total-CO_2 are combined with their immobile counterparts to yield concentrations of the total-component and total-kinetic species. These species concentrations are then passed to the batch chemistry solver, which uses an iterative search scheme to generate initial guess of the species concentrations for the Newton-Raphson iterative solver. Resulting salt concentrations are converted to equivalent NaCl concentrations that are used to compute phase properties and dissolution of CO_2 in the aqueous phase for the coupled flow and transport solver.

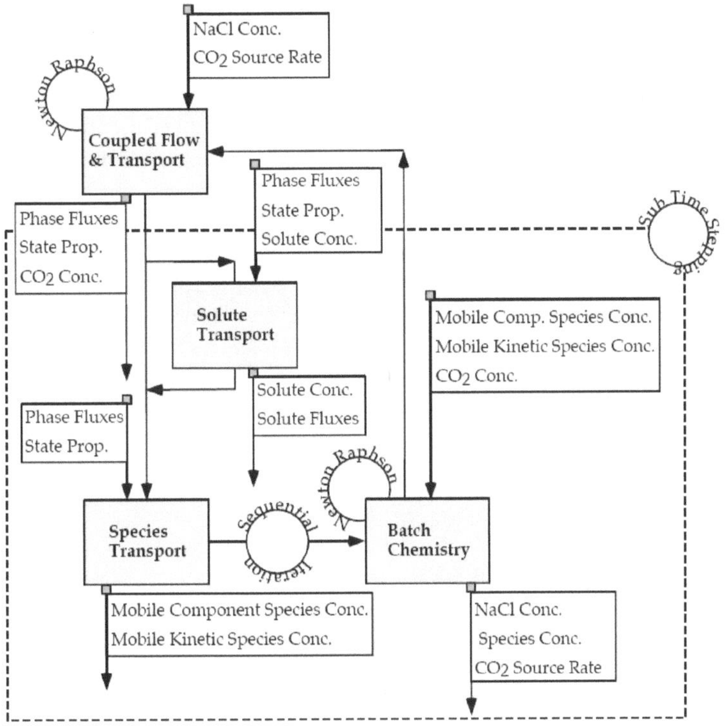

Figure 1: Illustration of STOMP-CO2/ECKEChem solution flow path, shaded boxes are processes, unshaded boxes are input/output for each process, circles are approaches used or connections between processes.

Numerical Solution Improvements

Large, complex geochemical systems have demonstrated the need for more numerically efficient solution approaches in ECKEChem. Two additional solution improvements have been implemented in ECKEChem and one additional option currently is being implemented. The original solution scheme includes all three types of equations (*i.e.*, equilibrium, conservation, and kinetic) into the Jacobian matrix of the Newton-Raphson iteration; where the unknowns are corrections to the individual species concentrations. This approach requires a linear-system solve of order equal to the number of species. The first solution-improvement option involves only solving for corrections to the total-component and total-kinetic species and then converting those corrections to corrections to the individual species concentrations *via* the system of equilibrium equations. This option maintains the general Newton-Raphson iteration scheme, except that the equilibrium equations are incorporated after the linear-system solution for the conservation and kinetic equations. For some problems this approach can significantly reduce the number of unknowns in the linear system solution. The second option is directed at improving the numerical solution efficiency for problems involving large numbers of single-species ordinary differential equations, such as in the demonstration problem in the following section. This option removes the kinetic equations with single-species from the set of equations solved in the Jacobian matrix of the Newton-Raphson linearization. These equations are solved using successive substitution for the single-species kinetic equations with an inner Newton-Raphson iterative loop for the equilibrium, conservation, and multi-species kinetic equations. Species concentrations are resolved when both the inner and outer iteration schemes have converged. The decoupled equilibrium equation solution option can be applied to the inner Newton-

Raphson iterative loop. In spite of the nested iterative looping schemes, this numerical solution approach improves the numerical efficiency by reducing the order of the linear system. The final solution option, which is now under development, is the log-concentration scheme. The original solution scheme solves for the actual species concentrations. The log-concentration solution option will allow the user to solve for the natural logarithm of species concentration; which effectively reduces the number of iterations for convergence for some geochemical systems.

4. DEMONSTRATION PROBLEM

Liu *et al.* [25] have conducted a series of column experiments and numerical investigations to understand the desorption of uranium from contaminated coarse-textured subsurface sediments with large fractions of river cobbles. At this site, uranium is predominately sorbed onto the finer grained sediments (*i.e.*, mineral surfaces, secondary grain-coating materials, intra-grain micro-factures, making its desorption and transport behavior through the subsurface sediments complex. These experimental and numerical investigations demonstrated that uranyl desorption from the fine-grained sediments was controlled by kinetic processes that exhibited multi-rate behavior. Liu *et al.* further demonstrated that a multicomponent kinetic model that integrated uranyl surface complexation reactions and distributed rate expressions could describe the reactive transport processes observed in the column experiments. Specifically the research team determined that uranyl desorption and reactive transport could be described by integrating the geochemical reaction network and kinetic parameters derived from laboratory investigations of the reactive mineral or size fraction and the physical transport properties of the whole sediment from independent tracer experiments. The objective of this demonstration is to reproduce the numerical investigations of Liu *et al.* [25] using STOMP-W/ECKEChem for a reactive transport problem, involving complex multi-rate uranyl desorption. The simulator version applied included the numerical solution improvements described in the previous section.

Speciation Reactions

Liu *et al.* [25] considered eight chemical components in their numerical simulations: UO_2^{2+}, CO_3^{2-}, Na^+, Ca^{2+}, Mg^{2+}, NO_3^-, H^+, >SOH, along with 43 relevant aqueous species; where >SOH represents a surface site for uranyl adsorption. The equilibrium aqueous speciation reactions and their equilibrium constants are shown in Table **1**.

Table 1: Aqueous speciation reactions.

Speciation Reaction	**log K_{eq}**
$UO_2^{2+} + H_2O = UO_2\,OH^+ + H^+$	-5.25
$UO_2^{2+} + 2H_2O = UO_2\,(OH)_2(aq) + 2H^+$	-12.15
$UO_2^{2+} + 3H_2O = UO_2\,(OH)_3^- + 3H^+$	-20.25
$UO_2^{2+} + 4H_2O = UO_2\,(OH)_4^{2-} + 4H^+$	-32.4
$2UO_2^{2+} + H_2O = (UO_2)_2\,OH^{3+} + H^+$	-2.7
$2UO_2^{2+} + 2H_2O = (UO_2)_2\,(OH)_2^{2+} + 2H^+$	-5.62
$3UO_2^{2+} + 4H_2O = (UO_2)_3\,(OH)_4^{2+} + 4H^+$	-11.90
$3UO_2^{2+} + 5H_2O = (UO_2)_3\,(OH)_5^+ + 5H^+$	-15.55
$3UO_2^{2+} + 7H_2O = (UO_2)_3\,(OH)_7^- + 7H^+$	-32.2
$4UO_2^{2+} + 7H_2O = (UO_2)_4\,(OH)_7^+ + 7H^+$	-21.9
$UO_2^{2+} + CO_3^{2-} = UO_2CO_3(aq)$	9.94
$UO_2^{2+} + 2CO_3^{2-} = UO_2(CO_3)_2^{2-}$	16.61
$UO_2^{2+} + 3CO_3^{2-} = UO_2(CO_3)_3^{4-}$	21.84
$3UO_2^{2+} + 6CO_3^{2-} = (UO_2)_3(CO_3)_6^{6-}$	54.0
$2UO_2^{2+}+CO_3^{2-}+3H_2O=(UO_2)_2CO_3(OH)_3^-+3H^+$	-0.855
$3UO_2^{2+}+CO_3^{2-}+3H_2O = (UO_2)_3O(OH)_2HCO_3^+ + 3H^+$	0.655
$11UO_2^{2+}+6CO_3^{2-}+12H_2O= (UO_2)_{11}(CO_3)_6(OH)_{12}^{2-} + 12H^+$	36.43

Table 1: cont….

$2Ca^{2+} + UO_2^{2+} + 3CO_3^{2-} = Ca_2UO_2(CO_3)_3$	30.7
$Ca^{2+} + UO_2^{2+} + 3CO_3^{2-} = CaUO_2(CO_3)_{32}^{-}$	27.18
$Mg^{2+} + UO_2^{2+} + 3CO_3^{2-} = MgUO_2(CO_3)_3^{2-}$	26.11
$UO_2^{2+} + NO_3^{-} = UO_2NO_3^{+}$	0.30
$Ca^{2+} + CO_3^{2-} = CaCO_3$	3.22
$Ca^{2+} + H^{+} + CO_3^{2-} = CaHCO_3^{+}$	11.434
$Ca^{2+} + NO_3^{-} = CaNO_3^{+}$	0.5
$Ca^{2+} + H_2O = H^{+} + CaOH^{+}$	-12.697
$2H^{+} + CO_3^{2-} = H_2CO_3$	16.681
$H^{+} + CO_3^{2-} = HCO_3^{-}$	10.329
$2Mg^{2+} + CO_3^{2-} = Mg_2CO_3^{2+}$	3.59
$Mg^{2+} + CO_3^{2-} = MgCO_3$	2.92
$H^{+} + Mg^{2+} + CO_3^{2-} = MgHCO_3^{+}$	11.34
$Mg^{2+} + H_2O = H^{+} + MgOH^{+}$	-11.417
$Na^{+} + CO_3^{2-} = NaCO_3^{-}$	1.27
$Na^{+} + H^{+} + CO_3^{2-} = NaHCO_3$	10.029
$Na^{+} + NO_3^{-} = NaNO_3$	-0.55
$Na^{+} + H_2O = H^{+} + NaOH$	-13.897
$H_2O = H^{+} + OH^{-}$	-13.997

These speciation reactions will ultimately appear as equilibrium equations in the ECKEChem form, but the reaction equilibrium constants generally are altered by the preprocessor, depending on the components specified by the user.

Surface Complexation Reactions

Liu *et al.* [25] used the work of Bond *et al.* [34] that suggested that equilibrium partitioning of U(VI) between aqueous and solid phases could be used to model the surface complexation reactions of either uranyl carbonate or uranyl hydroxyl species, or their combination for the site sediments. From this work two U(VI) surface complexation reactions were proposed, as shown in Table **2** with their equilibrium constants. These equilibrium reactions were used to calculate the adsorption extent of the two components across all sorption sites. The kinetic sorption reactions described are based on the deviation of the sorbed concentrations from the local equilibrium state, defined by the adsorption extent. Using surface complexation reactions to define the adsorption extent makes it a function of all the aqueous components (*e.g.*, pH, CO_3^{2-}, U(VI), *etc.*) and the sorption site properties.

Table 2: Surface complexation reactions.

Speciation Reaction	**log K_{eq}**
$\text{>SOH} + UO_2^{2+} + H_2O = \text{>SOUO}_2\text{OH} + 2H^{+}$	-4.42
$\text{>SOH} + UO_2^{2+} + CO_3^{2-} = \text{>SOUO}_2\text{HCO}_3$	16.53

Kinetic Sorption Reactions

The column experiments of Liu *et al.* [25] and previous uranyl adsorption investigations by Qafoku *et al.* [35] indicated that the U(VI) sorption reactions were occurring across multiple sites. These observations were consistent with the microscopic and spectroscopic characterizations of the site sediments that revealed sorbed U(IV) were associated with mineral surfaces, dispersed in grain coating materials, and/or distributed within intra-aggregates. Without more specific site information, the multisite sorption system was conceptualized with a distributed rate model, as previously used to describe multisite, time-variable mass transfer rates of uranyl adsorption to other similar sediments [35]. When applied to a kinetic sorption reaction model the distributed rate approach results in a unique sorption rate equation for each sorption site, where the sorbed concentration, adsorption extent and kinetic rate constant are specific to the site:

$$\frac{\partial S_i^m}{\partial t} = \alpha^m \left(S_i^{m*} - S_i^m \right) \; for \; i = 1,2; \; m = 1,2,\ldots 50 \tag{51}$$

where, i is the sorbed component (*i.e.*, $>SOUO_2OH$, $>SOUO_2HCO_3$) and m is site index. Inherent in the above equation is that 50 sorption sites are sufficient to model the sediment. To minimize the total number of parameters needed to define the multisite kinetic sorption model, the reaction rate constants were assumed to follow a lognormal probability distribution:

$$p(\alpha) = \frac{1}{\alpha \sigma \sqrt{2\pi}} \exp\left[-\frac{1}{2\sigma^2} \left\{ \ln(\alpha) - \mu \right\}^2 \right] \tag{52}$$

where p is the probability of a site with a corresponding rate constant of α; and μ and σ define the probability distribution function. The reaction rate constant for a particular site, m, was determined from the ratio of the sorption site density for the site to the total sorption site density for the sediment:

$$f^m = S_T \int_{\alpha^m - \frac{\Delta\alpha}{2}}^{\alpha^m + \frac{\Delta\alpha}{2}} p(\alpha) d\alpha \tag{53}$$

Dual-Domain Kinetic Reactions

Liu *et al.* [25] conducted uranium desorption experiments in two columns; fine-grain sediment (< 2mm) and field-textured sediment from the capillary fringe beneath Hanford's historic 300 Area North Process Pond. The field-textured sediment was a coarse-textured, alluvial flood deposit containing river cobble, gravel, sand, silt and clay. Kinetic-reaction parameters, including those for the lognormal probability distribution, and hydraulic-transport parameters were determined from parameter estimation techniques using the uranyl desorption and tracer transport experiments and numerical simulation. Hydraulic transport parameters were also determined from tracer transport experiments conducted on the field-textured sediments. The uranium effluent profile from the field-texture sediment desorption experiments were similar to those observed for the fine-grained sediments, but the concentrations were consistently lower. Numerical simulations using the reactive transport scheme described above with 1) the kinetic-reaction parameters determined from the fine-grained-sediment experiments, 2) the sorbed U(VI) concentration and sorption sites in the fine-grained sediments diluted by the cobble fraction in the field-textured sediments, and 3) hydraulic-transport parameters determined from the tracer experiments with the field-textured sediments, over predicted the effluent U(VI) concentrations. Liu *et al.* [25] concluded that this over prediction suggested that the field-textured sediment was behaving as a dual-domain system having an immobile and mobile fractions. To simulate the field-textured sediment a dual domain model was developed that included kinetic exchange of the mobile and immobile fractions of the reactive species:

$$\frac{\partial \{C_i\}^{im}}{\partial t} + \sum_{m=1}^{50} \alpha^m \left(\{S_i^{m*}\}^{im} - \{S_i^m\}^{im} \right) = \omega \left(\{C_i\}^{mo} - \{C_i\}^{im} \right) \tag{54}$$

The developed numerical model that included the described aqueous speciation reactions, surface complexation reactions, distributed rate sorption kinetics, and dual-domain properties successfully reproduced the uranyl desorption column experiments in the field-textured sediments.

ECKEChem Approach and Solution

A founding concept for ECKEChem was that it provides robust reactive transport capabilities for geologic media through a mathematical framework that solved a system of nonlinear equations. To further demonstrate this conceptual approach to reactive transport, the complex dual-domain simulations of U(VI) desorption and transport in field-textured sediments will be executed using STOMP-W/ECKEChem. The starting point for developing reactive transport system using ECKEChem is to create a reaction network. The reaction network preprocessor [17] accepts reactions in either equilibrium or kinetic reaction types. Both reaction types are written in stochiometric form, and the equilibrium reactions include the log of the equilibrium constant. A total of 297 reactions were used to model the dual-domain system of Liu *et al.* [25]:

- 36 equilibrium aqueous speciation reactions for the mobile-fraction species (Table **1**).
- 36 equilibrium aqueous speciation reactions for the immobile-fraction species (Table **1**).
- 2 equilibrium surface complexation reactions for the mobile-fraction species (Table **2**).
- 2 equilibrium surface complexation reactions for the immobile-fraction species (Table **2**).
- 50 kinetic uranyl-hydroxyl species sorption site reactions for the mobile-fraction.
- 50 kinetic uranyl-hydroxyl species sorption site reactions for the immobile-fraction.
- 50 kinetic uranyl-carbonate species sorption site reactions for the mobile-fraction.
- 50 kinetic uranyl-carbonate species sorption site reactions for the immobile-fraction.
- 21 kinetic mass exchange reactions between mobile and immobile domains (Table **3**), where an immobile to mobile porosity ratio of 0.2 is used as stoichiometry in the exchange reactions for unit consistency.

Table 3: Kinetic mobile-immobile exchange reactions.

Exchange Species	Porosity Fraction
$(UO_2)_{11}(CO_3)_6(OH)_{12}^{2-}$	0.2
$(UO_2)_2(OH)_2^{2+}$	0.2
$(UO_2)_2CO_3(OH)_3^-$	0.2
$(UO_2)_2OH^{3+}$	0.2
$(UO_2)_3(CO_3)_6^{6-}$	0.2
$(UO_2)_3(OH)_5^+$	0.2
$(UO_2)_3(OH)_7^-$	0.2
$(UO_2)_3O(OH)_2HCO_3^+$	0.2
$(UO_2)_4(OH)_7^+$	0.2
$MgUO_2(CO_3)_3^{2-}$	0.2
$UO_2(CO_3)_2^{2-}$	0.2
$UO_2(CO_3)_3^{4-}$	0.2
$UO_2(OH)_2(aq)$	0.2
$UO_2(OH)_3^-$	0.2
$UO_2(OH)_4^{2-}$	0.2
UO_2CO_3	0.2
$UO_2NO_3^+$	0.2
UO_2OH^+	0.2
UO_2^{2+}	0.2

Table 3: cont....

$Ca_2UO_2(CO_3)_3$	0.2
$CaUO_2(CO_3)_3^{2-}$	0.2

The output from the preprocessor [17] is a series of equations and species:

- 76 equilibrium equations.
- 0 conservation equations.
- 216 kinetic equations.
- 43 mobile species (aqueous species in the mobile domain).
- 249 immobile species, including sorbed species residing in mobile and immobile domains, and aqueous species in the immobile domain.

The STOMP-W/ECKEChem input is structured around "cards"; where input cards should be thought as a collection of like input data, not the infamous paper punch cards of yesterday.

The Aqueous Species Card identifies mobile aqueous species and assigns associate parameters, and was used to specify the 43 aqueous species in mobile domain. The Aqueous Species Card is additionally used to set aqueous molecular diffusion coefficient for all aqueous species, activity coefficient model option, species charge, species diameter, and species molecular weight. The Solid Species Card identifies immobile solid species and assigns associate parameters, and was used to specify all 146 aqueous and sorbed species in immobile domain and 103 sorbed species in mobile domain. The Gas Species Card, not used for this problem, is used to identify gas species and specify gas molecular diffusion coefficient for all species and gas-aqueous partitioning parameters, and define an associate aqueous species.

The Equilibrium Equation Card identifies the species in the equilibrium equation and assigns the stochiometric parameters, and was used to specify the 72 mobile- and immobile-fraction speciation reactions and the 4 mobile- and immobile-fraction surface complexation reactions. The Equilibrium Reaction Card assigns the parameters used in the temperature dependent equations for equilibrium constants and was used to specify the equilibrium constants. The equilibrium constant used in an equilibrium equation is identified in the Equilibrium Equation Card by referencing a specified equilibrium reaction in the Equilibrium Reaction Card. As the 38 mobile fraction and 38 immobile-fraction equilibrium equations use identical equilibrium constants, only 38 Equilibrium Reaction Card inputs are required.

The Kinetic Equation Card identifies the species in the kinetic equation, the stochiometric parameters, and the kinetic reaction rates and associated coefficients, and was used to specify the 216 kinetic equations. The Kinetic Reaction Card identifies the reaction rate form, the dependent species, and assigns reaction parameters, and was used to specify the 221 kinetic reactions. The Kinetic Equations Card entries vary in complexity. The species and reactions involved in each kinetic equation can vary from only 1 species and 0 kinetic reactions to 23 species and 121 kinetic reactions.

The Conservation Equation Card identifies the species in the conservation equation and the stochiometric parameters, but was not used to specify conservation equations as no conservation equations resulted from the preprocessing stage. The Species Link Card is used to link reactive species with coupled equation components and to identify the hydrogen ion species for output of pH. The Lithology Card, not used for this problem, specifies the mineral specific areas and species volume fractions for every rock/soil type defined on the Rock/Soil Zonation Card. The Solute/Porous Media Interaction Card is used to specify hydraulic dispersion parameters for all reactive species and tracers for every rock/soil type defined on the Rock/Soil Zonation Card. In addition to the ECKEChem specific input cards, reactive species input can additionally be specified on the Initial Conditions Card, Boundary Conditions Card, Output Control Card, Surface Flux Card and Source Card.

Comparison Against SC-DR-DP Model

The numerical investigation of Liu *et al.* [25] demonstrated the need for a surface complexation-distributed rate-dual porosity (SC-DR-DP) model to match the effluent concentrations of U(VI) from the uranium desorption column experiments in the field-textured sediments. Eliminating the surface complexation and distributed rate components of the model, resulted in poor agreement with experimental results for the fine-grained sediments; and eliminating the dual porosity component of the model, resulted in over–prediction of the effluent U(VI) concentrations. To compare the STOMP-W/ECKEChem simulator against the SC-DR-DP modeling of Liu *et al.* [25], two simulations were conducted with two of the solute transport options in STOMP-W/ECKEChem: 1) Patankar scheme and 2) Total-Variational-Diminishing scheme, using the geochemical model described above. Simulation results for the SC-DR-DP Model and the two transport schemes for STOMP-W/ECKEChem are shown in Fig. **2**.

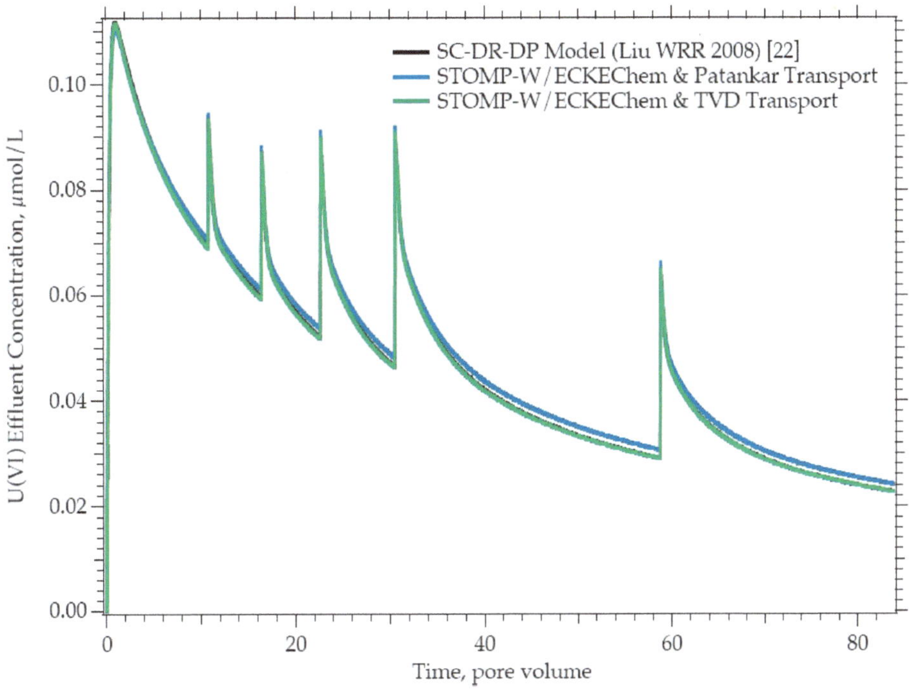

Figure 2: Numerical simulations of U(VI) effluent concentrations with desorption from a field-textured sediment. Black curve is simulation results from the SC-DR-DP model of Liu *et al.* [25]; blue curve is simulation results from STOMP-W/ECKEChem using Patankar solute transport; and green curve is simulation results from STOMP-W/ECKEChem using total-variational-dimensioning solute transport.

5. CONCLUSIONS

Engineering problems frequently require solutions to nonlinear equations or systems of nonlinear equations, and numerical solution techniques for these classes of problems are common in undergraduate and graduate engineering texts [36]. In developing a reactive transport capability for the STOMP simulator, an engineering perspective was taken by reducing the geochemistry to a system of nonlinear equations of three types: equilibrium, conservation, and kinetic. The approach is attractive to engineers faced with resolving complex geochemical systems. The concept for this engineering approach is founded in the general paradigm for batch biogeochemical systems developed by Fang *et al.* [17]. The conceptual approach was first demonstrated with the dissolution of basalt and precipitation of carbonates with CO_2 injection into a deep saline basalt formation for geologic sequestration. The application of this conceptual approach has required the implementation of several numerical solution options to improve the computational efficiency of the solution scheme. Whereas engineers often power their way through to problem solutions, subtle adjustments (*i.e.*, the numerical solution scheme improvements) to the original engineering approach by astute geochemists have significantly improved the numerical efficiency of the solution, as required to solve a demonstration problem involving uranyl reactive transport.

LIST OF ABBREVIATIONS

Roman

a_i	stochiometric coefficient of species *i*
A_m	specific reactive surface area, m^2/kg aqu.
A_s	area of grid surface *s*, m^2
b	first-order biomass decay coefficient, 1/s
b_i	stochiometric coefficient of species *i*
c_i	reaction rate coefficient of reaction *i*
C	species conc., mol/m^3 aqu.
C_a	acceptor species conc., mol/m^3 aqu.
C_{aq}	aqueous-species conc., mol/m^3 aqu.
C_d	donor-species conc., mol/m^3 aqu.
C_i	species conc. of species *i*, mol/m^3 aqu.
(C_i)	aqueous activity of species *i*, mol/m^3 aqu.
C_i	species conc. of species *j*, mol/m^3 aqu.
(C_j)	aqueous activity of species *j*, mol/m^3 aqu.
C_j^{tc}	total-comp. conc. of species *j*, mol/m^3 aqu.
$\{C_j^{tc}\}_\gamma$	tc conc. of species *j* in phase γ mol/m^3 aqu.
C_j^{tk}	total-kinetic conc. of species *j*, mol/m^3 aqu.
$\{C_j^{tk}\}_\gamma$	tk conc. of species *j* in phase γ mol/m^3 aqu.
C_{oil}	mobile oil conc., kg oil/m^3 aqu
d_c	equivalent collector diameter, m
D	molecular diffusion coefficient, m^2/s
$\mathbf{D}_h$	hydraulic dispersion tensor, m^2/s
e_i	stochiometric exponent of species *i*
eq	equilibrium
E_a	activation energy, J/mol
f^m	site density for site *m*, mol/kg soil
F	thermal diffusive flux, W/m^2
g	acceleration of gravity, m/s^2
h	enthalpy, J/kg
J	diffusive flux vector, kg/m^2 s
k	thermal conductivity, W/m C
k	reaction rate constant, mol/m^2 s

k	intrinsic permeability tensor, m^2
k_b	backward reaction rate constant, 1/s
k_d	disolution reaction rate constant, 1/s
k_f	forward reaction rate constant, 1/s
k_m	mass transfer coefficient, 1/s
$k_{r\gamma}$	relative permeability of phase γ
k_{ref}	rate const. at ref. temperature, mol/m^2 s
k_s	sorption reaction rate constant, 1/s
K_a	half-sat. const. for acceptor, mol/m^3
K_d	half-sat. const. for donor, mol/m^3
K_d	sorption coefficient, m^3 aqu./kg soil
K_{eq}	equilibrium constant
$\{K_{eq}\}_j$	equilibrium constant of equil. equation j
m	mass rate, kg/s
M	molecular weight, kg/kmol
n	unit surface normal vector
N_C	num. of components
$N_{eqn}{}^{cn}$	num. of conservation eqns.
$N_{eqn}{}^{eq}$	num. of equilibrium eqns.
$N_{eqn}{}^{kn}$	num. of kinetic eqns.
N_F	num. of degrees of freedom
N_P	num. of phases
$N_{products}$	num. of products
$\{N_r^{tk}\}_j$	num. of reactions in total-kinetic species j
$N_{reactants}$	num. of reactants
$\{N_s^{eq}\}_j$	num. of species in equilibrium eqn. j
$\{N_s^{kn}\}_j$	num. of species in kinetic eqn. j
$\{N_s^{tc}\}_j$	num. of species in total-comp. species j
$\{N_s^{tk}\}_j$	number of species in total-kinetic species j
p	probability
P_γ	pressure of phase γ, Pa
q	power, W
q_i	source of species i, mol/s m^3 aqu.
q_m	max. rate of substrate util., mol d/mol cell/s
Q	ion activity product
R	ideal gas constant, J/mol K

R_k	kinetic reaction rate, mol/s m^3 aqu.
$\{R_k\}_j$	kinetic reaction rate *i*, mol/s m^3 aqu.
R_k^m	kin. reaction rate for site *m*, mol/s m^3 aqu.
$\{R_k^{tc}\}_j$	tc reaction rate: sp. *j*, reac. *k*, mol/s m^3 aqu
$\{R_k^{tk}\}_j$	tk reaction rate: sp. *j*, reac. *k*, mol/s m^3 aqu
s	saturation
S	sorbed species conc., kg/kg soil
S^*	max. sorbed species conc., kg/kg soil
S_i	sorbed species *i* conc., kg/kg soil
S^*_i	max. sorbed species *i* conc., kg/kg soil
S_{oil}	sorbed oil conc., kg oil/kg soil
S_{oil}^*	max. sorbed oil conc., kg oil/kg soil
S_i^m	sorbed site *m* species *i* conc., kg/kg soil
S_i^{m*}	max. sorbed site *m*, sp. *i* conc., kg/kg soil
S_T	total sorption site density, mol/kg soil
t	time, s
tc	total-component
tk	total-kinetic
T	temperature, K
T_{ref}	reference temperature, K
V	volume, m^3
V_l	aqueous volumetric flux, m/s
V_p	volume of node *p*, m^3
$\{V_s\}_\gamma$	vol. flux across surface *s* of phase γ, m/s
$\mathbf{V}_\gamma$	volumetric flux vector of phase γ, m/s
X_m	biomass, mol cells/m^3 aqu.
Y	microbial yield coef., mol cells/mol donor
$\mathbf{z}$	gravitational unit vector

Greek

α	empty bed collision efficiency
α	kin.-reac. rate constant, 1/s
α^m	kin.-reac. rate constant for site *m*, 1/s
Γ	surface, m^2
Γ_p	surface surrounding node *p*, m^2
Δt	time step, s
η	single collector efficiency
μ	viscosity, Pa s

μ	probability distribution func. parameter
π	pi (*i.e.*, 3.14159265)
ρ	density, kg/m^3
ρ_b	bulk density, kg/m^3
ϕ	porosity
σ	probability distribution func. parameter
τ	tortuosity factor
χ	mole fraction
ω	mass fraction

Roman Subscripts

b	bulk
g	gas phase
h	hydrate phase
i	ice phase
i	reactive species i
im	immobile fraction
j	reactive species j
l	aqueous phase
m	sorption site m
mo	mobile fraction
mp	mineral phase
n	nonaqueous phase liquid
p	precipitated salt
r	reaction
s	geologic media
s	species
ss	sorbed species

Greek Subscripts

γ	phase

Roman Superscripts

cn	conservation
eq	equilibrium
kn	kinetic
t^o	current time step level
t^+	new time step level

t old time step level

tk total kinetic

tc total component

ACKNOWLEDGEMENTS

The principal author (engineer) of this chapter would like to thank Dr. Yilin Fang, the second author (geochemist), for her guidance in the subtleties of geologic reactive transport. Jointly the authors would like to thank Dr. Chongxuan Liu for his collaboration in comparing numerical modeling results. The development of STOMP-CO2/ECKEChem was funded by the ZERT Project, Assistant Secretary for Fossil Energy, Office of Sequestration, Hydrogen, and Clean Coal Fuels, NETL, of the U.S. Dept. of Energy under Contract No. DE-AC05-76RL01830. Funding for the preparation of this chapter was provided by the Carbon Sequestration Initiative within the Laboratory Directed Research and Development program at the Pacific Northwest National Laboratory.

REFERENCES

[1] M. D. White, M. Oostrom, "*STOMP Subsurface Transport Over Multiple Phases, Version 2.0, Theory Guide*", PNNL-12030, UC-2010, Pacific Northwest National Laboratory, Richland, WA, 2000.

[2] M. D. White, M. Oostrom, "*STOMP Subsurface Transport Over Multiple Phases, Version 4.0, User's Guide*", PNNL-15782, Pacific Northwest National Laboratory, Richland, WA, 2006.

[3] M. Rosing, J. Nieplocha and S. B. Yabusaki, "Toward efficient compilation of user-defined extensible Fortran directives", In Proceeding Ninth International Workshop *on High-Level Parallel Programming Models and Supportive Environments*, 9th, Santa Fe, NM. IEEE, New York. pp. 61–69. 2004.

[4] M. D. White, M. Oostrom, M. L. Rockhold, and M. Rosing, "Scalable modeling of carbon tetrachloride migration at the Hanford site using the STOMP simulator". *Journal of Vadose Zone* vol. 7, pp. 654–666, 2008.

[5] J. Nieplocha, B. Palmer and V. Tipparaju, "Advances, Applications and Performance of the Global Arrays Shared Memory Programming Toolkit", *International Journal of High Performance Computing and Applications*, vol. 20, pp. 203-231, 2006.

[6] M. D. White, R. J. Lenhard, W. A. Perkins, K. R. Roberson, "*Arid-ID Engineering Simulator Design Document*", PNL-8448, Pacific Northwest Laboratory, Richland, WA, 1992.

[7] M. D. White, W. E. Nichols, "*MSTS Multiphase Subsurface Transport Simulator Theory Manual*", PNL-8636, Pacific Northwest Laboratory, Richland, WA, 1992.

[8] W. E. Nichols, M. D. White, "*MSTS Multiphase Subsurface Transport Simulator User's Guide and Reference*", PNL-8637, Pacific Northwest Laboratory, Richland, WA, 1992.

[9] D. H. Bacon, M. D. White, B. P. McGrail, "*Subsurface Transport Over Reactive Multiphases (STORM): A General Coupled, Nonisothermal Multiphase Flow, Reactive Transport, and Porous Medium Alteration Simulator, Version 2. User's Guide*", PNNL-13108, Pacific Northwest National Laboratory, Richland, WA, 2000.

[10] A. Chilakapati, "*RAFT: A simulator for ReActive Flow and Transport of groundwater contaminants*", PNL-10636, Pacific Northwest National Laboratory, Richland, WA, 1995.

[11] C. I. Steefel, S. B. Yabusaki, "*OS3D/GIMRT, Software for multicomponent-multidimensional reactive transport: User's Manual and Programmer's Guide*", PNL-11166, Pacific Northwest National Laboratory, Richland, Washington, 1996.

[12] C. I. Steefel, "*GIMRT, version 1.2: Software for modeling multicomponent, multidimensional reactive transport. User's Guide*", UCRL-MA-143182, Lawrence Livermore National Laboratory, Livermore, California, 2001.

[13] D. A. Dzombak, F. M. M. Morel, "Surface Complexation Modeling: Hydrous Ferric Oxide", *Wiley-Interscience*, 393 p, 1990.

[14] M. D. White, B. P. McGrail, "*STOMP Subsurface Transport Over Multiple Phases, Version 1.0, Addendum: ECKEChem Equilibrium-Conservation-Kinetic Equation Chemistry and Reactive Transport*", PNNL-15482, Pacific Northwest National Laboratory, Richland, WA, 2005.

[15] M. D. White, B. P. McGrail, H. T. Schaef, D. H. Bacon, "Numerically simulating carbonate mineralization of basalt with injection of carbon dioxide into deep saline formations", In Proceedings of CMWR XVI - *Computational Methods in Water Resources XVI International Conference*, Copenhagen, Denmark, June 19-22, 2006.

[16] J. Bear, and J. J. Nitao, "On equilibrium and primary variables in transport in porous media", *Transport in Porous Media*, vol. 18, no. 2, pp. 151-184, 1995.

[17] Y. Fang, G. T. Yeh, WD Burgos, "A general paradigm to model reaction-based biogeochemical processes in batch systems", *Water Resources Research*, vol. 39, no. 4, pp.1083-1107, 2003.

[18] P.W. Atkins, J. de Paula, *Physical Chemistry* (7th ed.), Oxford University Press, Oxford, UK, 2002.

[19] C. I. Steefel and A. C. Lasaga, "A coupled model for transport of multiple chemical species and kinetic precipitation/dissolution reactions with application to reactive flow in single phase hydrothermal systems", *American Journal of Science*, vol. 294, pp. 529-592, 1994.

[20] A. C. Lasaga, "Chemical kinetics of water-rock interactions", *Journal of Geophysical Research*, vol. 89, pp. 4009-4025, 1984.

[21] P. Aagaard, H. C. Helgeson, "Thermodynamic and kinetic constraints on reaction rates among minerals and aqueous solution. I. Theorical considerations", *American Journal of Science*, vol. 282, pp. 237-285, 1983.

[22] K. Pruess, J. Garcia, T. Kovscek, C. Oldenburg, J. Rutqvist, C. Steefel, T. Xu, "*Intercomparison of Numerical Simulation Codes for Geologic Disposal of CO_2*", Lawrence Berkeley National Laboratory, LBNL-51813, Berkeley, California, 2002.

[23] C. Tebes-Stevens, A. J. Valocchi, "*Numerical Solution Techniques for Reaction Parameter Sensitivity Coefficients in Multicomponent Subsurface Transport Models*". Hydraulic Engineering Series No. 59. UILI-ENG-98-2009. Department of Civil and Environmental Engineering, University of Illinois at Urbana-Champaign, 1998.

[24] C. Tebes-Stevens, A. J. Valocchi, van Briesen, B. E. Rittmann, "Multicomponent transport with coupled geochemical and microbiological reactions – model description and example simulations", *Journal of Hydrology*, vol. 209, pp. 8-26, 1998.

[25] C. Liu, J. M. Zachara, N. P. Qafoku, Z. Wang, "Scale-dependent desorption of uranium from contaminated subsurface sediments", *Water Resources Research*, vol. 44, W08413, 2008.

[26] I. Langmuir, "The constitution and fundamental properties of solids and liquids. part i. solids", *Journal of American Chemical Society*, vol. 38, pp. 2221-2295, 1916.

[27] R. C. Borden, "Effective distribution of emulsified edible oil for enhanced anaerobic bioremediation", *Journal of Contaminant Hydrology*, vol. 94, pp. 1-12, 2007.

[28] T. Xu, F. Gerard, K. Pruess, G. Brimhall, "*Modeling non-isothermal multiphase multi-species reactive chemical transport in geologic media*", Lawrence Berkeley National Laboratory, LBNL-40504, UC-400, Berkeley, California. 1997.

[29] C. I. Steefel, S. B. Yabusaki, "*OS3D/GIMRT, Software for modeling multicomponent-multidimensional reactive transport, User's manual and programmer's guide*", Pacific Northwest Laboratory, PNL-11166, Richland, Washington, 1996.

[30] G. T. Yeh, V. S. Tripathi, "A critical evaluation of recent developments in hydrogeochemical transport models of reactive multichemical components", *Water Resources Research*, vol. 25, no. 1, pp. 93-108, 1989.

[31] G. T. Yeh, V. S. Tripathi, "A model for simulating transport of reactive multispecies components: Model development and demonstration", *Water Resources Research*, vol. 27, pp. 3075-3094, 1991.

[32] M. W. Saaltink, J. Carrera, C. Ayora, "On the behavior of approaches to simulate reactive transport", *Journal of Contaminant Hydrology*, vol. 48, pp. 213-235, 2001.

[33] G. T. Yeh, Y. Fang, W. D. Burgos, "*BIOGEOCHEM 1.0: A numerical model to simulate BIOGEOCHEMical reactions under multiple phases systems in batches*", Department of Civil and Environmental Engineering, technical report, University of Central Florida, Orlando, Florida, 2005.

[34] D. L. Bond, J. A. Davis, J. M. Zachara, "Uranium (VI) release from contaminated vadose zone sediments: Estimation of potential contributions from dissolution and desorption", in *Adsorption of Metals by Geomedia II: Variables, Mechanisms, and Model Applications*, edited by M.O. Barnett and D.B. Kent, chap. 14, Elsevier, Amsterdam, Netherlands. pp. 375-416, 2008.

[35] N. P. Qafoku, J. M. Zachara, C. Liu, P. L. Gassman, O. Qafoku, and S. C. Smith, "Kinetic desorption and sorption of U(VI) during reactive transport in contaminated Hanford sediment", *Environmental Science and Technology*, vol. 39, pp. 3157-3165, 2005.

[36] W. F. Stoecker, *Design of Thermal Systems*, Third Edition, McGraw-Hill, Inc. 1989.

CHAPTER 6

PFLOTRAN: Reactive Flow & Transport Code for Use on Laptops to Leadership-Class Supercomputers

G. E. Hammond[1], P. C. Lichtner[2*], C. Lu[3] and R. T. Mills[4]

[1]Pacific Northwest National Laboratory, Richland, WA, USA; [2]Los Alamos National Laboratory, Los Alamos, NM, USA; [3]Energy and Geoscience Institute, University of Utah, Salt Lake City, UT, USA and [4]Oak Ridge National Laboratory, Oak Ridge, TN, USA

Abstract: PFLOTRAN, a next-generation reactive flow and transport code for modeling subsurface processes, has been designed from the ground up to run efficiently on machines ranging from leadership-class supercomputers to laptops. Based on an object-oriented design, the code is easily extensible to incorporate additional processes. It can interface seamlessly with Fortran 9X, C and C++ codes. Domain decomposition parallelism is employed, with the PETSc parallel framework used to manage parallel solvers, data structures and communication. Features of the code include a modular input file, implementation of high-performance I/O using parallel HDF5, ability to perform multiple realization simulations with multiple processors per realization in a seamless manner, and multiple modes for multiphase flow and multicomponent geochemical transport. Chemical reactions currently implemented in the code include homogeneous aqueous complexing reactions and heterogeneous mineral precipitation/dissolution, ion exchange, surface complexation and a multirate kinetic sorption model. PFLOTRAN has demonstrated petascale performance using 2^{17} processor cores on problems composed of over 2 billion degrees of freedom. The code is currently being applied to simulate uranium transport at the Hanford 300 Area and CO_2 sequestration in deep geologic formations.

Keywords: High performance computing, reactive transport, carbon sequestration, multiple realizations, multiphase flow and transport, Richards equation, domain decomposition.

1. INTRODUCTION

Over the past several decades, subsurface flow and transport models have become vital tools for the U.S. Department of Energy (DOE) in its environmental stewardship mission. These models have been employed to evaluate the impact of fossil energy resources, such as CO_2 sequestration in deep geologic formations, on the environment, and the efficacy of proposed remediation strategies for legacy waste sites.

For years, traditional models—simulating groundwater flow and solute transport, with basic chemical reactions such as aqueous complexing, mineral precipitation/dissolution, sorption to rock/soil surfaces and radioactive decay—have been employed in 1D or 2D systems [1]. Although these simplified groundwater models are still in wide use, advances in subsurface science have enabled the development of more sophisticated models that employ multiple fluid phases and chemical components coupled through a network of biological and geochemical reactions at multiple scales. With this increased complexity, however, comes the need for increased computing power, typically far beyond that of the average desktop computer. This is especially true when applying these models to large-scale three-dimensional problem domains. This paper gives a brief description of PFLOTRAN—a next-generation, highly-scalable code for simulations of reactive flows in geologic media—and discusses some of the challenges encountered and the progress made in scaling PFLOTRAN simulations to the petascale on Cray XT4 and XT5 architectures.

PFLOTRAN is a subsurface multiphase, multicomponent reactive flow and transport code intended for use on a variety of computer architectures ranging from laptops to leadership-class supercomputers (see Section 4). It is founded upon established frameworks for high-performance computing [*i.e.* HDF5 (Hierarchical

[*]**Address correspondence to P.C. Lichtner:** Los Alamos National Laboratory, Los Alamos, NM, USA; Tel: 1-505-667-3420; E-mail: lichtner@lanl.gov

Fan Zhang, Gour-Tsyh (George) Yeh, Jack C. Parker and Xiaonan Shi (Eds)

Data Format 5), MPI (Message Passing Interface), PETSc (Parallel Extensible Toolkit for Scientific computing), SAMRAI (Structured Adaptive Mesh Refinement Application Interface)], and supports seamless integration of Fortran 9X, C and C^{++} programming languages. PFLOTRAN is licensed under an open source GNU Lesser General Public License (LGPL).

2. GOVERNING EQUATIONS

The governing equations employed in PFLOTRAN to model subsurface flow depend on the physical and chemical processes simulated. Thus, the code is divided up into several flow modes including multiphase CO_2-H_2O, air-liquid water, Thermal-Hydrologic-Chemical (THC), and Richards' equation for variably saturated porous media. The flow modes are coupled to a multicomponent geochemical transport mode through temperature, pressure, flow velocity, and phase saturation state. Likewise, the geochemical transport mode may alter the flow field through changes in porosity, permeability and tortuosity caused by chemical reactions.

2.1. Multiphase Flow

Local equilibrium is assumed between phases for modeling multiphase systems with PFLOTRAN. The multiphase partial differenttial equations for mass and energy conservation solved by PFLOTRAN have the general form [2]:

$$\frac{\partial}{\partial t}\left(\varphi\sum_{\alpha} s_{\alpha}\rho_{\alpha}X_i^{\alpha}\right)+\nabla\cdot\sum_{\alpha}\left[q_{\alpha}\rho_{\alpha}X_i^{\alpha}-\varphi s_{\alpha}D_{\alpha}\rho_{\alpha}\nabla X_i^{\alpha}\right]=Q_i \tag{1a}$$

for the i th component, and

$$\frac{\partial}{\partial t}\left(\varphi\sum_{\alpha} s_{\alpha}\rho_{\alpha}U_{\alpha}+(1-\varphi)\rho_r c_r T\right)+\nabla\cdot\sum_{\alpha}\left[q_{\alpha}\rho_{\alpha}H_{\alpha}-\kappa\nabla T\right]=Q_e \tag{1b}$$

for energy. In these equations α designates a fluid phase (*e.g.* H_2O, supercritical CO_2) at temperature T and pressure P_α with the sums over all fluid phases present in the system; species are designated by the subscript i (*e.g.* w = H_2O, c = CO_2); φ denotes the porosity of the geologic formation; s_α denotes the phase saturation state; X_i^α denotes the mole fraction of species $i\left(\sum_i X_i^\alpha=1\right)$; ρ_α, H_α, U_α refer to the molar density, enthalpy, and internal energy of each fluid phase, respectively; and q_α denotes the Darcy flow rate defined by

$$q_\alpha = -\frac{kk_\alpha}{\mu_\alpha}\nabla(P_\alpha - W_\alpha\rho_\alpha g z), \tag{2}$$

where k refers to the intrinsic permeability, k_α denotes the relative permeability, μ_α denotes the fluid viscosity, W_α denotes the formula weight, g denotes the acceleration of gravity, and z designates the vertical of the position vector. The source/sink terms, Q_i and Q_e, describe injection and extraction of mass and heat at wells, respectively. The quantities ρ_r, c_r, and κ refer to the density, heat capacity, and thermal conductivity of the porous rock.

Additional constitutive relations are needed to account for capillary pressure, and changes in phase which are not discussed in detail here (see [3]). In PFLOTRAN a variable switching approach is used to account for phase changes enforcing local equilibrium. According to the Gibbs phase rule there are a total of N_C +1 degrees of freedom where N_C denotes the number of independent components. This can be seen by noting that the intensive degrees of freedom are equal to $N_{\text{int}} = N_C - N_P + 2$, where N_P denotes the number of phases. The extensive degrees of freedom equals $N_{ext} = N_P - 1$. This gives a total number of degrees of freedom $N_{dof} = N_{int} + N_{int} = N_C + 1$, independent of the number of phases N_P in the system.

2.2. Richards' Equation

The governing mass conservation equation for PFLOTRAN's variably-saturated single phase flow mode is given by

$$\frac{\partial}{\partial t}(\varphi s \rho) + \nabla \cdot (\rho \mathbf{q}) = Q_w, \tag{3}$$

and

$$q = -\frac{k k_r}{\mu} \nabla (P - \rho g z). \tag{4}$$

Here, φ denotes porosity, s saturation, ρ water density, q Darcy flux, k intrinsic permeability, k_r relative permeability, μ viscosity, P pressure, g gravity, and z the vertical component of the position vector. Supported relative permeability functions for Richards' equation include van Genuchten, Books-Corey and Thomeer-Corey, while the saturation functions include Burdine and Mualem. Water density and viscosity are computed as a function of temperature and pressure through an equation of state for water.

2.3. Geochemical Transport

In PFLOTRAN, the geochemical transport equations may be coupled to the flow equations or run in standalone mode. In coupled mode, the flow equations provide the pressure, temperature, Darcy flow velocity, and saturation as functions of time and position. In standalone mode, these quantities are given constant values. Chemical reactions currently implemented in the code are listed in Table **1** and consist of homogeneous aqueous complexing reactions, and heterogeneous mineral dissolution/precipitation, ion exchange and surface complexation reactions. Thermodynamic data are read from an extensive database for equilibrium constants over a range of temperatures from 0–300°C and fixed pressure at 1 bar for temperatures below 100°C and along the saturation curve for pure water for higher temperatures, reaction stoichiometry, mineral molar volumes, species valence, and Debye-Hückel parameters. The user may also use a customized database for higher temperatures, although pressure must be fixed in the current implementation. Surface complexation reactions may be treated either as intrinsically fast reactions in local chemical equilibrium or through a kinetic multirate model defined below. Reactions are transformed to canonical form [4] using a basis set of N_c aqueous primary species that may differ from the species used to construct the database.

The governing mass conservation equation for PFLOTRAN's geochemical transport mode for a multiphase system written in terms of a set of independent aqueous primary or basis species has the form.

$$\frac{\partial}{\partial t}\left(\varphi \sum_{\alpha} s_{\alpha} \Psi_j^{\alpha}\right) + \nabla \cdot \sum_{\alpha} \left(q_{\alpha} - \varphi s_{\alpha} D_{\alpha} \nabla\right) \Psi_j^{\alpha} = Q_j - \sum_{m} v_{jm} I_m - \frac{\partial S_j}{\partial t} \tag{5}$$

where the sums over α are over all fluid phases in the system, and where Ψ_j^{α} denotes the total concentration in the α th fluid phase for primary species $\mathcal{A}_j^{\mathrm{pri}}$ defined by:

$$\Psi_j^{\alpha} = \delta_{l\alpha} C_j^l + \sum_{i=1}^{N_{\mathrm{sec}}} \nu_{ji} C_i^{\alpha}. \tag{6}$$

Table 1: Chemical reactions implemented in PFLOTRAN written in terms of primary species $\mathcal{A}_j^{\mathrm{pri}}$, secondary species $\mathcal{A}_j^{\mathrm{sec}}$, minerals $\mathcal{M}_m$, gaseous species $\mathcal{A}_l^g$, sorbed species $X_{z_k}\mathcal{A}_k$, surface complexes $\Xi\mathcal{A}_k^\sigma$ and empty surface sites ΞX_σ, with corresponding stoichiometric coefficients ν_{jl}, ν_{jm}, ν_{jl}^g, and ν_k^σ. Reaction rates are based on local equilibrium (LEQ) or kinetic rate laws. Primary and secondary aqueous species may be interchanged provided the resulting reactions are linearly independent.

Type	Reaction	Rate
	Homogeneous	
Aqueous	$\sum_j \nu_{ji}\mathcal{A}_j^{\mathrm{pri}} \rightleftharpoons \mathcal{A}_i^{\mathrm{sec}}$	LEQ
	Heterogeneous	
Mineral	$\sum_j \nu_{jm}\mathcal{A}_j^{\mathrm{pri}} \rightleftharpoons \mathcal{M}_m$	Kinetic
Gaseous	$\sum_j \nu_{jl}^g\mathcal{A}_j^{\mathrm{pri}} \rightleftharpoons \mathcal{A}_l^g$	LEQ
Ion Exchange	$z_k\mathcal{A}_j^{z_j+} + z_jX_{z_k}\mathcal{A}_k \rightleftharpoons z_j\mathcal{A}_k^{z_k+} + z_kX_{z_j}\mathcal{A}_j$	LEQ
Surface Complexation	$\nu_k^\sigma \Xi\mathcal{X}_\sigma + \sum_j \nu_{jk}^\sigma\mathcal{A}_j^{\mathrm{pri}} \rightleftharpoons \Xi\mathcal{A}_k^\sigma$	LEQ/Kinetic Multirate

The superscript l denotes the aqueous phase and C_j^l represents the concentration of primary species $\mathcal{A}_j^{\mathrm{pri}}$ assumed to be chosen from the set of aqueous species. The secondary species concentration C_i^α is computed in terms of the primary species from the mass action relation

$$C_i^\alpha = (\gamma_i^\alpha)^{-1} K_i^\alpha \prod_{j=1}^{N_c} (\gamma_j^l C_j^l)^{\nu_{ji}}, \tag{7}$$

with equilibrium constant K_i^α, and activity coefficients γ_j^α, γ_i^α. The activity coefficients are currently computed using the Debye-Hückel equation. Both kinetic and equilibrium sorption are described through the quantity S_j which denotes the concentration of sorbed species in units of mol per bulk volume (see [5] for more details). The quantity Q_j denotes a source/sink term. The coefficient D_α represents hydrodynamic dispersion and molecular diffusion as a diagonal tensor. A full dispersion tensor is not currently supported, as the approach presents the potential for oscillatory behavior and negative concentrations that are physically impossible and cause the geochemical algorithms to fail. Although other codes may truncate minimum concentrations to resolve this problem, cutoffs for concentrations are arbitrary and can introduce error themselves. Diffusion is considered to be species-independent in the current formulation, which greatly simplifies the flux term in the transport equations. Mineral reactions are described by the reaction rate I_m for mineral $\mathcal{M}_m$ with the mineral concentration computed from the equation

$$\frac{\partial \varphi_m}{\partial t} = \overline{V}_m I_m, \tag{8}$$

with molar volume $\overline{V}_m$. The reaction rate I_m is based on transition state theory with the form

$$I_m = -k_m a_m \mathcal{P}_m (1 - K_m Q_m)\, \zeta_m, \tag{9}$$

using the sign convention of a positive value for precipitation and negative for dissolution, where k_m and K_m denote the kinetic rate constant and equilibrium constant, respectively, both functions of temperature and pressure, and a_m refers to the mineral specific surface area. The ion activity product Q_m is defined by

$$Q_m = \prod_k \left(\gamma_k^l C_k^l\right)^{\nu_{km}}. \tag{10}$$

The quantity $\mathcal{P}_m$ is a prefactor with accounts for the pH dependence and other attenuation factors on the rate. The factor ζ_m takes on the values zero or one to ensure that if a mineral is not present at some particular point in space, it cannot dissolve:

$$\zeta_m = \begin{cases} 1, & \varphi_m > 0 \text{ or } K_m Q_m > 1, \\ 0, & \varphi_m = 0 \text{ and } K_m Q_m \leq 1 \end{cases} \tag{11}$$

Porosity and permeability may be altered through chemical reactions and coupled back to the flow equations. Porosity is obtained from the relation

$$\varphi = 1 - \sum_m \varphi_m, \tag{12}$$

which presumes that the total porosity of the porous medium is connected. From this relation permeability may be computed using a phenomenological relation between permeability and porosity, for example, a power law relation of the form

$$k(\varphi) = k_0 \left(\frac{\varphi}{\varphi_0}\right)^n, \tag{13}$$

for some power *n*, with initial porosity and permeability φ_0, k_0, respectively. Likewise, mineral surface area can vary according to a power law relation of the form

$$a_m = a_m^0 \left(\frac{\varphi_m}{\varphi_m^0}\right)^{n_m}, \tag{14}$$

where typically n_m=2/3, and a_m^0 and φ_m^0 denote the initial mineral surface area and volume fraction. This relation only applies to primary minerals with $\varphi_m^0 \neq 0$.

3. PFLOTRAN FRAMEWORK

PFLOTRAN is developed for use with leadership class, high-performance computing architectures. The code is designed from the ground up to efficiently scale on the latest ultrascale supercomputers. It can also run efficiently on desktops and notebook computers and is developed on the latter by most of the development team. The code is written in a modular manner facilitating the incorporation of new computational algorithms and scientific processes. This section describes several concepts or features that are considered to be important for modern code development and key to attaining scalability on the latest leadership-class supercomputers. These key aspects of PFLOTRAN's development include the integration of computational science within an object-oriented coding paradigm, PFLOTRAN's founding upon the robust and sophisticated PETSc library, and the incorporation of scalable I/O through parallel HDF5.

3.1. Object-Oriented Fortran 9X

In order to facilitate code development and modularity in PFLOTRAN, an object-oriented coding paradigm is employed through the judicious use of Fortran9X features. The nature of Fortran9X dictates that the object-oriented paradigm employed within PFLOTRAN differ somewhat from traditional object-oriented languages such as C^{++} and Java in that no attempt is made to explicitly leverage polymorphism or inheritance within the framework, though some use is coincidental. Bea *et al.* [6] explains that Fortran9X polymorphism can be achieved through the use of appropriate interfaces within the code. Although their statement is technically correct, the user must still provide a unique argument list to enable the interface to select the correct subroutine. Within PFLOTRAN, most science subroutines that could exhibit polymorphism pass identical arguments, and thus the interface cannot differentiate between routines. Besides, it can be argued that the user might as well explicitly declare the name of the subroutine called for improved clarity and readability within the code. That being said, interfaces are used widely throughout the code, but not with the explicit intent of enabling polymorphism.

It is our view that the most important aspects of the object-oriented paradigm are data encapsulation and modularity within the code. Since data abstraction, polymorphism and inheritance are not necessarily naturally implemented within Fortran9X, it can be argued that one should use a truly object-oriented language such as C^{++} or Java for those purposes. As of late, Fortran 2003/2008 has increased the number of object-oriented features within Fortran (*e.g.* type extension and inheritance, polymerphism, dynamic type allocation and type-bound procedures) [7]. However, in PFLOTRAN extensive utilization of Fortran 2003/2008 paradigms has been avoided due to inconsistent implementation across compiler platforms, the use of which would greatly limit the platform independence of the code.

As would be expected with any object-oriented code, the numerical algorithms, data structures, and scientific processes within PFLOTRAN all revolve around *classes* and their instantiations, *objects*. A PFLOTRAN class is defined in Fortran through derived data types composed of standard Fortran data structures (*i.e.* dynamically allocated arrays of characters, integers, reals) and pointers to other lower-level derived types/classes; it is not an actual Fortran 2003/2008 CLASS statement described by [7]. Within the code a hierarchy of classes exist ranging from lower-level ones such as auxiliary data structures that contain raw data such as pressures, temperatures, or concentrations, to mid-level ones such as the realization class, which encompasses all low-level data sets and objects, or the timestepper class that controls the procedural workflow to the high-level simulation class that encompasses all other objects in a PFLOTRAN simulation.

Although a comprehensive description of PFLOTRAN's object-oriented paradigm is beyond the scope of this paper, a brief descryption of several mid-to high-level classes is provided. For instance, PFLOTRAN's highest level object is the simulation object (or multi-realization simulation object if run in stochastic mode). The simulation object contains pointers to two types of objects, the realization and the timesteppers for flow and transport, as shown in Fig. **1**.

```
type :: simulation_type
  type(realization_type), pointer :: realization
  type(timestepper_type), pointer :: flow_stepper
  type(timestepper_type), pointer :: tran_stepper
end type simulation_type
```

Figure 1: Fortran9X data type for PFLOTRAN simulation class.

The realization object contains a hierarchy of data structures and objects that define the problem statement and scientific algorithms being employed to solve the mathematical/scientific equations that govern the problem statement. The timestepper encompasses the nonlinear and linear solvers utilized to solve the systems of PDEs either for steady-state conditions or for an increment in time (*i.e.* a time step). The timestepper essentially utilizes the realization object to first determine which scientific processes are being simulated and then populate the system of equations being solved through residual and Jacobian function evaluations based on realization parameters/properties (*e.g.* rock/soil permeabilities, reaction rate constants, time step size, *etc.*) and processes (Richards' equation, saturation functions, geochemical reaction equations, *etc.*).

Fig. **2** illustrates a lower-level class for defining an aqueous species using linked lists. Within the aqueous species object, primary and secondary aqueous species are assigned an id, name, Debye-Hückel ion size parameter (a0), molecular weight (molar_weight), valence (Z) and flag (print_me) for I/O purposes. For the case where the species is a secondary aqueous complex, an equilibrium reaction object (*i.e.* equilibrium_rxn_type) is allocated, which contains parameters for calculating the complex concentration (*i.e.* primary species ids, stoichiometries, equilibrium constant). As PFLOTRAN reads each species from the prescribed input file, an aqueous species object is created and appended to a linked list of species. Parameters from these species lists are later parsed into compact arrays for efficient lookup during PFLOTRAN execution.

```
type, public :: aq_species_type
  PetscInt :: id
  character(len=MAXWORDLENGTH) :: name
  PetscReal :: a0
  PetscReal :: molar_weight
  PetscReal :: Z
  PetscTruth :: print_me
  type(equilibrium_rxn_type), pointer :: eqrxn
  type(aq_species_type), pointer :: next
end type aq_species_type
```

Figure 2: Fortran9X data type for PFLOTRAN aqueous species class.

Object-oriented Fortran9X syntax can appear quite different in comparison to traditional Fortran programming. Fig. **3** illustrates two *do* loops for setting total component concentrations to a value using object-oriented Fortran9X with PFLOTRAN objects and traditional Fortran77. Notice that in the case of object-oriented Fortran9X, the total component concentration array total, of size reaction%ncomp, is embedded within each element of the rt_aux_vars array, which is an array of reactive transport auxiliary objects of size grid%nlmax. In this case, the grid object owns the size of the local on-processor portion of the computational grid (*i.e.* nlmax) and the reaction object owns the number of primary chemical degrees of freedom (*i.e.* ncomp). All of these objects (grid, reaction, rt_aux_vars) are compartmentalized as descendants of the high-level realization object. The arrays and parameters are accessed by referencing descendants of the realization object instead of using a common statement. This object-oriented approach improves the modularity of the code and better ensures the correct data locality (*i.e.* there is no need for common statements throughout the code).

```
grid => realization%patch%grid
reaction => realization%reaction
rt_aux_vars => &
realization%patch%aux%Global%aux_vars

do n = 1, grid%nlmax
   do j = 1, reaction%ncomp
      rt_aux_vars(n)%total(j) =..
   enddo
enddo
```

OO Fortran9X

```
common/array/a(nlmax,ncomp)

do n = 1, nlmax
   do j = 1, ncomp
      a(n,j) =..
   enddo
enddo
```

Traditional Fortran77

Figure 3: Comparison of object-oriented Fortran9x and traditional Fortran loops.

Constructor and destructor routines exist for all classes employed within PFLOTRAN. Objects are created as needed and initialized to default values in their respective constructor routines. The first object to be created is the simulation object (*i.e.* SimulationCreate()), after which all underlying objects are created. In certain cases, objects can be replicated in the construction process by passing the original object to the constructor interface. The use of constructors guarantees that the initialization of data types within PFLOTRAN objects is uniform throughout the code. At the end of a simulation, the destructor routine is called for the highest simulation object (*i.e.* SimulationDestroy()). This routine in turn calls the destructors for its realization and timestepper objects, each of which do the same for underlying objects and data structures. In the case of standard Fortran data structures (*e.g.* arrays of integers, reals), deallocate is called instead of the destructor. This process propagates to the lowest-level objects until all objects are destroyed and memory is freed.

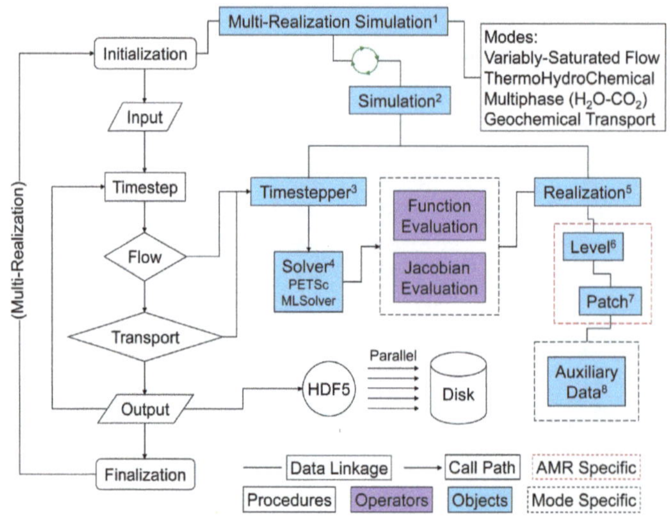

Figure 4: Schematic of PFLOTRAN workflow and data dependence illustrating the use of procedures, operators, and objects within the code (see text for explanation of numbers).

Fig. **4** illustrates PFLOTRAN's workflow and data dependence. Explanation of numbers appearing in the figure are:

1. Multi-Realization Simulation object: Highest level data structure providing all information for running simulations composed of multiple realizations.
2. Simulation object: Data structure providing all information for running a single simulation.
3. Timestepper object: Pointer to Newton-Krylov solver and tolerances associated with time stepping.
4. Solver object: Pointer to nonlinear Newton and linear Krylov solvers (PETSc) along with associated convergence criteria.
5. Realization object: Pointer to all discretization and field variables associated with a single realization of a simulation.
6. Level object: Pointer to discretization and field variables associated with a single level of grid refinement within a realization.
7. Patch object: Pointer to discretization and field variables associated with a subset of grid cells within a level.
8. Auxiliary Data object: Pointer to auxiliary data within a realization/patch.

A comprehensive list of PFLOTRAN objects and their linkage to the processes in the workflow would be difficult to include in a single schematic, and thus, only higher-level objects/processes are shown. To the left is the standard flowchart for flow and transport simulators where the model initializes, reads the input deck, and steps through time computing flow and transport while writing results at select times. In addition to this traditional sequence, a multi-realization loop has been added to signify the simulation of more than one realization. To the right, higher-level data objects are illustrated along with their connectivity within the code's data space. The flowchart illustrates that the flow and transport algorithms call a solver within the timestepper. The solver then calls for a (residual) function evaluation and the calculation of the associated Jacobian matrix. Both solver function calls require auxiliary data provided by the realization object to complete the task (*e.g.* permeability, reaction rates, *etc.*), and thus, the realization object possessing this data is passed to the routines.

For Adaptive Mesh Refinement (AMR), auxiliary data is associated with levels of grid refinement and patches of cells within each level; thus, the linkage through the level and patch objects. The hierarchy of objects is designed to minimize the amount of data explicitly passed to procedures (without resorting to global variables) and maximize compartmentalization of the parameters and processes. In doing so, parameters and processes may be altered and customized with minimal impact to the remainder of the code (*i.e.* procedure argument lists), thus greatly facilitating code development and maintenance. The object-oriented paradigm is also vital in the implementation of the AMR framework within PFLOTRAN as the physical domain must be divided among grid levels and patches, which are treated as lists of objects and are easily accommodated within the PFLOTRAN framework.

Critics of object-oriented programming and the extensive use of pointers within Fortran codes may argue that computational performance degrades due to limitations in the compilers' ability to optimize the code. In the case of PFLOTRAN, the bulk of computation time is spent in isolated "kernels", and it has been demonstrated that the object-oriented design of the code has minimal, if any, negative impact on the code's execution time. Considering that a large fraction of the code's execution time is spent within the PETSc nonlinear solvers for larger parallel runs (*e.g.* 95+% for large-scale flow simulations and 60+% for large-scale reactive transport simulations) and that the residual function evaluations scale exceptionally well, one could argue that the ease of programming and exceptional modularity of the code far outweigh any existing degradation in performance.

3.2. Parallel Implementation

PFLOTRAN is founded upon parallel data structures and solvers provided by PETSc [8]. The code employs domain decomposition through PETSc DA (Distributed Array) or DM objects (The DM object is a generalization of the DA object for managing an abstract grid object.) to partition a physical gridded domain across processor cores, depending on whether structured or unstructured grids are employed (Note: Unstructured grids are currently in the development phase, yet to be completed). With this approach, each processor core possesses locally the data necessary to calculate its portion of the global problem being solved, regardless of the algorithm employed. With the decomposition in place, PETSc provides the necessary index sets or mappings for passing data between processors during the course of the simulation (*e.g.* updating ghost cells, checkpointing vectors, *etc.*). In fact, PETSc actually performs the necessary vector gather/scatters when requested, masking the details of communication.

PFLOTRAN accesses PETSc linear and nonlinear solvers through the SNES (Scalable Nonlinear Equation Solvers) component which provides an interface to various methods (primarily Newton-based) for solving systems of nonlinear equations. Through the SNES and the underlying linear equation solver (KSP) and preconditioner (PC) components, the user may specify which solver algorithms and associated parameters (*e.g.* tolerances, maximum iteration counts, *etc.*) are to be employed to solve the system PDEs for flow and transport. From the "physics" or "science" side, PFLOTRAN provides to SNES two types of function evaluations for solving the Newton-Raphson method: one that computes the residual vector on the right hand side and another that calculates the coefficients for the Jacobian matrix. For each nonlinear solver iteration, SNES calls the function evaluation routines on each processor core to compute the local entries in the residual vector and coefficients for the local rows of the Jacobian. The SNES solver then employs the selected parallel KSP solver to solve the global, linear system of equations and update the nonlinear solution on each processor. This iterative process continues until the solution converges as dictated by the tolerances specified. From the outside, the SNES solve appears no different in parallel than in serial. This is the beauty of parallel domain decomposition within the PETSc framework: the framework masks the intricate details of parallel computing (data decomposition, message passing, *etc.*) at the upper-level user interfaces, exposing them only if the user so desires.

In order to obtain scalable solutions on massively-parallel leadership class machines, the developer must take care to prevent unnecessary bottlenecks from impeding efficient parallel performance. The developer must understand the data requirements and communication patterns of the parallel algorithms employed, including underlying numerical libraries, and resolve parallel inefficiencies. The better the developer understands the parallel implementation of the entire code, the easier it is to resolve parallel bottlenecks. For this reason, the use of poorly understood, *black box* algorithms within parallel codes increases the likelihood of inefficient scalability. If the user introduces algorithms that continually require off-processor data, a potential exists for excessive communication, which may limit parallel scalability. Often, such bottle-necks are not so evident on tens to hundreds of processor cores. However, as the number of cores increases, the inefficiencies become more evident.

Take for instance, parallel I/O through processor zero with a round-robin approach where data is passed to processor zero on a processor by processor basis (*i.e.* a global vector is not required). The model may scale well up to a few hundred processor cores even through such an I/O algorithm is a serial bottleneck; the bottleneck has just not yet become obvious. However, as the number of processor cores grows into the thousands, I/O through processor zero is a known bottleneck, especially for ASCII I/O. At that point, the developer must improvise and employ a more novel approach to I/O such as parallel HDF5, as is employed in PFLOTRAN. At tens to hundreds of thousands of processor cores, the algorithms that limit PFLOTRAN's scalability are in the parallel solvers (*i.e.* global reductions or dot products) and I/O.

3.3. Multiple Realization Simulations

Perhaps one of the most unique and innovative features of PFLOTRAN is its ability to launch multiple simulations of different realizations simultaneously, each realization being executed across multiple processor cores. Although the embarrassingly parallel execution of multiple simultaneous realizations is common these days, the discharge of each in parallel through domain decomposition within a processor subcommunicator group is a novel feat within subsurface simulation. This ability will revolutionize Monte-Carlo style analyses used to better quantify uncertainty in the subsurface.

For the user, the implementation of a stochastic multi-realization simulation is quite straightforward. For example, suppose multiple realizations of permeability are to be employed in a Monte-Carlo fashion. Assuming the correlated random fields have been generated beforehand, a simple script is utilized (*e.g.* a Python script using h5py and numpy libraries) to load these datasets into an HDF5 formatted file under a dataset with a name describing the dataset and the realization *id* (*e.g.* Permeability1 – permeability for realization #1). At the prompt or within the job script, the user enters command line arguments that specify that the simulation be run in stochastic mode with a specified number of realizations and processor groups (Note: the number of processor groups must be less than or equal to the number of processor cores). An example of the command line arguments follows:

```
mpirun -np 10000 pflotran
   -stochastic
   -num realizations 1000
   -num_groups 100
```

Upon execution, the realizations and parallel job's processor cores are divided as evenly as possible among the processor groups. In this case, 1000 realizations will be run on 100 processor groups using 10,000 processor cores.

Each processor group will utilize 100 processor cores (np/num groups) and run 10 realizations apiece (num realizations/num groups), one after another. Thus, only 100 realizations may be executed simultaneously as each processor group may only simulate a single realization on 100 processor cores at a time. Each processor group continues to run realizations until its allocation of 10 has completed.

An alternative approach would be a masterslave paradigm where the root processor core assigns realizations to sub-communicator groups on a one by one basis. This approach would prove beneficial should significant load imbalance exist between sub-communicator groups. Either way, the implementation of the algorithm is straightforward and embarrassingly parallel.

Output for the stochastic simulation is written to files labeled by the realization id and/or processor group id. The user then employs scripts or codes to postprocess the results, computing statistical averages, sampling data, *etc.* To date, this approach has been successfully demonstrated with PFLOTRAN on stochastic simulations composed of hundreds of thousands of realizations and utilizing thousands of processor cores.

3.4. Parallel HDF5 I/O

Parallel I/O plays an important role in enabling, or perhaps better stated, not degrading parallel scalability within a code. PFLOTRAN employs scalable I/O through the parallel HDF5 [9] which leverages collective MPI-IO operations across high-performance file systems. The HDF5 data model provides a sophisticated set of data objects and associated metadata for archiving virtually any combination of data within a single binary file. The HDF5 API provides the programmer with flexible interfaces for reading/writing data from/to a file with relative ease. Through this interface, one specifies the attributes, properties, and data types associated with the data set, all of which can be compartmentalized within a hierarchy of groups within the file. In fact, data sets within an HDF5 file are arranged in a manner similar to the file and directory structure of a standard file system, but within a single file. That is not to say that the file can be navigated in the same manner.

HDF5 files are platform-independent, and data can be accessed through high-level API interfaces written in C, C^{++}, Fortran 90, Java, and Python regardless of the native datatypes on a particular machine (*e.g.* 32-bit vs 64-bit, bigendian vs little-endian, C-specific vs Fortran-specific). For instance, one may write an HDF5 file using Python on a 32-bit, Windows-based laptop computer and read it in parallel across 100,000 processor cores using Fortran90 on a 64-bit Linux-based supercomputer. The library also provides linkage to external libraries such as zlib and szip for compressed storage. All data are stored in an optimal binary format for rapid access.

Critical to the scalability of its parallel I/O, HDF5 employs MPI-IO to enable super-computers to write data either independently or collectively to the same file across hundreds of thousands of processor cores. Our profiling has demonstrated that parallel HDF5's collective writes (all processor cores within an MPI communicator writing in unison) are much more scalable than independent writes (processor cores writing independently), as one would expect. For data sets associated with structured grids, PFLOTRAN scalable parallel I/O relies heavily on the HDF5 hyperslab data layout (*i.e.* up to 3D in memory and file space) to efficiently write data in parallel.

All input data files are read as one-dimensional data in file space and mapped accordingly to memory space, using hash tables to sample for local on-processor values. This approach has reduced initialization time by orders of magnitude for highly parameterized and heterogeneous problem domains. The one-dimensional file space enables PFLOTRAN to read a wider variety of grid structures, as opposed to solely structured grids. Another enhancement to the HDF5 reading algorithms within PFLOTRAN is the use of subsets of processors through MPI sub-communicators to read data and distribute it to the remainder of the processor cores (*e.g.* within the global communicator): this results in more optimal communication patterns and reduces contention at file metadata servers.

4. PARALLEL PERFORMANCE

We have put significant focus on achieving parallel scalability on leadership-class high-performance computing architectures as part of a SciDAC-2 groundwater project (http://ees.lanl.gov/pflotran/). Designed from the ground up for parallel efficiency, PFLOTRAN delivers to the end user the ability to simulate real-world problems at scales limited by only the scalability of linear/nonlinear system solvers. The code runs on machines ranging from the laptop to the largest massively-parallel computer architectures. In fact, core development of the code generally takes place on lap-tops and desktops (Mac, Linux, Windows) with testing on smaller 1D, 2D or 3D problems sized to fit within the machine's memory limits. Testing at scale is carried out on a supercomputer. It should be noted that on any specified architecture, a single PFLOTRAN executable can be run utilizing any of the code's flow or geochemical transport modes in both serial and parallel without modifying the executable (assuming that the job fits into the available memory). In other words, changes to the number of processor cores, parallel decomposition, simulated physical and chemical processes, problem size, *etc.* do not require recompiling the code, since PFLOTRAN utilizes dynamic memory and processor allocation. Production jobs using PFLOTRAN can be run in serial or parallel (provided MPI is installed) on desktop or laptop computers for smaller problem sizes (*e.g.* 1D or small 2D or 3D problems), whereas to simulate large 3D field-scale problems generally much larger supercomputing resources are required.

Perhaps the most significant factor in achieving PFLOTRAN's scalability is the consistent use of PETSc data structures and solvers. Because PFLOTRAN uses PETSc's parallel data structures and associated routines to manage parallel domain decomposition, details such as MPI communication patterns are handled by highly efficient algorithms in PETSc. Furthermore, the large number of solver and preconditioner algorithms made available within PETSc and through PETSc interfacing allows the user to choose the most suitable and scalable solver algorithms for a particular application. For instance, for steady-state flow in a saturated, heterogeneous porous medium, one might choose PETSc's stabilized biconjugate gradient algorithm (Bi-CGStab) coupled with a multilevel preconditioner such as Hypre's PFMG solver [10]. Whereas, for multicomponent biogeochemical transport, PETSc's block Jacobi algorithm may adequately precondition Bi-CGStab for most problem scenarios. Thus through use of PETSc, PFLOTRAN has access to a wide range of algorithms for testing and comparison purposes. In addition, PETSc provides shells for research and development of user-defined solvers/preconditioners (*e.g.* [11]).

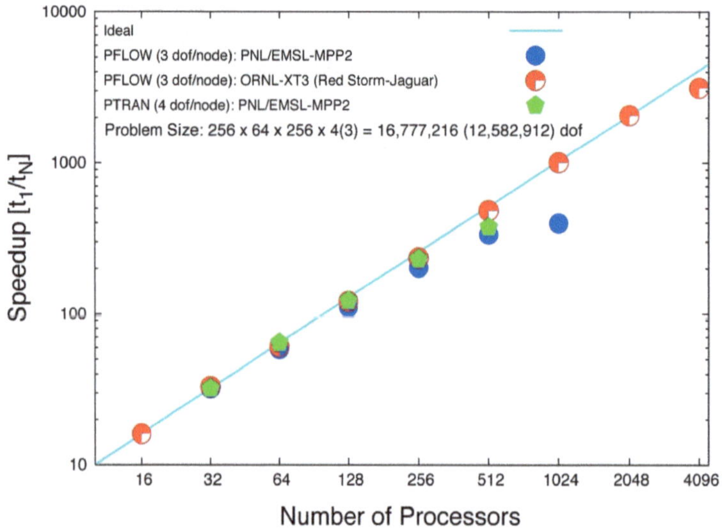

Figure 5: Performance of PFLOTRAN (PFLOW refers to flow and PTRAN to reactive transport) running a single phase thermohydrologic benchmark problem on a 256×64×256 grid with three and four degrees of freedom per node, respectively (approximately 12.6 and 16.8 million degrees of freedom total), on the now-defunct Cray XT3 incarnation of Jaguar at ORNL and the Itanium2-based MPP2 cluster at PNNL.

Prior to being funded by SciDAC, PFLOTRAN demonstrated efficient scalability as shown in Fig. **5** for a moderately-sized, single-phase, thermo-hydrologic benchmark problem composed of 12 to 16 million degrees of freedom (dofs) executed on 16 to 4096 processors on Oak Ridge National Laboratory's (ORNL) Cray XT3 supercomputer. Since SciDAC-funded development, PFLOTRAN has been run on problems composed of up to two-billion degrees of freedom and utilizing up to 131,072 (2^{17}) processor cores on ORNL's Jaguar Cray XT5, the world's fastest open science supercomputer. These large-scale problems are based on real-world variably-saturated flow and geochemical transport modeling of uranium at the Hanford 300 Area in Washington State (see Section 5.1). At this site the migration of a Cold-War era uranium plume is being modeled to better quantify the mass flux of uranium leaching into the neighboring Columbia River, a quantity crucial to environmental decision making and policy at the site.

Fig. **6** demonstrates the strong scaling performance of PFLOTRAN while simulating variably-saturated flow at the Hanford 300 Area using Richards' equation on a problem composed of 270 million degrees of freedom on up to 27,580 processor cores on ORNL's Jaguar XT4 supercomputer. The scenario was run for 50 time steps with I/O turned off. Although good, the scalability is not perfect as the performance deviates from ideal as the processor core count grows. There are several factors contributing to this behavior. First, at 27,580 cores, a 10% increase in the number of linear Bi-CGStab solver iterations was observed relative to the number in the 1024 core run. This increase in linear solver iterations is attributable to an expected and well-understood loss of effectiveness in single-level domain decomposition preconditioners such as block Jacobi. Block-Jacobi preconditioning approximates the inversion of the local on-processor portion of the Jacobian matrix within the linear solver (in this case, through ILU[0] factorization). As the number of cores grows, each processor core owns a smaller portion of the global matrix, resulting in decreased coupling of matrix coefficients (fewer rows and columns of coefficients are coupled), and thus a less-accurate approximate inverse of the matrix.

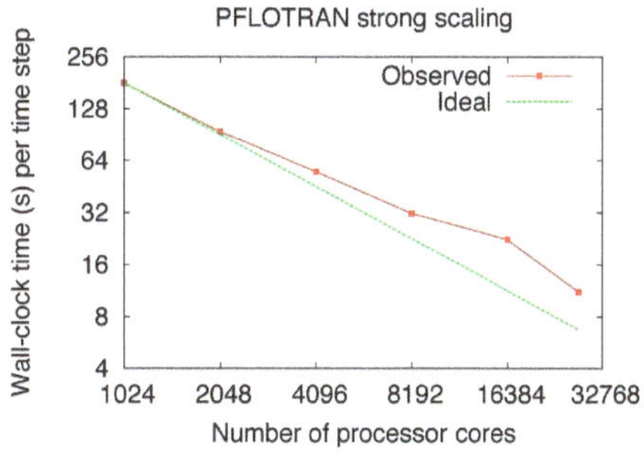

Figure 6: Strong scaling performance on the Cray XT4 partition of ORNL's Jaguar for a variably-saturated flow problem at the Hanford 300 Area with 270 million degrees of freedom (no I/O).

A perhaps more important factor in the departure from linear scaling is the large growth in the relative cost of vector norms and inner products performed within the linear solver as more cores are employed to solve the problem. The cost of these operations is dominated by the cost of global reduction operations, which increases substantially as the number of cores approaches extreme scales. For instance, the time spent in *MPI Allreduce*() calls grew from 10% of the total wall clock time on 1024 cores to up to 56% on 16,384 cores. (We note that we have since worked with the PETSc development team to implement a restructured version of Bi-CGStab that requires only one *MPI Allreduce*() operation per iteration.) Finally, with the current structured-grid formulation of the Hanford 300 Area problem, inactive grid cells exist on the eastern edge of the problem domain where grid cells lie above the river bank and river bottom (see Section 5.1). As the number of processor cores grows, a load imbalance develops, the impact of which is not necessarily easy to quantify when superimposed on the inefficiencies in the preconditioner and increased communication costs.

Aside from the communication issue, which is likely hardware bound, there are several ongoing efforts to further improve PFLOTRAN performance, such as the development of novel solver and preconditioner algorithms to improve both strong (fixed problem size) and weak (fixed problem size per processor core) scalability of the code. Preliminary studies have demonstrated that for steady-state problems, where the Jacobian is less diagonally dominant, multi-level solvers such as multigrid improve weak scalability with much success [12]. With regard to the inactive grid cells, the addition of unstructured gridding is currently underway and will enable a more optimal domain decomposition of the structured grid outside the PETSc Distributed Array (DA) through mesh partitioning software such as ParMETIS, an MPI-based parallel library that implements a variety of algorithms for partitioning unstructured graphs, meshes, and for computing fill-reducing orderings of sparse matrices. The PETSc DA currently does not handle the removal of inactive grid cells, which requires an unstructured grid formulation. However, parallel efficiency is not expected to be nearly as good with unstructured grids as that obtained from structured grids due to such issues as the difficulties in generating optimal mesh partitioning (load balancing), and the lack of geometric multigrid solvers/preconditioners.

Fig. **7** further illustrates the strong scalability of PFLOTRAN for the simulation of uranium transport (see Section 5.1) depicted in Fig. **8** which utilizes up to 65,536 cores on ORNL's Cray XT5 petaflop incarnation of Jaguar. Here, 15 chemical components depicting a subset of geochemical components coupled to uranium mobility through changes in pH, carbonate and calcium concentrations as groundwater and river water mix, are transported within a grid composed of 136 million cells or spatial degrees of freedom. Thus, for flow the nonlinear system is composed of 136 million dofs while geochemical transport consists of 2 billion dofs. Not surprisingly, these performance results reflect relatively poor scalability for flow at larger processor core counts. This is expected given the increased size of the utilized interconnect (increasing cost of global communications) and the small number of flow degrees of freedom per processor core at near 2000—as a general rule of thumb a minimum of 10,000 dof per core is needed to obtain good scaling performance—which results in a very poor ratio of computation to communication. On the other hand for transport, the number of unknowns per core remains above 30,000 using 64,536 cores, and thus it scales rather well considering that only a conventional PETSc solver and preconditioner are being employed (*i.e.* Bi-CGStab with block-Jacobi ILU[0]). For situations such as this where the size of the flow problem is much smaller than the transport problem, it may make sense to add the ability for PFLOTRAN to solve the flow problem redundantly on disjoint groups of MPI processes, much as parallel multigrid solvers do for coarse-grid problems. This may prove unnecessary, however, as in many cases the cost of the transport solve dominates.

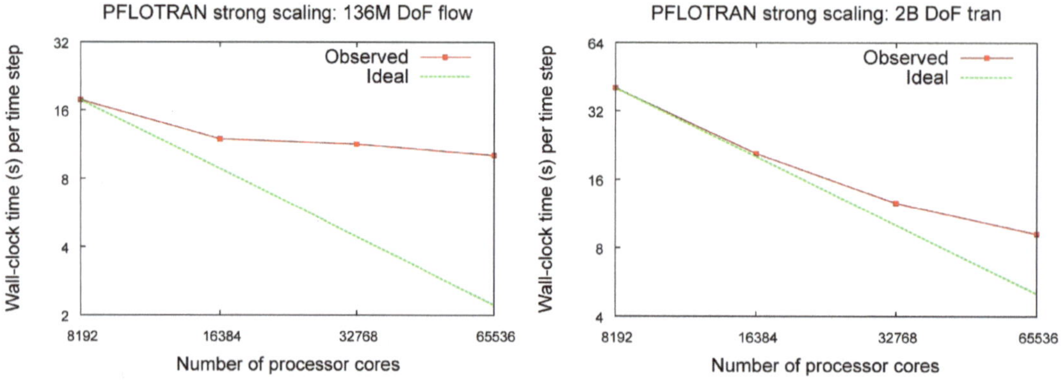

Figure 7: Strong scalability of flow (left) and geochemical transport (right) portions of PFLOTRAN for the Hanford 300 Area 136M dof and 2B dof problems, respectively.

5. APPLICATIONS

In this section two applications of PFLOTRAN are described briefly: (i) uranium migration at the Hanford 300 Area, and (ii) CO_2 sequestration in a deep geologic formation. These examples illustrate, respectively, the use of complex time-dependent boundary conditions and multi-component chemistry coupled to variably saturated flow, and two-phase simulation of H_2O and supercritical CO_2.

5.1. Hanford 300 Area

The Hanford 300 Area is located along the Columbia River in southeastern Washington State. The site was part of plutonium production beginning in 1943. Significant quantities of uranium, copper and other contaminants were disposed at the site in liquid waste streams. Waste was disposed of in two ponds which lie at a distance of approximately 100 m west of the river referred to as the North (NPP) and South (SPP) Processing Ponds. In addition, waste was also placed in nearby trenches. Today a uranium plume persists at the site with concentrations exceeding the EPA 30 μg/L maximum contaminant level [13] despite excavation of contaminated sediments. Particularly perplexing is the persistence of the uranium plume and the factors controlling uranium mobility. Simulations using a K_d model predicted that uranium would be removed by ambient groundwater flow within a decade. Fifteen years later the uranium plume remains where it was with no apparent measurable degradation [14]. Presumably, this is a consequence of slow leaching of uranium from millimeter or smaller scales which causes the uranium source to persist over long time spans.

In a preliminary effort, PFLOTRAN has been applied to model the migration of uranium at the Hanford 300 Area taking into account multicomponent chemistry with a realistic description of uranium sorption through surface complexation [5, 15]. This work differs from other attempts to model the site (*e.g.* [16]) in that the evolution of the uranium plume is divided into three distinct phases. These correspond to: (I) the disposal of uranium and other contaminants and their subsequent migration into the Columbia River, (II) present-day conditions in which non-labile forms of uranium are released at slow rates characterized by diffusion-limited mass transfer from various source regions providing an approximate steady release of uranium into the river, and finally (III) the period in which all non-labile forms of uranium have been removed and multirate sorption governs the behavior of the uranium plume [5]. It is extremely difficult, if not impossible, to model Phase I because of the lack of historical data for the waste stream. Therefore, in Phase II it is assumed that uranium is present in both labile and non-labile forms with sorbed uranium extending from the source regions to the river.

The Hanford 300 Area model consists of a three-dimensional domain measuring 900×1300×20 meters (x, y, z) in size with orientation aligned with the Columbia River at 14° west of north. The base of the model domain lies at 90 meters elevation above sea level. The computational grid is composed of 1,872,000 grid cells with 5-meter horizontal and 0.5-meter vertical grid spacing. Aquifer material properties (*e.g.* permeability, porosity, *etc.*) are assigned to grid cells based on hydrostratigraphic data available at the site [17]. The predominant hydrologic unit at the Hanford 300 Area is the highly-permeable Hanford Unit, which is underlain by several less transmissive Ringold Units. Hydraulic conductivities in the Hanford Unit are on the order of thousands of meters per day, while those of the Ringold Units are hundreds of meters or less *per* day.

Model geochemistry consists of 15 primary aqueous species (H^+, Ca^{2+}, Cu^{2+}, Mg^{2+}, UO_2^{2+} K^+, Na^+, HCO_3^-, Cl^-, F^-, HPO_4^{2-}, NO_3^-, SO_4^{2-} and 2 tracers), 88 secondary aqueous complexes, 2 kinetically-formulated minerals (Calcite and Metatorbernite), 2 surface complexes ($>SOUO_2OH$ and $>SOHOU_2CO_3$) and 1 surface site ($>SOH$). Important geochemical reactions for equilibrium and multirate surface complexation and mineral dissolution include:

$$>SOH + UO_2^{2+} - 2H^+ + H_2O \rightleftharpoons >SOUO_2OH, \tag{15a}$$

$$>SOH + UO_2^{2+} - 2H^+ + H_2O + CO_2 \rightleftharpoons >SOHUO_2CO_3, \tag{15b}$$

$$2\,UO_2^{2+} + Cu^{2+} + 2\,HPO_4^{2-} + 8H_2O - 2H^+ \rightleftharpoons \text{Metatorbernite (s)}, \tag{15c}$$

$$Ca^{2+} + CO_2 - 2H^+ + H_2O \rightleftharpoons \text{Calcite (s)}. \tag{15d}$$

Mobility of uranium is also affected by aqueous complexation reactions with the two dominant species $CaUO_2(CO_3)_3^{2-}$ and $Ca_2UO_2(CO_3)_{3(aq)}$. It should be noted that Metatorbernite serves as surrogate source of non-labile U(VI), and although identified at the site in small quantities, the precise form of non-labile U(VI) is still unknown. Overall, 28,080,000 geochemical degrees of freedom are represented in the model.

The tight coupling of Hanford 300 Area hydrology to rapid fluctuations in the Columbia River stage necessitates the use of transient boundary conditions. For variably saturated flow, transient hydrostatic and seepage face/conductance boundary conditions are assigned to the western inland and river boundaries, respectively, with time-varying datums and gradients on a 1-hour time interval. The seepage face boundary condition calculates the boundary flux for river bank cells above the water surface as a function of internal cell and external atmospheric pressures. The conductance boundary condition replicates the lower permeability of the river hyporheic zone sediments for submerged cells. A constant specified surface recharge is assigned to the top of the model domain, while the north and south boundaries are no flow.

For geochemical transport, chemical compositions of background groundwater and river water are obtained from Hanford 300 Area IFRC and USGS data, respectively. The background groundwater concentrations are assigned to the inland boundary and river chemistry to the river seepage face/conductance boundary. Note that a slightly higher pH and lower carbonate concentration in the river relative to the groundwater significantly impact the sorption of U(VI) to Hanford Unit media.

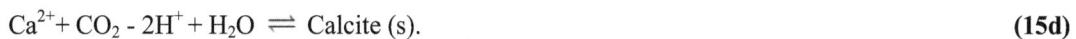

Initial geochemical conditions are set by running a non-sorbing simulation of U(VI) transport based on the boundary conditions discussed above for several years to generate an initial U(VI) plume that reaches to the river. An intermediate initial condition is derived from a snapshot of the flow and transport solution at that time. Sorbed phase U(VI) concentrations are then equilibrated with the aqueous phase to generate the final initial condition.

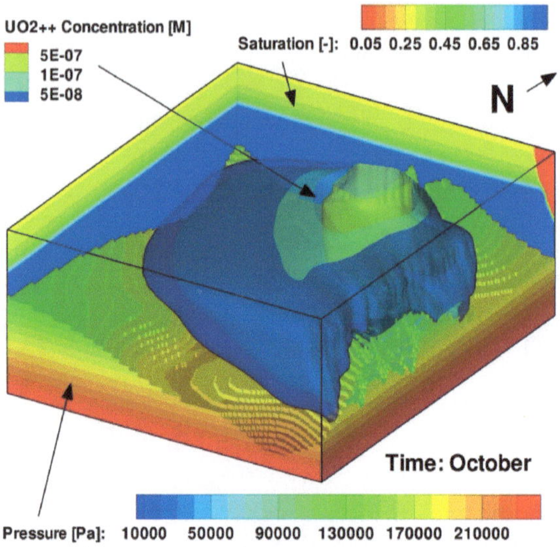

Figure 8: Isosurfaces of total aqueous U(VI) concentration. The Hanford Unit is hidden to better illustrate the U(VI) plume that resides above the Ringold Units. Piezometric pressure is draped on the top of the Ringold Units while water saturation is contoured on the posterior side surfaces of the domain.

Central to this work is the use of high performance computing carried out on the world's fastest open science computer, Jaguar, the Cray XT4/5 at ORNL. Use of HPC made possible the ability to capture the rapidly fluctuating Columbia River stage and multi-component U(VI) chemistry at a sufficiently fine 3D grid resolution. Simulations of the approximately 1.8 million flow and 28 million geochemical transport degrees of freedom were run on over 4k processor cores—runs that would have required several years of computation time on a conventional single processor work-station. Fig. **8** illustrates simulated concentration isosurfaces of a U(VI) plume at the Hanford 300 Area during the October time frame when the river stage is relatively low and groundwater flow is primarily toward the river. Several conceptual models based on this real-world application of PFLOTRAN to radionuclide migration have served as test problems for assessing the parallel performance of PFLOTRAN at the extreme scale. More details can be found in [5].

5.2. Supercritical CO_2 Sequestration

An example of modeling sequestration of supercritical CO_2 using PFLOTRAN is discussed in this section. The code is based on a variable switching approach to account for phase changes between CO_2 and H_2O. The Span and Wagner [18] equation of state is used to compute the density of supercritical CO_2 and correlations for the solubility of CO_2 in brine are taken from [19]. The density of the brine-CO_2 mixture is taken from [20]. To illustrate the code a 3D simulation was carried out for a sandstone host rock containing calcite cement [3].

An isotropic permeability of 2×10^{-12} m^2 is used in the simulation with a porosity of 15%. The nominal temperature and pressure is 50°C and 200 bars. The computational domain is 250 m thick and 7×7 km in lateral extent. CO_2 is injected at a depth of 50 m below the top of the domain. No flow boundary conditions are imposed at the top, bottom, front, and back of the domain with constant pressure at the left and right sides. An injection rate of 1 Mt/y for 20 years was used in the simulations. This corresponds to roughly 75% of the CO_2 produced by a 1000 MW gas-fired power plant in 20 years. Calculations were carried out with 256 processor cores for $160 \times 160 \times 25 \times 3 = 1,920,000$ degrees of freedom. A grid spacing of 47.35 m in the x-and y-directions and 10 m in the z-direction is used. The strong fingering observed in the figures is

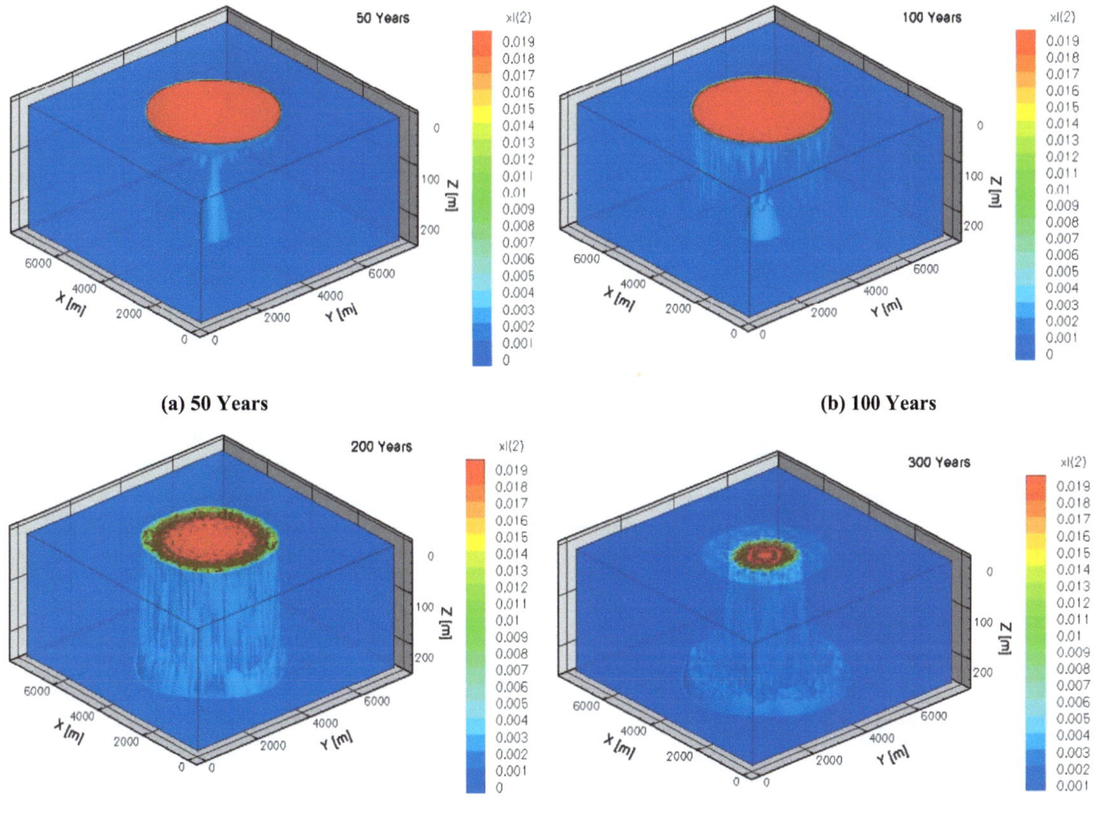

(a) 50 Years **(b) 100 Years**

(c) 200 Years **(d) 300 Years**

Figure 9: Mole fraction of dissolved CO_2 plotted at the indicated times. Fingering is caused by a density instability as CO_2 dissolves into the aqueous phase.

a result of the high permeability of the formation and is caused by a density instability as CO_2 dissolves into the brine. The result is sensitive to the mixing rule used for the density of the CO_2-brine system. Ideal mixing, for example, results in the mixture density being lighter than the original brine causing the CO_2-brine mixture to move upwards rather than downwards. As can be seen in Fig. **9(d)**, grid orientation effects appear in the solution as a result of the simple 7-point stencil employed. To remove these effects higher order discretization methods are being considered which honor the positive definiteness of the solution. This latter property is especially important to maintain in the reactive transport algorithms which solve for the logarithm of the concentration.

Further development of the CO_2 algorithm will involve adding methane and oil fluid phases, implementing higher order discretization methods to avoid grid orientation effects, and refining the time stepping algorithm to allow for larger time steps as phase changes take place. One possibility being investigated is the use of a kinetic formulation based on persistent variables, rather than the current basis switching algorithm. This formulation would also have the advantage of avoiding the problem of solving for different independent variables on different discretization levels when using multilevel solvers.

6. CONCLUSION & FUTURE DEVELOPMENT

A flexible, extensible, parallel computer code, PFLOTRAN, has been developed for modeling subsurface groundwater processes. The code has achieved petascale performance on ORNL's quad-core Cray XT5, Jaguar. PFLOTRAN is based on an object-oriented framework implemented in Fortran 9X, but it provides seamless integration with C and C^{++} packages such as SAMRAI for adaptive mesh refinement. Rapid development has been enabled through the use of PETSc as the parallel framework providing data structures, message passing and efficient solvers. A novel approach for running multiple realizations with each realization run on multiple processor cores was implemented. The code can be applied to various scenarios ranging from variably saturated media to CO_2 sequestration in deep geologic formations.

Future work will add capabilities for unstructured grids, Adaptive Mesh Refinement (AMR) based on the SAMRAI package, colloid-facilitated transport, Pitzer activity coefficient model, operator splitting algorithms, and multiple continua, among others. In addition, more work remains to be done to improve preconditioners and solvers at the extreme limit of hundreds of thousands of processor cores.

ACKNOWLEDGEMENTS

We would like to acknowledge the contributions of the PETSc development team and especially Barry Smith, Satish Balay, and Matt Knepley, without whose help PFLOTRAN would not have achieved its high degree of scalability. We are also indebted to G. (Kumar) Mahinthakumar and Vamsi Sripathi for their enhancement of PFLOTRAN's parallel I/O scalability. This research is supported under the U.S. Department of Energy SciDAC-2 program with funding provided by DOE Offices of Biological & Environmental Research (BER) and Advanced Scientific Computing Research (ASCR). Supercomputing resources were provided by the DOE Office of Science Innovative and Novel Computational Impact on Theory and Experiment (INCITE) program with allocations on NCCS Jaguar at Oak Ridge National Laboratory.

REFERENCES

[1] C. I. Steefel, D. DePaolo, P. C. Lichtner, "Reactive transport modeling: An essential tool and a new research approach for the Earth sciences", *Earth and Planetary Science Letters*, vol. 240, no. 3-4, pp. 539-558, 2005.

[2] P. C. Lichtner, "Continuum formulation of multicomponent-multiphase reactive transport", in P. Lichtner, C. Steefel, E. Oelkers (Eds.), *Reactive Transport in Porous Media*, Vol. 34 of Reviews in Mineralogy, Mineralogical Society of America, pp. 1-81, 1996.

[3] C. Lu, P. C. Lichtner, "High resolution numerical investigation on the effect of convective instability on long term CO_2 storage in saline aquifers", in D. Keyes (Ed.), SciDAC 2007 *Scientific Discovery through Advanced Computing*, Vol. 78 of Journal of Physics: Conference Series, IOP Publishing, Boston, Massachusetts, p. 012042, 2007.

[4] P. Lichtner, "Continuum model for simultaneous chemical reactions and mass transport in hydrothermal systems", *Geochimica et Cosmochimica Acta*, vol. 49, pp. 779-880, 1985.

[5] G. E. Hammond, P.C. Lichtner, "Field-scale model for the natural attenuation of uranium at the Hanford 300 Area using high performance computing", *Water Resources Research*, (ISSN 0043-1397).

[6] S. Bea, J. Carrera, C. Ayora, F. Batlle, M. Saaltink, "Cheproo: A Fortran 90 object-oriented module to solve chemical processes in earth science models", *Computers & Geosciences*, vol. 35, pp. 1098-1112, 2009.

[7] J. C. Adams, W. S. Brainerd, R. A. Hendrickson, R. E. Maine, J. T. Martin, B. T. Smith, *The Fortran 2003 Handbook: The Complete Syntax, Features and Procedures*, Springer-Verlag, London, 2009.

[8] S. Balay, K. Buschelman, V. Eijkhout, W.D. Gropp, D. Kaushik, M.G. Knepley, L.C. Melnes, B.F. Smith and H. Zhang. *PETSc users manual*, Tech. Rep. ANL95/11 -Revision 3.0.0, Argonne National Laboratory, 2009.

[9] The HDF Group, *HDF5 User's Guide: HDF5 Release 1.8.3*, NCSA, May 2009.

[10] J. E. Jones, C. S. Woodward, "Newton-Krylov-multigrid solvers for large-scale highly heterogeneous, variably saturated flow problems", *Advances in Water Resources*, vol. 24, pp. 763-774, 2001.

[11] G. E. Hammond, A. J. Valocchi, P.C. Lichtner, "Application of Jacobian-free Newton-Krylov with physics-based preconditioning to biogeochemical transport", *Advances in Water Resources*, vol. 28, no. 4, pp. 359-376, 2005.

[12] B. Lee, G. E. Hammond, "Parallel performance of preconditioned Krylov solvers for Richards equation", in preparation.

[13] EPA, *U. S. environmental protection agency soil cleanup criteria in 40 CFR part 192*, Tech. rep. 1998.

[14] J. M. Zachara, J. A. Davis, C. Liu, J. P. McKinley, N. Qafoku, D. M. Wellman, S. Yabusaki, "*Uranimum geochemistry in vadose zone and aquifer sediments from the 300 Area uranium plume*", Report PNNL-15121, Pacific Northwest National Laboratory, Richland, WA, 2005.

[15] D. Bond, J. Davis, J. Zachara, "Uranium(VI) release from contaminated vadose zone sediments: estimation of potential contributions from dissolution and desorption", *Adsorption of Metals by Geomedia II: Variables, mechanisms, and model applications*, pp. 375-416, 2008.

[16] S. B. Yabusaki, Y. Fang, S. R. Waichler, "Building conceptual models of field-scale uranium reactive transport in a dynamic vadose zone-aquifer-river system", *Water Resources Research*, vol. 44, pp. 1-24, 2008.

[17] M. D. Williams, M. L. Rockhold, P.D. Thorne, Y. Chen, "*Three-dimensional groundwater models of the 300 area at the hanford site, washington state*", Report PNNL-17708, Pacific Northwest National Laboratory, 2008.

[18] R. Span, W. Wagner, A new equation of state for carbon dioxide covering the fluid region from the triple-point temperature to 1100 K at pressures up to 800 MPa, *Journal of Physical and Chemical Reference Data* 25, pp. 1509-1596, 1996.

[19] Z. Duan, R. Sun, "An improved model calculating CO_2 solubility in pure water and aqueous NaCl solutions from 273 to 533 K and from 0 to 2000 bar", *Chemical Geology*, vol. 193, pp. 257-271, 2003.

[20] Z. Duan, J. Hu, D. Li, S. Mao, "Densities of the CO_2-H_2O and CO_2-H_2O-NaCl systems up to 647 k and 100 MPa", *Energy & Fuels*, vol. 22, no. 3, pp. 1666-1674, 2008.

CHAPTER 7

CORE^{2D} V4: A Code for Water Flow, Heat and Solute Transport, Geochemical Reactions, and Microbial Processes

J. Samper[1*], C. Yang[2], L. Zheng[3], L. Montenegro[1], T. Xu[3], Z. Dai[4], G. Zhang[5], C. Lu[6] and S. Moreira[1]

[1]Civil Engineering School, University of A Coruña, Campus de Elviña s/n, 15192-A Coruña, Spain; [2]Bureau of Economic Geology, University of Texas, University Station, Box X, Austin, TX, USA; [3]Lawrence Berkeley National Lab, One Cyclotron Road, Berkeley, CA, USA; [4]Earth and Environmental Sciences Division, Los Alamos National Laboratory, NM, USA; [5]200 N. Dairy Ashford Rd, Houston, TX , USA and [6]Lawrence Livermore National Lab, 7000 East Avenue, P.O.Box 808, Livermore, CA, USA

Abstract: Understanding natural groundwater quality patterns, quantifying groundwater pollution and assessing the performance of waste disposal facilities require modeling tools accounting for water flow, transport of heat and dissolved species as well as their complex interactions with solid and gaseous phases. Here we present, CORE^{2D} V4, a COde for modeling partly or fully saturated water flow, heat transport and multicomponent REactive solute transport under both local chemical equilibrium and kinetic conditions. It can handle abiotic reactions including acid-base, aqueous complexation, redox, mineral dissolution/precipitation, gas dissolution/exsolution, ion exchange and sorption reactions (linear K_d, Freundlich and Langmuir isotherms, and surface complexation using constant capacitance, diffuse layer and triple layer models) and microbial processes. Hydraulic parameters may change in time due to mineral precipitation/dissolution reactions. A sequential iterative approach is used for the numerical solution of coupled reactive transport equations. The capabilities of CORE^{2D} V4 are illustrated with six selected applications involving: 1) A laboratory concrete degradation experiment, 2) The long-term geochemical evolution of the near field of a High Level Radioactive Waste (HLW) repository in clay, 3) Cation exchange in a physically and geochemically heterogeneous medium, 4) An experiment of CO_2 injection in the vadose zone, 5) The prediction of the water quality of an open pit lake, and 6) Coupled thermo-hydro-chemical processes of compacted bentonite after FEBEX *in situ* test.

Keywords: Reactive transport, numerical model, CORE, HLW repository, pit lake, CO_2 injection.

1. INTRODUCTION

Progress in coupled thermal, hydrodynamic, geochemical and microbial models in porous and fracture media is essential to improve our understanding of how physical, geochemical and biological processes are coupled in groundwater and their effect on groundwater chemistry evolution and the reactive transport of contaminants and microorganisms. Numerical models have been increasingly used for this purpose, a trend that will continue because more sophisticated models and codes are being developed and computing speed keeps increasing. Significant efforts and attempts have been made during recent years towards the development of such tools [1-19]. A recent review on reactive transport modeling is presented by (Steefel EPSL 2005).

The group of University of La Coruña (UDC) has developed in the last 15 years a series of codes with the generic name of CORE (a COde for modeling partly or fully saturated water flow, heat transport and multi-component REactive solute transport under both local chemical equilibrium and kinetic conditions). The reference code, CORELE^{2D} [20] was an extended and improved version of a previous reactive transport code, TRANQUI [21]. CORE solves simultaneously for groundwater flow, heat transport and

*Address correspondence to J. Samper: Civil Engineering School, University of A Coruña, Campus de Elviña s/n, 15192-A Coruña, Spain; Tel: 34-981-167-000; Email: jsamper@udc.es

Fan Zhang, Gour-Tsyh (George) Yeh, Jack C. Parker and Xiaonan Shi (Eds)

multicomponent reactive solute transport under the following conditions 1) 2-D confined or unconfined, saturated or unsaturated, steady-state or transient groundwater flow with general boundary conditions; 2) chemical equilibrium including the following reactions: acid-base, redox, aqueous complexation, mineral dissolution/precipitation, cation exchange, surface complexation and gas dissolution/exsolution; and 3) transient heat transport considering conduction, heat dispersion and convection. The capabilities of CORE-LE^{2D} were later improved within several research projects funded by ENRESA and European Union.

CORE^{2D} V2 was released in 2000 [22]. Contrary to CORE-LE^{2D}, CORE^{2D} V2 handled kinetically-controlled dissolution-precipitation reactions. This version has been widely verified [22, 23]. CORE^{2D} V2 was the base for the development of: 1) INVERSE-CORE^{2D}, a code for automatic estimation of up to 16 different types of reactive transport parameters [24-26] and 2) BIOCORE^{2D} [27-29], a code which accounts for microbial processes in addition to geochemical reactions.

Here we present CORE^{2D} V4 [30], the most recent version of CORE^{2D} which was developed from CORE^{2D} V2 and shares the capabilities of BIOCORE^{2D} and INVERSE-CORE^{2D} such as automatic time stepping, kinetic aqueous complexation reactions, microbial processes, and inverse subroutines of INVERSE-CORE^{2D}.

The mathematical formulation of reactive transport is presented first. Then, the numerical methods for coupled flow, heat and reactive solute transport and geochemical reactions are described. Finally, several application examples are presented to illustrate the capabilities of CORE^{2D} V4 [30].

2. MATHEMATICAL FORMULATION OF CORE^{2D} V4

Based on mass conservation, reactive transport equations can be written in terms of total dissolved component concentrations, C_k, if diffusion and dispersion coefficients are the same for all aqueous species [11]:

$$\nabla\cdot(\theta\mathbf{D}\nabla C_k)-\mathbf{q}\nabla C_k+w(C_k^*-C_k)+\theta R_k=\theta\frac{\partial C_k}{\partial t} \qquad k=1,2,...,N_c \tag{1}$$

where θ is the volumetric water content, $\mathbf{D}$ is the dispersion tensor, $\mathbf{q}$ is the Darcy velocity, w is the external fluid sink/source term, C_k^* is the dissolved concentration of the external fluid source (generally C_k^* applies when w > 0 while $C_k^* = C_k$ for w < 0), k refers to the chemical component from 1 to N_c, and R_k is the chemical sink/source term which includes all the chemical interactions of the *k-th* component with solid species.

Chemical sink/source terms depend nonlinearly on concentrations of dissolved species, which makes Eq. (1) to become a set of coupled highly nonlinear partial differential equations. The Darcy velocity is derived from the solution of the flow equation. For $\mathbf{q} = 0$, Eq. (1) becomes a diffusion-reaction problem. Water flow through porous media is governed by Darcy's law which relates water flux $\mathbf{q}$ to the gradient of water pressure p and elevation z through the equation:

$$\mathbf{q}=-\frac{\mathbf{k}}{\mu}(\nabla p+\rho_w g\nabla z) \tag{2}$$

where ρ_w is the water density, μ is the dynamic viscosity, $\mathbf{k}$ is the intrinsic permeability tensor and g is the gravity acceleration. The governing equation for flow in variably saturated porous media can be derived from mass conservation and Darcy's Law [31]:

$$\nabla\cdot(k_r\mathbf{K}\nabla(p+z))+w=(\phi\frac{\partial S_w}{\partial p}+S_wS_s)\frac{\partial p}{\partial t} \tag{3}$$

where hydraulic conductivity is the product of relative conductivity k_r (a function of pressure head) and saturated conductivity $\mathbf{K}$, ϕ is the porosity, S_w is the saturation degree defined as the ratio of volumetric water content to porosity and S_s is the saturated storage coefficient.

Geochemical reactions can be divided into two classes: 1) Homogeneous reactions which occur in the liquid phase, such as aqueous complexation, acid-base and redox reactions and, 2) Heterogeneous reactions which involve mass transfer from the liquid to the solid/gas phases, and include mineral precipitation/dissolution, cation exchange, surface complexation, and gas dissolution/exsolution. The total dissolved concentration of a given component, C_k in Eq. (1) can be written in an explicit form as a function of the concentration of the N_C primary species:

$$C_k = c_k + \sum_{j=1}^{N_x} v_{jk} x_j = c_k + \sum_{j=1}^{N_x} v_{jk} \left(K_j^{-1} \gamma_j^{-1} \prod_{i=1}^{N_c} c_i^{v_{ji}} \gamma_i^{v_{ji}} \right) \tag{4}$$

where K_j is the equilibrium constant which depends on the pressure and temperature of the system, x_j and c_i are molar concentrations of the secondary and primary species, respectively, γ are activity coefficients, N_x is the number of secondary species, and v_{ji} is the stoichiometric coefficient of the *i-th* primary species in the aqueous complexation of the *j-th* secondary species.

Under equilibrium conditions, dissolution-precipitation reactions can be described by the mass action law which states that:

$$X_m \lambda_m K_m = \prod_{i=1}^{N_c} c_i^{v_{mi}^P} \gamma_i^{v_{mi}^P} \tag{5}$$

where X_m is the molar fraction of the *m-th* solid phase; λ_m is the thermodynamic activity coefficient (X_m and λ_m are taken equal to unity for pure phases), v_{mi}^P is the stoichiometric coefficient in the dissolution reaction of the *m-th* solid phase, and K_m is the corresponding equilibrium constant. The assumption of pure phases has been accepted for reactive transport problems involving cement-based materials [32]. The equilibrium condition provides a relationship between the concentrations of the involved aqueous species. The mass transfer needed to achieve this condition is not specified. In fact, Eq. (5) does not include the concentration of the *m-th* solid phase, and therefore the amount of dissolved/precipitated mineral cannot be computed explicitly.

The subsurface environment contains microbes which use organic and inorganic chemical species as substrates, electron acceptors and nutrients. Microbial growth is described by Monod kinetics laws such as:

$$\frac{\partial C_b^i}{\partial t}\Big|_{growth} = \sum_{p=1}^{N_s} \sum_{q=1}^{N_a} \sum_{r=1}^{N_n} \mu_i^{p,q,r} C_b^i G_i^{p,q,r} \frac{C_s^p}{K_s^p + C_s^p} \frac{C_a^q}{K_a^q + C_a^q} \frac{C_n^r}{K_n^r + C_n^r} \tag{6}$$

where C_s, C_a, and C_n are the concentrations of the substrates, electron acceptors and nutrients, respectively; K_s, K_a, and K_n are the half-saturation constants of the substrates, electron acceptors and nutrients, respectively; superscripts p, q and r refer to the order of substrates, electron acceptors and nutrients in the ecosystem; $\mu_i^{p,q,r}$ is the specific growth rate and $G_i^{p,q,r}$ is the lag coefficient of the ith microbe growing on the pth substrate and the qth electron acceptor; and N_s, N_a, and N_n are the total numbers of substrates, electron acceptors and nutrients, respectively. Expressions of other microbial processes, such as metabolic competition, decay, metabiosis, endogeneous respiration and attachment/detachment of micro-organisms on biofilms can be found in [27] and [28]. Consumption and yield rates of chemical species involved in microbiological processes are related to microbial growth rates by means of yield coefficients.

Energy is transported in the water-solid matrix system by groundwater flow and thermal conduction through the fluid and the solid. Conductive transport of heat is related to the temperature gradient in the fluid and solid while the convective transport of heat is related to the groundwater flow. The energy conservation equation is given by [22],

$$\nabla\cdot(\lambda\nabla T-\rho_w c_w qT)=\theta\rho_w c_w\frac{\partial T}{\partial t}+(1-\theta)\rho_s c_s\frac{\partial T_s}{\partial t}=\rho_m c_m\frac{\partial T}{\partial t} \quad \textbf{(7)}$$

where λ is the thermal conductivity tensor, T is the temperature, c_w is the specific heat of water, ρ_s and c_s are the density and the specific heat of the solid phase at a temperature $T_s = T$, and ρ_m and c_m are the density and specific heat of the bulk porous medium (water plus solid), so that:

$$\rho_m c_m=\theta\rho_w c_w+(1-\theta)\rho_s c_s \quad \textbf{(8)}$$

Chemical processes which involve mass transfer from the solid to the liquid phase can induce changes in physical and hydrodynamic properties of a porous medium [33, 34]. For instance, mineral dissolution (precipitation) can increase (decrease) the porosity. Such a change in porosity may, in turn, affect flow and transport properties (*e.g.*, diffusion coefficient and permeability). Changes in the porosity, $\Delta\phi_f$, can be evaluated from computed dissolution/precipitation rates according to the following equation:

$$\Delta\phi_f=\sum_{k=1}^{N_k}v_k\left(P_k^t\theta^t-P_k^{t-1}\theta^{t-1}\right) \quad \textbf{(9)}$$

where v_k is the molar volume (dm^3/mol) of the *k-th* mineral phase, P_k^t and P_k^{t-1} are concentrations (mol/L) of the *k-th* mineral phase at times *t* and (*t-1*), respectively, and θ^t and θ^{t-1} are volumetric water contents at times *t* and (*t–1*), respectively. A constitutive law is also required to account for the relationship between hydraulic conductivity and porosity. CORE^{2D} V4 incorporates the following Kozeny-Carman equation to relate *K* to Φ:

$$K=K_0\left(\frac{\phi}{\phi_0}\right)^3\frac{(1-\phi_0)^2}{(1-\phi)^2} \quad \textbf{(10)}$$

where K_0 and Φ_0 are reference values of *K* and Φ.

The effective diffusion coefficient may change with porosity of the porous medium due to mineral dissolution-precipitation. Different expressions of effective diffusion coefficient in terms of porosity have been implemented in CORE^{2D} V4. One of them is the following expression derived by [35] for concrete:

$$D_e=D_0\left[10^{-3}+0.07\phi^2+1.8(\phi-\phi_c)^2H(\phi-\phi_c)\right] \quad \textbf{(11)}$$

where D_e is the effective diffusion coefficient, D_0 is the diffusion coefficient in pure water, Φ is porosity, $\Phi_c = 0.18$ represents the critical capillary porosity of the material, and $H(\Phi-\Phi_c)$ is a Heavyside step function which is equal to 1 for $\Phi-\Phi_c \geq 0$ and zero otherwise.

3. NUMERICAL METHODS

A Galerkin finite element method is used in CORE^{2D} V4 to solve the groundwater flow, solute and heat transport equations. At each time step the coupled reactive transport equations are solved by the following steps:

Solution of the groundwater flow equation. It is solved in terms of hydraulic heads for flow in saturated media (confined or unconfined aquifers). The unconfined aquifer flow equation is solved iteratively using a predictor-corrector scheme. Flow in variably saturated media is solved in terms of pressure heads by a Newton-Raphson iterative method. For steady-state flow, the equation is solved only once at the first time step.

Computation of the groundwater velocity. Water velocities, which are needed to evaluate the advective and dispersive solute and heat fluxes, are computed from nodal head values by direct application of Darcy's law to the finite element solution.

Solution of the heat transport. Thermal conduction and advection are considered. The solution of the heat transport equation provides nodal temperatures which are used to update temperature-dependent chemical parameters such as chemical equilibrium constants and activity coefficients. The solution of the heat transport equation shares the subroutines used for solving the solute transport equation, because the structure of both equations is similar.

Solution of the solute transport equations. Each chemical component has a transport equation in terms of its total dissolved concentration, C. Since chemical sink/source terms are evaluated at the previous iteration, each transport equation can be solved separately. Transport matrices may be different for each component when the diffusion coefficient and the effective porosity is different for each chemical species. In such cases, transport equations must be solved separately for each chemical component.

Solution of the biogeochemical equations. Temperature-dependent chemical constants are updated in non-isothermal conditions. A Newton-Raphson method is used to solve the geochemical equations. Biogeochemical reactions are solved in a node-wise manner.

The time step can be specified in advance by the user or derived from an automatic time stepping algorithm [27]. Two types of stabilization methods have been implemented in CORE^{2D} V4 to reduce the numerical oscillations produced by classical finite element methods when solving advection-dominated problems. They include the Stream Upwinding Petrov-Galerkin (SUPG) method and the subgrid scale stabilized method [36].

A Sequential Partly Iterative Approach (SPIA) has been reported by [37] which improves the accuracy of the traditional Sequential Non-Iterative Approach (SNIA) and is more efficient than the general sequential iterative approach (SIA). While SNIA leads to a substantial saving of computing time, it introduces numerical errors which are especially large for cation exchange reactions. SPIA improves the efficiency of SIA because the iteration between transport and chemical equations is only performed in nodes with a large mass transfer between the solid and liquid phases. SPIA is found to be as accurate as SIA while requiring significantly less CPU time. In addition, SPIA is much more accurate than SNIA with only a minor increase in computing time.

CORE^{2D} V4 can cope with heterogeneous systems having irregular internal and external boundaries. The code can also handle heterogeneous and anisotropic media. It is not restricted to specific chemical species, and therefore can consider any number of aqueous, exchanged and sorbed species, minerals and gases. Thermo-dynamic data and stoichiometric coefficients of chemical equilibrium reactions are read directly from a database which is modified from the EQ3NR database [38]. A pre-processor is provided to convert the contents of other thermodynamic databases, for chemical reactions at chemical equilibrium, such as PHREEQE, MINTEQ, NEA, CHEMVAL and HATCHES into the format of CORE databases. It is worth noting that CORE^{2D} V4 can handle with: 1) Anisotropic diffusion to deal with diffusion anisotropy in clay media [39], 2) Isotopic transport coupled with chemical reactions for the purpose of simulating radionuclide release from a HLW repository, and 3) Automatic estimation of flow, solute transport, chemical and biological parameters [40].

4. VERIFICATION

CORE^{2D} V4 has been extensively verified against analytical solutions. The conservative solute and heat transport subroutines of CORE^{2D} V4 have been verified for one-dimensional conditions. The 1-D test case corresponds to the time evolution of concentrations in a semi-infinite confined aquifer under steady-state uniform flow. Reactive transport with kinetic dissolution-precipitation reactions and kinetic aqueous complexation has been verified against analytical solutions for 1-D problems in saturated media [36]. Multicomponent reactive transport coupled with cation exchange reactions has been verified against analytical solutions derived by [41].

Reactive transport with kinetic rate laws was verified for calcite and smectite dissolution against 1DREACT [42]. The capabilities of CORE^{2D} V4 to deal with redox reactions were verified against DYNAMIX with a case of uranium migration through a column [43]. Reactive transport with cation exchange was verified against a test case reported in PHREEQM user's manual problem [44] which involves a column initially filled with 1 mM $NaNO_3$ and 0.2 mM KNO_3 which is flushed by a 0.6 mM $CaCl_2$ solution. Subroutines for solving microbial processes have been verified against analytical solutions derived by (Salvage JH 1998) [5] and against other codes such as BIOCLOG^{3D} [45] and FEREACT [7].

5. APPLICATIONS

CORE^{2D} V4 and other CORE codes have been extensively used for the interpretation of several *in situ* experiments performed at Underground Rock Laboratories (URL) in DIB, DR and VE experiments at Mont Terri (Switzerland) [39, 46], CERBERUS experiment at Mol (Belgium) [28, 29], DIR experiments at Bure (France) [47], Redox Zone II and REX experiments at Äspö (Sweden) [48-50], and FEBEX experiment at Grimsel (Switzerland) [51, 52].

CORE^{2D} V4 has been also used to simulate the long-term geochemical evolution of HLW repositories in clay [19] and granite [18, 53] in integrated performance assessment projects such as BENIPA, NFPRO and PAMINA.

CORE^{2D} V4 has been used to simulate: 1) Solute transport in aquifers including uranium transport in the Andújar aquifer (Spain), the geochemistry of the Aquia aquifer (USA) [54] and salt water flushing in the Llobregat Delta aquitard, 2) Drainage of civil engineering works, flow into tunnels and underground excavations and 3) Groundwater flow and heat transport in hydrothermal systems.

The current capabilities of CORE^{2D} V4 are illustrated with some selected applications involving: 1) A laboratory concrete degradation experiment, 2) The simulation of the long-term geochemical evolution of the near field of a High Level Radioactive Waste (HLW) repository in clay, 3) The reactive transport in a physically and geochemically heterogeneous medium, 4) An experiment of CO_2 injection in the vadose zone, 5) The prediction of the water quality of a pit lake, and 6) Coupled thermo-hydro-chemical reactions of compacted bentonite after the FEBEX *in situ* test.

5.1. Case 1: Concrete Degradation

This case corresponds to the model of a laboratory concrete degradation experiment reported by [55]. A cylindrical cement specimen with a length of 1.02 cm and 13.2 cm^2 of cross-sectional area was flushed with pure water (Fig. **1**). Constant water heads of 1.7 and 0 m were imposed at the left and right ends, respectively. Since the cement specimen was leached by pure water, porosity and hydraulic conductivity of cylindrical cement specimen increase due to dissolution of cementitious minerals. Concentrations of Na^+, K^+, and Ca^{2+} were measured in the outflow water. The model of this experiment using CORE^{2D} V4 has been reported by [56].

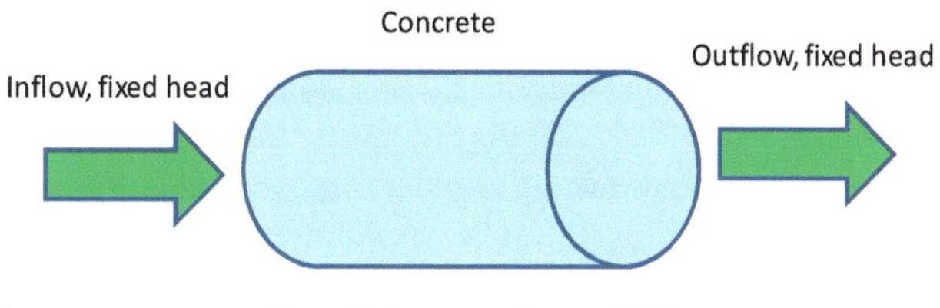

Figure 1: Scheme of the laboratory concrete degradation experiment [56].

The cement column was simulated with a 1-D model and using Dirichlet conditions for water flow at both boundaries. The geochemical model accounts for 21 aqueous complexation reactions and 4 mineral dissolution-precipitation reactions which are listed in Table **1**. The experiment was run for almost a year. The time evolution of modeled and measured concentrations of Na^+, K^+ and Ca^{2+} in the discharge water are shown in Fig. **2** and **3**.

The concentrations of both Na^+ and K^+ in the discharge water (see Fig. **2**) decrease and reach values much smaller than their initial concentrations as more and more pure water is flushed through the cement specimen. The concentration of dissolved Ca^{2+} increases with time due to the dissolution of Ca-minerals, mainly portlandite and C-S-H, reaches a maximum value and then decreases slightly (see Fig. **3**). This indicates that the overall dissolution rates of Ca-minerals may reach a maximum value and then slowly decrease with time. The pH of the discharge water decreases from its initial value of 13.5 to approximately 12.5. One can see from Fig. **2** and **3** that the model captures the overall trends of the time evolution of Na^+, K^+ and Ca^{2+} concentrations in the discharge water. However, the model underestimates slightly the concentrations of Na^+ and K^+ and overestimates that of Ca^{2+}. Such deviations may be caused by the fact that effective diffusion coefficients computed with Eq. (11) may overestimate the actual diffusion coefficients.

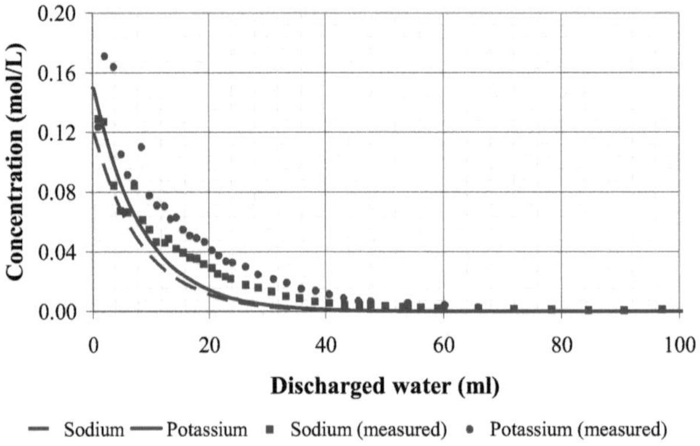

Figure 2: Comparison of computed and measured concentrations of Na^+ and K^+ in the outflow water [56].

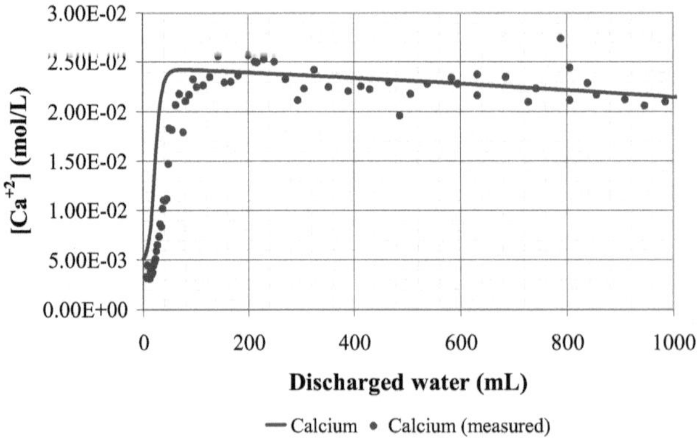

Figure 3: Comparison of computed and measured concentrations of Ca^{2+} in the outflow water [56].

Portlandite dissolves much faster than the C-S-H minerals in the numerical model. This is consistent with experimental observations indicating that portlandite is the first cementitious mineral to be exhausted when a cement specimen is exposed to water flow. Therefore, portlandite dissolution dominates the change in porosity in the cement specimen before portlandite is exhausted. Although there are no data on changes in porosity in the cementitious column, average hydraulic conductivity of the flushed cement specimen can be calculated from the outflow rate measured and the hydraulic gradient imposed upon the cement specimen according to Darcy's law.

Table 1: Hydrogeochemical reactions considered in the reactive transport model of the cement experiment [56].

Aqueous complexes	**Log K**
$H_3SiO_4^- + H^+ \leftrightarrow SiO_2(aq) + 2H_2O$	+9.812
$H_2SiO_4^{2-} + 2H^+ \leftrightarrow SiO_2(aq) + 2H_2O$	+22.912
$NaHCO_3(aq) \leftrightarrow HCO_3^- + Na^+$	−0.154
$NaCO_3^- \leftrightarrow CO_3^{2-} + Na^+$	−0.514
$NaOH(aq) + H^+ \leftrightarrow H_2O + Na^+$	+14.180
$NaHSiO_3(aq) + H^+ \leftrightarrow H_2O + Na^+ + SiO_2(aq)$	+8.304
$NaSO_4^- \leftrightarrow Na^+ + SO_4^{2-}$	−0.820
$KOH(aq) + H^+ \leftrightarrow H_2O + K^+$	+14.460
$KSO_4^- \leftrightarrow K^+ + SO_4^{2-}$	−0.880
$CaH_2SiO_4(aq) + 2H^+ \leftrightarrow Ca^{2+} + SiO_2(aq) + 2H_2O$	+18.562
$CaH_3SiO_4^+ + H^+ \leftrightarrow Ca^{2+} + SiO_2(aq) + 2H_2O$	+8.792
$Ca(H_3SiO_4)_2(aq) + 2H^+ \leftrightarrow Ca^{2+} + 2SiO_2(aq) + 4H_2O$	+15.053
$CaHCO_3^+ \leftrightarrow Ca^{2+} + HCO_3^-$	−1.047
$CaCO_3(aq) + H^+ \leftrightarrow Ca^{2+} + HCO_3^-$	+7.002
$CaOH^+ + H^+ \leftrightarrow Ca^{2+} + H_2O$	+12.850
$CaSO_4(aq) \leftrightarrow Ca^{2+} + SO_4^{2-}$	−2.111
$CO_3^{2-} + H^+ \leftrightarrow HCO_3^-$	+10.329
$CO_2(aq) + H_2O \leftrightarrow H^+ + HCO_3^-$	−6.345
$H^+ + OH^- \leftrightarrow H_2O$	+13.995
$H_2SO_4(aq) \leftrightarrow SO_4^{2-} + 2H^+$	+1.020
$HSO_4^- \leftrightarrow H^+ + SO_4^{2-}$	−1.979
Minerals	**Log K**
$Ca(OH)_2(s) + 2H^+ \leftrightarrow Ca^{2+} + 2H_2O$	+22.800
$SiO_2(s) \leftrightarrow SiO_2(aq)$	−3.999
$C_{1.65}\text{-S-}H_{2.45} + 3.3H^+ \leftrightarrow 4H_2O + SiO_2(aq) + 1.65Ca^{2+}$	+29.285
$C_{1.10}\text{-S-}H_{1.9} + 2.2H^+ \leftrightarrow 3H_2O + SiO_2(aq) + 1.1Ca^{2+}$	+17.097

The time evolution of computed and measured hydraulic conductivities is shown in Fig. **4**. It can be seen clearly that the hydraulic conductivity increases with time due to mineral dissolution. Fig. **4** shows also the hydraulic conductivities calculated with the numerical model using Eq. (10). One can see that the hydraulic conductivity estimated from the measured outflow rate and the hydraulic gradients using Darcy's law are reproduced remarkably well by the model.

5.2. Case 2: Long-Term Geochemical Evolution of a Repository in Clay

This case corresponds to the model of the geochemical evolution in a multibarrier system for a High Level radioactive waste (HLW) disposal in a clay formation. The model includes a bentonite buffer having a thickness of 0.75 m, a concrete sustainment of 0.20 m of thickness and the argillaceous formation with a thickness of about 24 m (Fig. **5**). The coupled non-isothermal reactive transport model for the long-term geochemical evolution of a HLW repository in clay using CORE^{2D} V4 has been reported by [19].

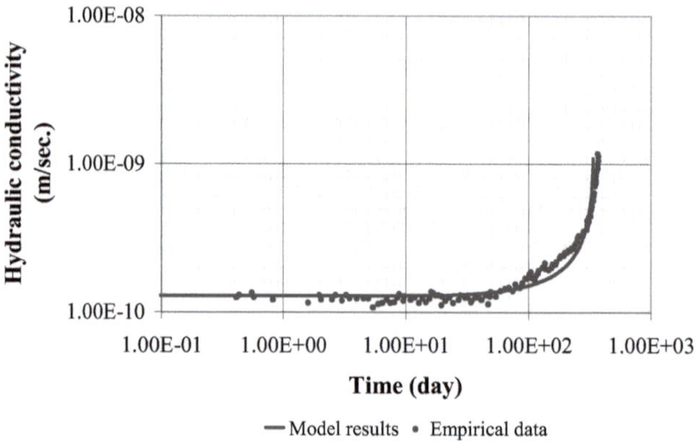

Figure 4: Comparison of computed hydraulic conductivities with those derived from measured outflows [56].

The porewater of the clay rock infiltrates into the concrete and interacts with the cementitious materials and produces a hyperalkaline porewater which then diffuses into the bentonite. The high pH fluid interacts with the bentonite minerals as it diffuses, inducing the dissolution of primary minerals and the neoformation of minerals by precipitation in the bentonite buffer and the concrete. Initially, the bentonite buffer is unsaturated with a volumetric water content of 14%. Water flow is negligible once the bentonite buffer is saturated. Then, solute transport in the domain occurs mainly by molecular diffusion.

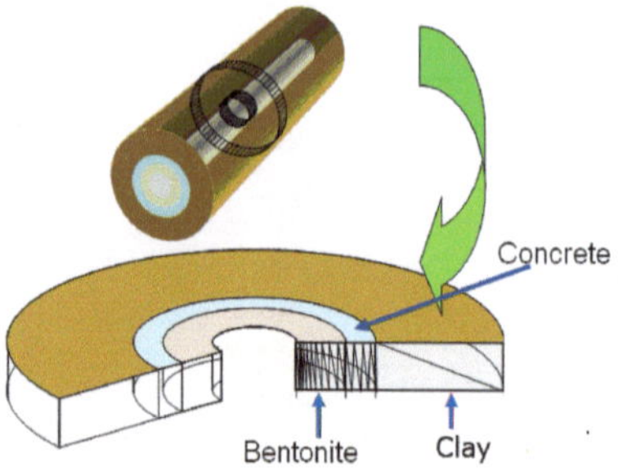

Figure 5: Scheme of a multibarrier system for a HLW repository in a clay formation. It includes: the containers, the bentonite, the concrete and the clay formation [19].

Table 2: Components, aqueous complexes, minerals and exchange cations considered in the coupled hydrogeochemical model of the long-term geochemical evolution of a HLW repository in clay.

Components
Ca^{2+}, Mg^{2+}, Na^{+}, K^{+}, H^{+}, Cl^{-}, HCO_3^{-}, SO_4^{2-}, $SiO_{2(aq)}$, H_2O
Aqueous complexes
CO_3^{2-}, $CaCO_3(aq)$, $CaHCO_3^{+}$, $CaSO_4(aq)$, $CaOH^{+}$, $CaCl^{+}$, $CaCl_2(aq)$, $Ca(H_3SiO_4)_2(aq)$, $CaH_2SiO_4(aq)$, $CaH_3SiO_4^{+}$, $MgCO_3(aq)$, $MgHCO_3^{+}$, $MgSO_4(aq)$, $MgCl^{+}$, $MgOH^{+}$, $MgH_2SiO_4(aq)$, $MgH_3SiO_4^{+}$, $NaOH(aq)$, $NaCO_3^{-}$, $NaHCO_3(aq)$, $NaCl(aq)$, $NaH_3Si_4O(aq)$, $NaHSiO_3(aq)$, $CO_2(aq)$, $KOH(aq)$, $KCl(aq)$, KSO_4^{-}, $KHSO_4(aq)$, $NaSO_4^{-}$, $H_3SIO_4^{-}$, $H_2SiO_4^{2-}$, $H_4(H_2SiO_4)_4^{4-}$, $HSiO_3^{-}$, $H_6(H_2SiO_4)_4^{2-}$, $HCl(aq)$, HSO_4^{-}, OH^{-}
Minerals
calcite, brucite, gyrolite, tobermorite, quartz, dolomite, portlandite, sepiolite, gypsum
Exchanged cations
Ca^{2+}, Mg^{2+}, Na^{+}, K^{+}, H^{+}

The geochemical model accounts for 38 aqueous complexation reactions, 9 minerals and 4 cation exchange reactions which are listed in Table **2**. All mineral reactions are assumed to proceed at local equilibrium.

Thermodynamic data for aqueous species are taken from EQ3/6 database [38]. The model takes into account the temperature dependence of reaction constants.

The numerical model aims at studying the evolution and propagation of high pH in the system. Fig. **6** shows the spatial distribution of the pH computed at different times. pH in the concrete increases first due to the dissolution of portlandite. It reaches a maximum after 10,000 years and then decreases slowly after 40,000 years when portlandite is completely exhausted. pH in the bentonite increases due to the penetration of the hyperalkaline plume from concrete into the bentonite. However, the precipitation of calcite, brucite and sepiolite buffers the hyperalkaline plume. pH reaches a peak value and then decreases to 9.5 after 1 Ma. The hyperalkaline plume penetrates slowly into the clay formation. The front of the hyperalkaline plume penetrates a radial distance of 0.7 m into the clay formation after 1 Ma.

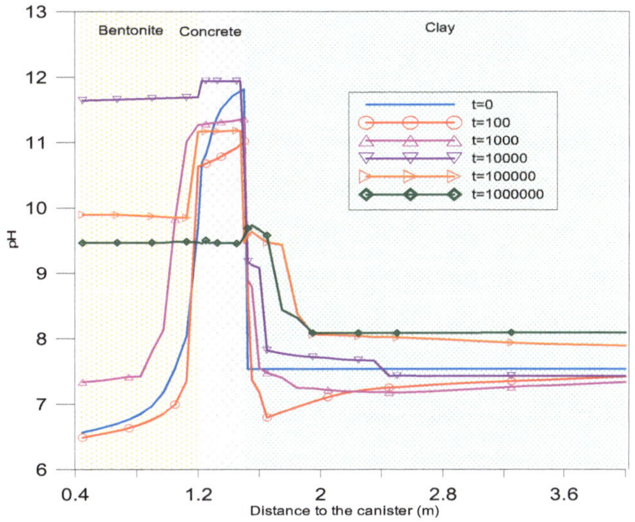

Figure 6: Radial distribution of computed pH at different times [19].

The model provides valuable insight into the mineral alterations induced by the diffusion of hyperalkaline fluids in the multibarrier system. Figs. **7** and **8** show the cumulative amount of dissolved/precipitated mineral at r = 1.125 m (in bentonite) and 1.35 m (in concrete), respectively. Positive (negative) values in both figures indicate precipitation (dissolution). The cumulative amount of mineral dissolution/precipitation is calculated as the difference between the concentrations of a mineral at the current time and the initial time. Brucite precipitates in the bentonite for the first 200,000 years (Fig. **7**). Gypsum precipitates until 300,000 years and then dissolves until the precipitated amount is exhausted. Quartz dissolves rapidly. Calcite precipitates during 1 Ma. Sepiolite precipitates slightly until 200,000 years.

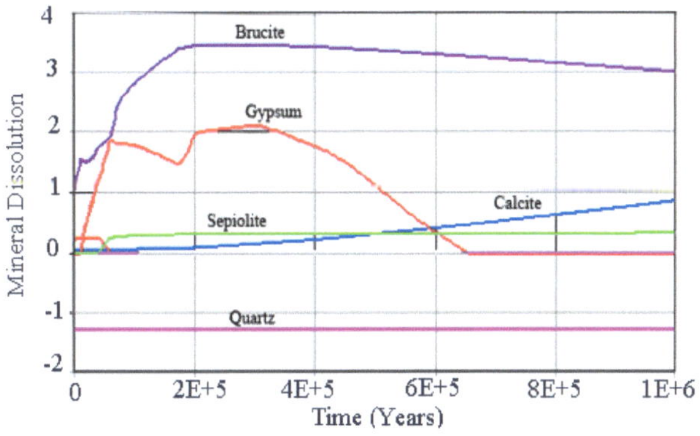

Figure 7: Computed cumulative amount of mineral dissolution/precipitation (in mol/l) in the bentonite at r = 1.125 m. The amount of dissolution-precipitation of portlandite, dolomite, gyrolite and tobermorite in bentonite is too small to be seen in this plot. Values are positive for precipitation and negative for dissolution [19].

Mineral dissolution/precipitation in concrete is shown in Fig. **8**. Portlandite is exhausted in the concrete after 40,000 years and then tobermorite starts dissolving. The dissolution of the portlandite leads to an increase in pH from 11.6 by 20 years to 12.5 by 1000 years. Gypsum and calcite precipitate due to portlandite dissolution. Gypsum precipitates as long as portlandite dissolves until it is exhausted. Precipitation of sepiolite after 60,000 years is caused by the dissolution of brucite. Sepiolite precipitation stops when the tobermorite is exhausted. It should be noted that the cumulative amounts of precipitated calcite and gypsum in the concrete are much greater than those in the bentonite.

The time evolution of the porosity in the system is shown in Fig. **9**. One can see that porosity of the clay formation decreases from its initial value of 0.37 to 0.18 before 10,000 years near the concrete-clay interface (r = 1.525 m). Later, it decreases slowly reaching a value of 0.16. Changes in clay porosity far from the concrete-clay interface are negligible. Porosity in the bentonite decreases from its initial value of 0.407 to 0.3 during the first 200,000 years due the precipitation of brucite and gypsum. After that, it increases slowly up to a value of 0.35. In the middle of the concrete, porosity decreases slightly from its initial value of 0.0805 to 0.045 before 350,000 years. After that time, it increases slightly.

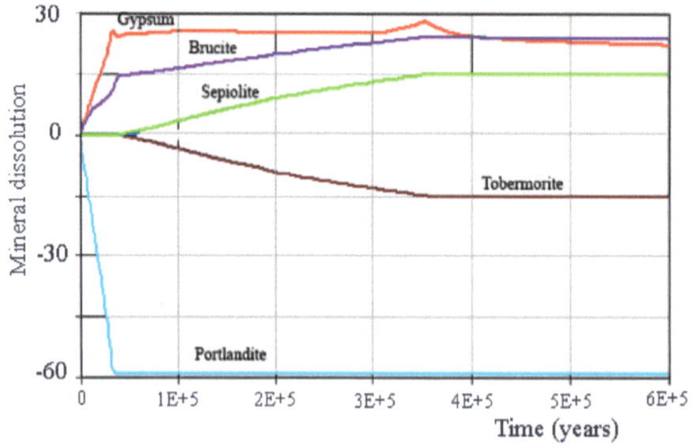

Figure 8: Computed cumulative amount of mineral dissolution/precipitation (in mol/l) in concrete at r = 1.35 m. The amount of dissolution-precipitation of calcite, dolomite, gyrolite and quartz is extremely small to be seen in this plot Values are positive for precipitation and negative for dissolution [19].

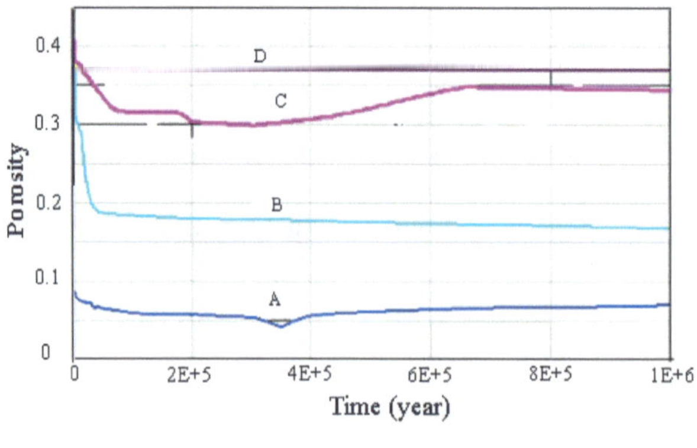

Figure 9: Computed time evolution of porosity in the bentonite (curve C, r = 1.125 m), concrete (curve A, r = 1.35 m) and clay (curves B, r = 1.525 m and D, r = 13.5 m) [19].

5.3. Case 3: Cation Exchange in a Physically and Geochemically Heterogeneous Aquiffer

This case corresponds to a set of chemical species which undergo cation exchange reactions in an aquifer having random Log K and Log CEC. Numerical simulations are performed for plumes moving through a vertical heterogeneous porous medium in the x-z plane with a size of 40 m × 10 m (Fig. **10**). Water heads are prescribed on the left and right boundaries, yielding a uniform mean horizontal field with a mean longitudinal hydraulic gradient J = - 0.1. Top and bottom boundaries of the domain are impervious.

Initially the system is filled with water having 1 mM $NaNO_3$ and 0.2 mM KNO_3. It is flushed by a 0.6 mM $CaCl_2$ solution from the left boundary. Na^+ is weakly adsorbed, and is eluted first. K^+ is more tenaciously held than Na^+, and it appears retarded. NO_3^- is used as anion and Cl^- is used as a conservative solute to display pure transport in the heterogeneous porous medium.

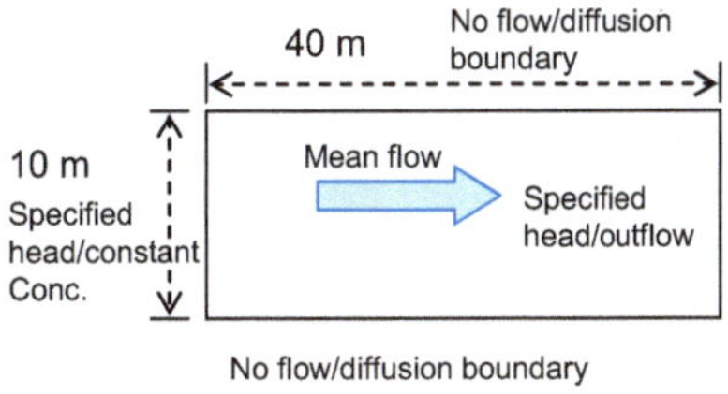

Figure 10: Model domain for the test case dealing with a physically and geochemically heterogeneous aquifer [57].

The numerical evaluation of multicomponent cation exchange reactive transport in physically and geochemically heterogeneous porous media using CORE^{2D} V4 has been reported by [57].

Log K and Log CEC are assumed to be random Gaussian functions with isotropic spherical semivariograms. Realizations of Log K and Log CEC were generated and used for Monte Carlo simulations. Five sets of numerical experiments were considered in terms of Log K and Log CEC and their correlation structures. The means of Log K and Log CEC were taken equal to 1.1 and -3.85, respectively. Their correlation lengths were equal to 10 m.

Fig. **11** shows the spatial distribution of a selected realization of Log K and Log CEC having means and variances equal to 1.099 and 0.5 for Log K and -3.85 and 1.0 for Log CEC, respectively. Log K and Log CEC are negatively correlated. It can be seen that zones with large permeability have low values of CEC and vice versa.

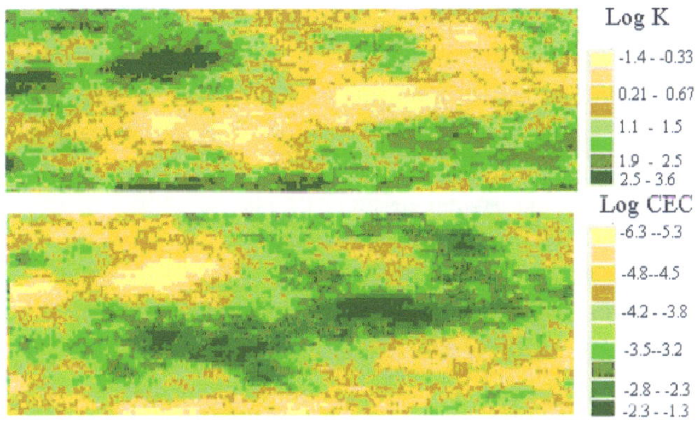

Figure 11: Spatial distributions of a selected realization of Log K and Log CEC [57].

Fig. **12** shows the plumes of Cl^-, Na^+, K^+ and Ca^{2+} in the aquifer for a selected realization. It can be seen clearly that the fronts of the plumes of the chemical species are irregular. In order to characterize the longitudinal behaviour of the conservative species (Cl^-) and the reactive species (Na^+, K^+ and Ca^{2+}), first and second order spatial moments are calculated. First-order spatial moments depend on the variances of Log K and Log CEC and their correlation structures. The greater the variance of Log K, the larger the displacements of the plume fronts for both non-reactive and reactive species. The effect of the variance of Log CEC on the first-order moments of reactive plumes is significant. The larger the variance of Log CEC, the smaller the first-order moments of the plume fronts. First-order moments of the reactive plume front depend on the correlation between Log CEC and Log K. A negative correlation structure leads to a displacement of the reactive plume larger than in the case of uncorrelated and positive correlation because for the case of negative correlation cations migrate easily through high permeability zones where cation exchange is mild (due to a small cation exchange capacity).

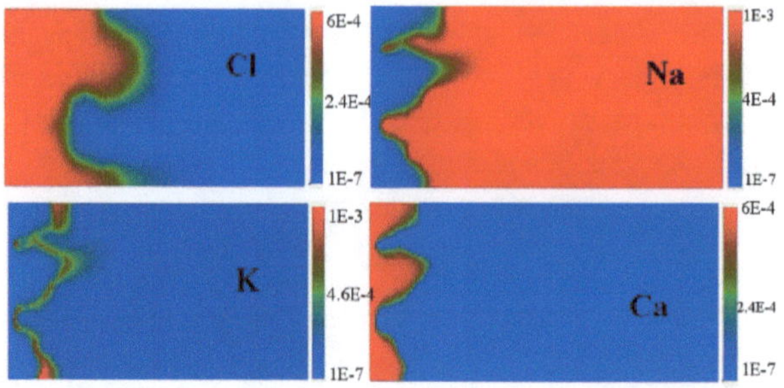

Figure 12: Simulated plumes of Cl^-, Na^+, K^+ and Ca^{2+} for the test case dealing with a physically and geochemically heterogeneous aquifer [57].

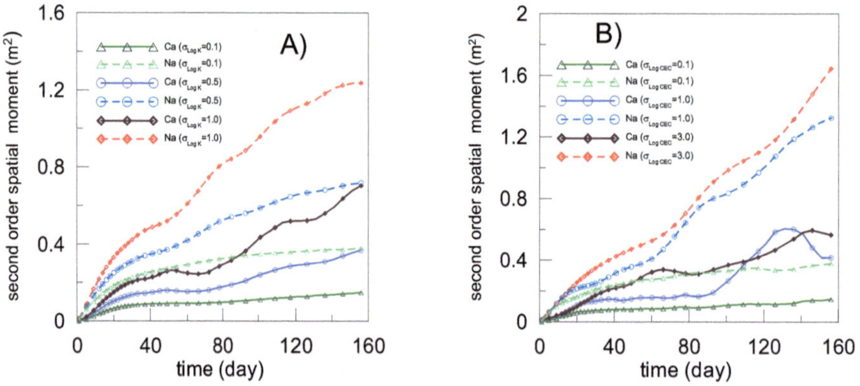

Figure 13: Second-order spatial moments of Na^+ and Ca^{2+} for A) random Log K and B) random Log CEC [57].

The second-order spatial moments of cations, which measure the extent of cation plumes, depend also on the variances of Log K and Log CEC and their correlation structures. Fig. **13A** shows the plots of second-order spatial moments of Na^+ and Ca^{2+} for different variances of Log K. One can see that the second-order spatial moments of the plume fronts of two reactive species increase with time. The larger the variance of Log K, the greater the second-order spatial moments of Na^+ and Ca^{2+}. Fig. **13B** shows the second-order spatial moments of Na^+ and Ca^{2+} as a function of time for different variances of Log CEC. These moments increase with the variance of Log CEC. For an uncorrelated structure they are greater than those for a positive correlation. The second-order spatial moment of Na^+ is almost 2 times that of Ca^{2+} when the variance of Log K is 0.1. Such a difference in second-order spatial moments increases with the increasing variance of Log K (see Fig. **13A**).

5.4. Case 4: CO_2 Injection in the Vadose Zone

A numerical model was constructed to simulate CO_2 injection in the vadose zone and to validate the sensors for measuring soil CO_2 concentrations. The model takes into account CO_2 diffusion and dissolution in the pore water. Air in the pore space is assumed stagnant. The transport of CO_2 in the vadose zone, therefore, obeys the following equations:

$$\frac{\partial R_f c_a}{\partial t} = \nabla \cdot \left(D_E \nabla c_a \right) - q_E \nabla c_a + r \quad \textbf{(12)}$$

$$R_f = \theta_a + k\theta_w \quad \textbf{(13)}$$

$$q_E = q_a + kq_w \quad \textbf{(14)}$$

$$\theta_a = \phi - \theta_w \quad \textbf{(15)}$$

where R_f is the apparent retardation factor, c_a is the concentration of CO_2 in the air space of the soil, D_E is the effective diffusion coefficient of CO_2 in the air space of the soil, q_E is the total advective flux of CO_2 in the soil, r is a sink/source term, θ_a is the air content, k is the Henry constant, θ_w is the water content, q_a is the advective velocity of CO_2, q_w is the advective velocity of water, and ϕ is the porosity. Microbial processes and root respiration are not accounted for.

CORE^{2D} V4 was updated and several new subroutines were added to solve the transport of CO_2 in the vadose zone. This new capability for simulating CO_2 transport was verified against an analytical solution for a case of CO_2 migration in a 1-D semi-domain with a fixed concentration boundary condition. Water and air in the 1-D domain are assumed stagnant. Two scenarios were simulated using the analytical solution and the updated numerical code: 1) The 1-D domain is dry with no water in the pore space (scenario 1); and 2) There is water in the pore space, and CO_2 may dissolve into the pore water (scenario 2). The results of the numerical simulations obtained with the updated CORE^{2D} V4 and the analytical solutions for the two scenarios are shown in Fig. **14** and Fig. **15**.

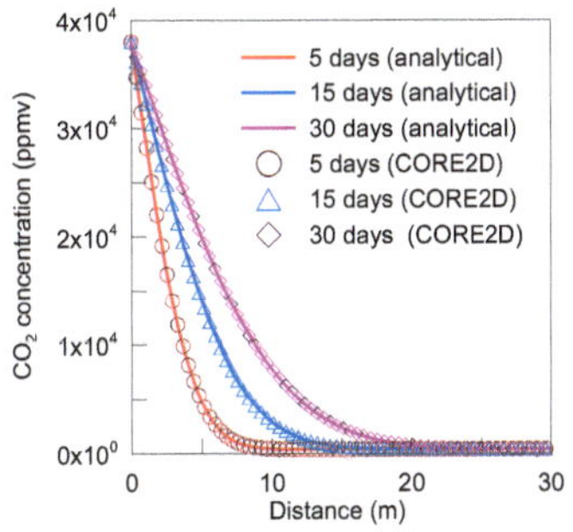

Figure 14: Verification of the updated version of CORE for simulating CO_2 transport against the analytical solution for the scenario 1 of the 1-D semi-domain [58].

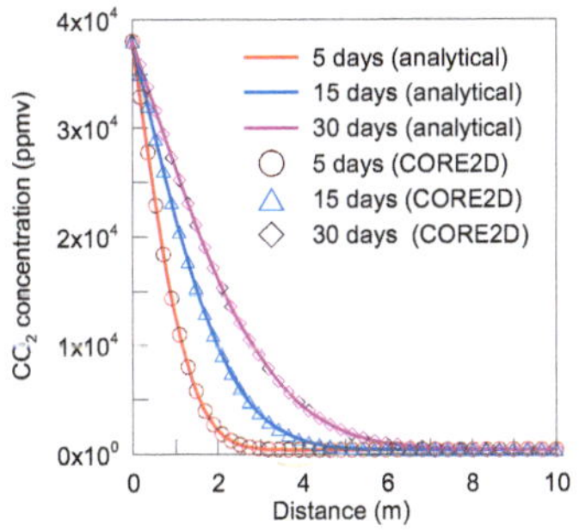

Figure 15: Verification of the updated version of CORE for simulating CO_2 transport against the analytical solution for the scenario 2 of the 1-D semi-domain [58].

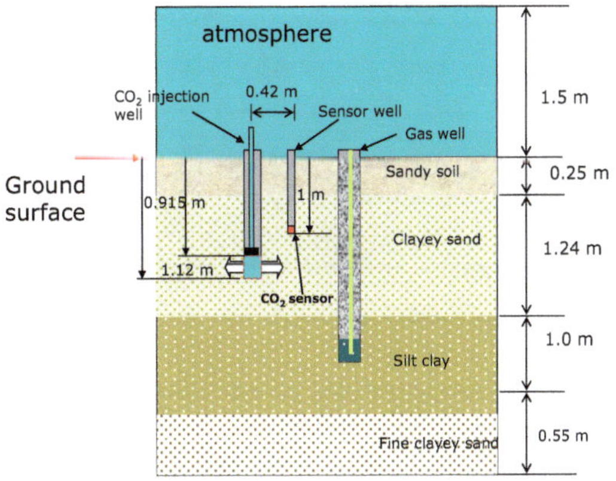

Figure 16: CO_2 injection setup at the Brackenridge site (injection well, sensor well and gas well station) [58].

A 2-D grid representing a 3-D axi-symmetric domain was used in the model (Fig. **17**). The left boundary coincides with the center of the injection well. A finite element mesh of 8,190 triangular elements and 4,224 nodes was used in the numerical simulation. The element size increases gradually outwards from the elements near the injection well. Porosities for the four soil layers downwards from the surface are 0.5, 0.41, 0.45 and 0.43, respectively. The atmospheric layer is assumed to have a porosity of 0.99 and a water content of 0.001 in the model. Hydraulic parameters were calibrated to fit the water content because no moisture and water potential were measured during the CO_2 injection experiment. Furthermore, water flow in the soil layers is assumed to be at steady state. The fit of the water content data measured at the lab is shown in Fig. **18**. Water content in the vertical profiles increases gradually downwards. It is important to point out that the model does not take into account root respiration, and therefore cannot reproduce daily CO_2 fluctuations in the soils.

Measured concentrations of CO_2 at the observation wells were fit by adjusting the diffusion coefficient and the injection rate. Calibrated effective diffusion coefficients are shown in Fig. **19** and range from $2.58 \cdot 10^{-8}$ m^2/s at the bottom of the domain to $2.45 \cdot 10^{-5}$ m^2/s near to the land surface.

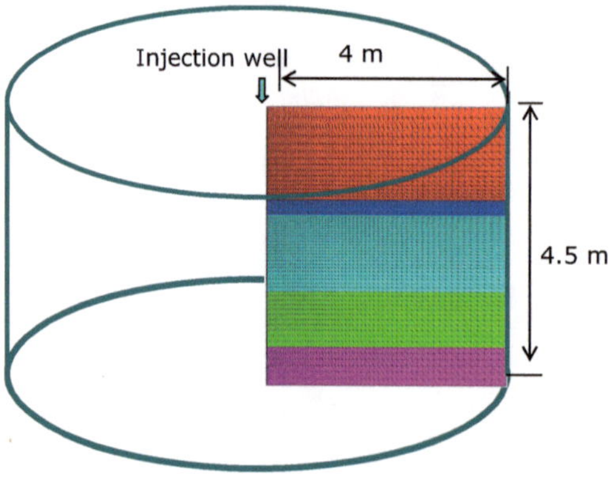

Figure 17: 2-D finite element grid considered in the model of the CO_2 injection test at the Brackenridge site [58].

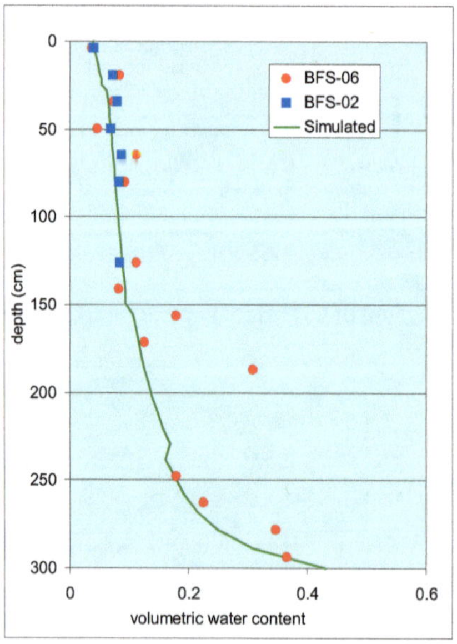

Figure 18: Measured (symbols) water contents at the lab for two boreholes and fit of the numerical model (line) [58].

This case illustrates the model of CO_2 injection at a gas well placed in the vadose zone at the Brackenridge site [58]. The model considers five layers. The upper layer simulates the atmosphere ant the other four, from ground surface downwards, are the following soil layers: sandy soil, clayey sand, silt clay and fine clayey sand. The atmospheric layer is included as a fixed CO_2 boundary concentration applied at the top of the layer. A scheme of the gas well with the thickness of each layer is shown in Fig. **16.** The water table is about 2.4 m below the ground surface. CO_2 tracer is injected at depths ranging from 0.915 to 1.12 m from the ground surface. Initial CO_2 concentrations are assumed to be 520 ppmv in the soil layers and 380 ppmv in the atmospheric layer. The soils were sampled and water content and porosity were measured in the lab.

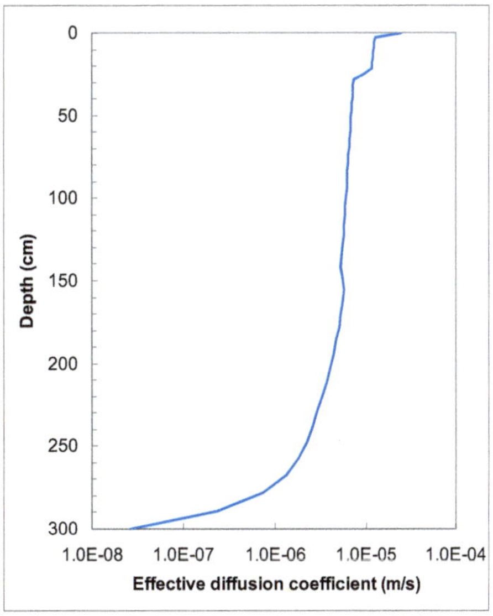

Figure 19: Calibrated effective diffusion coefficients of CO_2 in the air in the soil profile [58].

Using the calibrated effective diffusion coefficient, the best fit injection rate for the sensor data is 0.533 l/h. Fig. **20** shows the comparison of the final calibrated results with measured CO_2 concentrations at the observation wells. Model results agree well with measured CO_2 concentrations. The shape of the curves is diffusion-dominated. The model overestimates slightly the CO_2 measurements at the gas station well and there is no evidence that injected CO_2 was ever measured in this well.

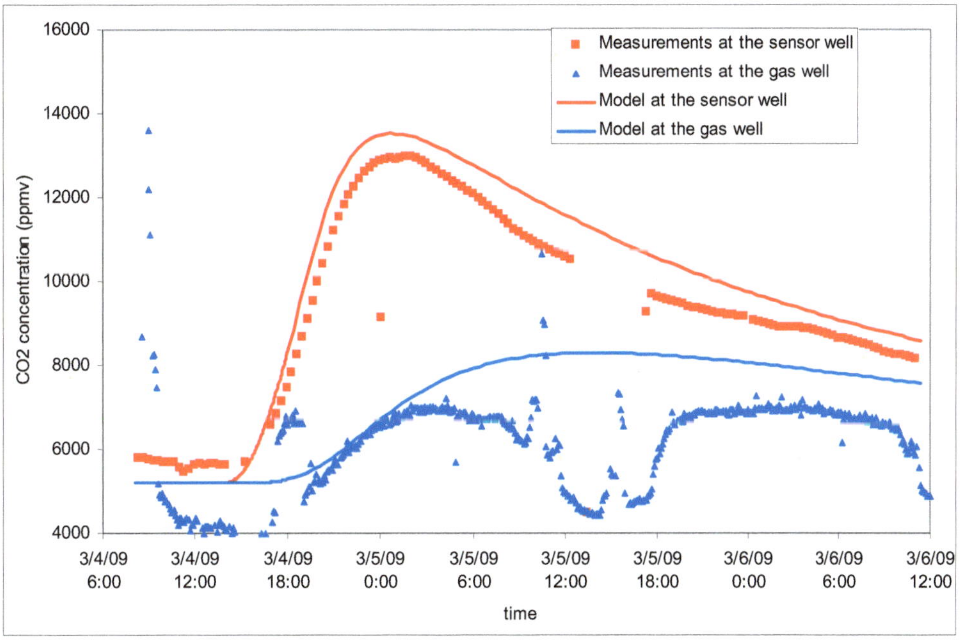

Figure 20: Comparison of model results (lines) to field measurements at the sensor well (red squares) and the gas well (blue triangles) [58]. Possible reasons that cause the CO_2 concentration drop measured at the gas well include: CO_2 dissolution and root respiration uptake.

5.5. Case 5: Water Quality in an Open Pit Lake

One of the most recent applications of CORE^{2D} V4 is the simulation of the water quality of pit lakes. Closure of the As Pontes coal mine in the Northwest of Spain will create an open pit lake. The future lake will have a surface of 8.1 km^2 and a maximum depth of 194 m. The total volume will be 547 hm^3. The lake will be filled with the runoff of the drainage basin of the mine, the runoff from an outside dump and the diversion of good-quality water from a nearby river (Fig. **21**).

Figure 21: Location of As Pontes mine in A Coruña, Spain. Also shown is the location of the outside dump [59].

A reactive transport model was constructed to predict the lake water quality during the 4-year period of lake filling by assuming first full mixing conditions. The geochemical conceptual model accounts for: aqueous complexation, acid-base, redox, mineral dissolution/precipitation, gas disolution/ex-solution and surface complexation reactions. The chemical system is defined in terms of the following primary species: H_2O, $O_2(aq)$, H^+, Ca^{2+}, Mg^{2+}, Na^+, K^+, Fe^{2+}, Mn^{2+}, Al^{3+}, Cl^-, SO_4^{2-}, HCO_3^-, $SiO_2(aq)$ and XOH (proton surface complexation). Ferryhydrite was selected as the main mineral phase for iron precipitation according to the following reaction: $Fe(OH)_3(s) + 3H^+ \Leftrightarrow Fe^{3+} + 3H_2O$, with a Log K = 4.9.

Based on the composition and volume of each one of the different inflows, CORE^{2D} V4 predicts the pH and the concentrations of the main dissolved species (SO_4^{2-}, Fe^{2+}, Al^{3+} and Mn^{2+}). CORE^{2D} V4 has been used also to evaluate the effect of different preventive and remedial actions such as covering acid-generating surfaces in the drainage basin of the mine lake, adding lime and limestone to the inflow water and adding fly ashes and adding the effluents of a nearby thermal power plant. Fig. **22** shows the sensitivity of pH to several remedial actions. The model has been used also to estimate the composition of the surface runoff water of the outside dumps which contributes significantly to the replenishment of the lake.

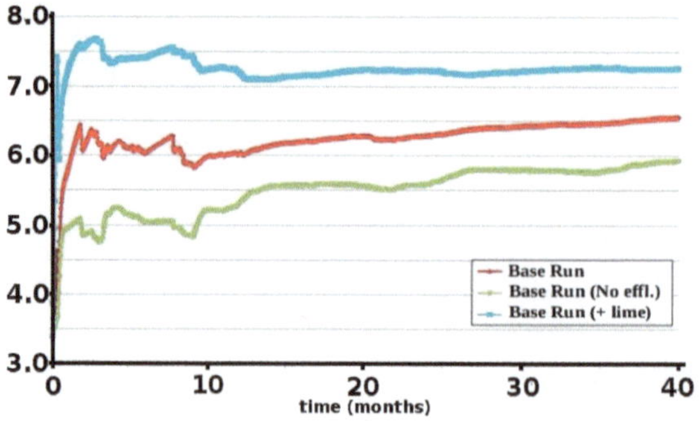

Figure 22: Full-mixing model predictions of pH in As Pontes pit lake considering: 1) base run (red line), 2) base run without effluents from the power plant, and 3) base run with the addition of 1,500 ton of lime (blue line) [59].

Most lakes in temperate climates undergo a cycle of stratification that affects the composition of the water column. Usually, less dense waters remain in the surface of the lake while saline acid waters mix and sink to the bottom of the lake. Thanks to the modularity of CORE^{2D} V4, it has been possible to couple its geochemical subroutines to a hydrodynamic code DYRESM [60, 61]. The result of such coupling is a new code called DYCD-CORE [59] which allows the modeling of the hydrodynamics, mixing phenomena and the stratification of mine lakes.

Lake inflows mix in the water column according to their densities and volumes. Energy sources for mixing coming mainly from solar radiation and wind shear stress are calculated by DYRESM. The net available energy is used for mixing different layers to get a stable density profile. Dissolved species and minerals are mixed during this process. When mixing finishes and a stable density profile is accomplished, chemical equilibrium and mineral dissolution/precipitation calculations are performed using the CORE^{2D} V4 subroutines.

Fig. **23** and Fig. **24** show the predictions of pH and sulfate in the water column of As Pontes pit lake using DYCD-CORE [59]. pH values in the surface are higher than in the bottom during all the filling period (see Fig. **23**). During the fall and the winter pH decreases in the lake surface due to the effect of the surface runoff coming from the outside dumps and mine walls. These waters content high concentrations of dissolved iron and when they get in contact with high pH waters in the lake, ferrihydrite precipitates and pH decreases slightly.

Large concentrations of sulfate are computed at the bottom of the lake, which are not renewed by mixing phenomena in the water column and create a stable bottom layer called monimolimion (Fig. **24**). Most of the non-reactive species behave in a similar manner to sulfate and they accumulate in the bottom waters. Natural waters from direct precipitation and Eume River remain in the surface of the lake.

Figure 23: Contour plots of pH in the water column of As Pontes pit lake predicted with DYCD-CORE. Depth (in m) is shown in the left y-axis, pH in the right y-axis and time in the x-axis [59].

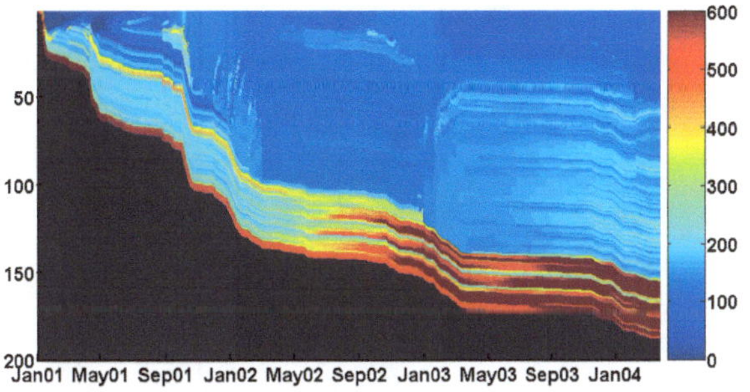

Figure 24: Contour plots of sulfate concentrations in the water column of As Pontes pit lake predicted with DYCD-CORE. Depth (in m) is shown in the left y-axis, sulfate concentrations (in mg/l) in the right y-axis and time in the x-axis [59].

5.6. Case 6: Coupled THC Models of Compacted Bentonite after FEBEX *in Situ* Test

This case corresponds to the coupled thermo-hydro-chemical models of compacted bentonite after an *in situ* full-scale engineered barrier experiment (FEBEX) reported by [51]. FEBEX is a demonstration and research project dealing with the bentonite engineered barrier designed for sealing and containment of a high-level radioactive waste repository [62]. The FEBEX *in situ* test is being performed in a gallery excavated in granite in the northern zone of the Grimsel underground laboratory operated by NAGRA in Switzerland. The test includes the heating system, the clay barrier (bentonite) and instrumentation, monitoring and control system. The drift is 70.4 m long and 2.28 m in diameter. The test zone is located at the last 17.4 m of the drift where heaters, bentonite and instrumentation were installed. The test zone was sealed with a concrete plug. The main elements of the heating system are two heaters (1 and 2), separated horizontally by 1 m, which simulate full-sized canisters. Heaters were placed inside cylindrical steel liners. Heaters are 4.54 m long, 0.9 m outer diameter and have a thickness of 0.1 m. Heaters are designed to maintain a maximum temperature of 100 ºC at the steel liner/bentonite interface.

The bentonite barrier is made of blocks of highly compacted bentonite. A layout of the *in situ* test is shown in Fig. 25. Weighted average values of dry density and water content of bentonite blocks are 1.70 g/cm3 and 14.4%, respectively [62]. The main mineral phase (90-92 wt.%) in the FEBEX bentonite is montmorillonite [63].

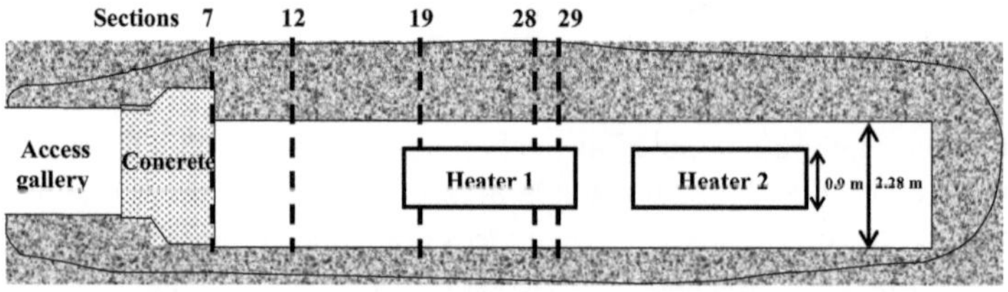

Figure 25: Layout of the FEBEX *in situ* test. Vertical lines show the location of the different sampling sections [51].

The *in situ* test began in February 27th, 1997. Heater 1 was switched-off in February 2002. Dismantling of heater 1 was performed from May to September 2002. A comprehensive post-mortem bentonite sampling and analysis program was designed to characterize solid and liquid phases, measure physical and chemical changes induced by the combined effect of heating and hydration; and test THM and THC models [64-67].

The TH conceptual model takes into account the following multiphase flow processes: 1) Advective flow of liquid water, 2) Advective and diffusive flow of water vapor, 3) Advective and diffusive flow of "dry" air, 4) Advective flow of air dissolved in water, 5) Heat conduction through solid, liquid and gas phases; and 6) Heat convection through liquid and gaseous phases. Solute transport processes include: advection, molecular diffusion, and mechanical dispersion.

Different Conceptual Geochemical Models (CGM) have been proposed for FEBEX bentonite. They differ depending on the type of test used to obtain chemical data: squeezing and Aqueous Extracts Tests (AET). Conceptual model CGM-0 was derived from a squeezing test performed on a bentonite sample having a gravimetric water content of 26.5% [62]. Geochemical models were used to derive the chemical composition at 14% water content. Later, Samper [68] presented a model derived from aqueous extract tests (model CGM-1) which assumes gypsum at equilibrium, calcite and dolomite with kinetic mineral dissolution, and a prescribed pressure of $10^{-3.5}$ bar for CO_2(g). This model reproduces adequately measured aqueous extract data performed at different durations and S/L ratios. However, it fails to reproduce bicarbonate and pH data, possibly due to changes in CO_2(g) pressure which are not considered in the model. In general, calculated results with model CGM-1 could not fit sulfate squeezing data. Samper [68] presented a modified model (CGM-2) based on data from both aqueous extract and squeezing tests. This model does not consider an initial amount of gypsum in the system but allows for its precipitation. Models CGM-1 and CGM-2 fit simultaneously both squeezing and aqueous extract data for most chemical species, although they present discrepancies for sulfate and bicarbonate data.

Models CGM-0, CGM-1 and CGM-2 account for aqueous complexation, acid-base, mineral dissolution/precipitation, gas dissolution/exsolution and cation exchange. These reactions are assumed at local equilibrium except for calcite and dolomite which may dissolve or precipitate under kinetic conditions in models CGM-1 and CGM-2 [51]. Equilibrium constants at 25 °C for aqueous complexes, minerals and gases were taken from EQ3/6 database [38]. The Gaines-Thomas convention is used for cation exchange. Kinetic laws and rate constants for calcite were taken from [69] while those of disordered dolomite were derived from [70]. Values of specific surface for calcite and dolomite were derived by [68] from calibration of AET. Bentonite chemical composition data at the initial water content of bentonite buffer (14.4%) for models CGM-0, CGM-1 and CGM-2 are reported in [51].

The model assumes axial symmetry. Spatial domain was discretized with a 1-D mesh of rectangular finite elements which contains 614 nodes and 307 elements (Fig. **26**). Axial symmetry conditions are simulated by taking into account that the thickness at each element node is assigned differently according to the radial location. Two material zones were considered in the mesh: bentonite and granite. The first two nodes are located at the heater-bentonite interface (r = 0.45 m). Bentonite which lies at the interval 0.45 m < r < 1.135 m is discretized with 220 nodes. Granite is simulated with 392 nodes and extends from 1.135 m to 50 m. The mesh was refined near the bentonite-granite interface because large concentration gradients are expected to take place in this zone. Model was run for 1930 days, which corresponds to the time of dismantling of heater 1.

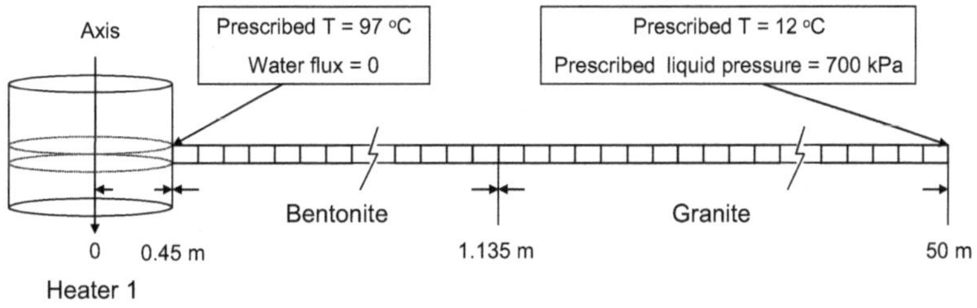

Figure 26: Scheme of finite element rectangular grid and thermal and flow boundary conditions used for THC model of FEBEX *in situ* test [51].

The initial temperature is uniform and equal to 12 °C. A constant temperature of 97 °C is prescribed at the heater-bentonite interface (r = 0.45 m) while temperature is assumed to remain constant and equal to its initial value of 12 °C at the external boundary (r = 50 m) because the thermal perturbation induced by the heaters does not extend to that boundary. Boundary conditions for flow include: 1) No flow at r = 0.45 m and 2) A prescribed liquid pressure of 700 kPa at r = 50 m (see Fig. **26**).

Since the FEBEX *in situ* test was not designed to be air-tight, nothing prevented gas from flowing in or out the test section at early stages of the test. However, it is believed that bentonite swelling sealed most of buffer gaps and voids, making difficult for gas to flow across the concrete plug. It is assumed in the model that the system is closed for gaseous phases. This means that there are neither gas flux across external boundaries nor prescribed gas pressures. Gas in granite and bentonite is simulated as a trapped phase overlapping the liquid phase.

Predictions performed with the three conceptual geochemical models are compared to measured temperature, water content and geochemical data obtained from AET of samples collected after dismantling of heater 1. These geochemical data require the use of inverse hydrogeochemical models for their interpretation in order to account for the geochemical reactions suffered by bentonite samples during aqueous extraction [52]. Inverse geochemical modeling is performed for all AET data of section 29, a section located in the heater zone (see Fig. **25**).

The THC Model reproduces properly the radial distribution of computed temperatures in bentonite and granite just when heater 1 was switched off at t = 1827 days (Fig. **27**). The model captures the general trend of measured gravimetric water contents at the bentonite barrier after 1930 days. However, it under-predicts measured water contents near granite-bentonite interface and over-predicts them near the heater (Fig. **28**). It should be noticed that measured water contents near granite-bentonite interface are larger than the water content corresponding to full saturation at the original dry density of the buffer (w = 25%) due to bentonite swelling, a process not taken into account in our model. Over-prediction of measured water contents near the heater is partly due to the fact that our model considers at the heater a fixed temperature of 97 °C which is equal to the average of measured temperatures at heater surface which range from 75 to 100 °C. Model fit to measured water contents near the heater would improve if a temperature larger than 97 °C had been prescribed at heater-bentonite interface. Another reason for overpredicting measured water contents near the heater is bentonite shrinkage which is not considered in the model.

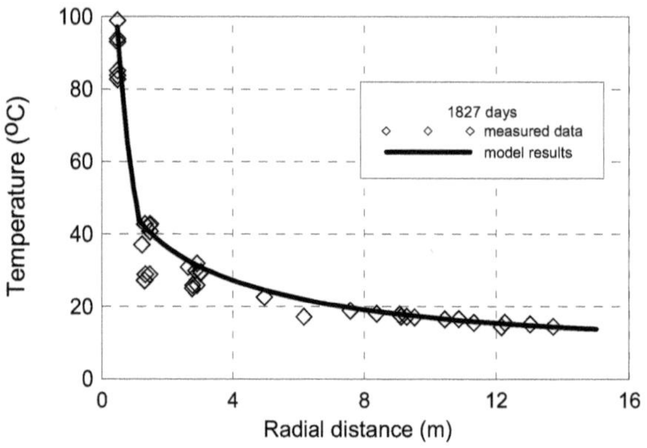

Figure 27: Radial distribution of computed and measured temperatures after 1827 days of heating, just the day in which heater 1 was switched off [51].

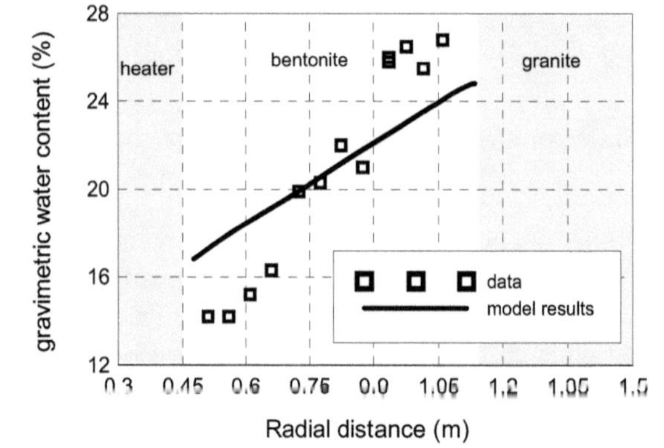

Figure 28: Radial distribution of computed and measured gravimetric water content after 1930 days of heating [51].

THC predictions of *in situ* test performed with CGM-0, CGM-1 and CGM-2 are compared to inferred AET geochemical data from section 29 at t = 1930 days. Such comparison provides a unique opportunity to test our THC model of the engineered barrier system. Spatial distribution of the concentration of a conservative species such as chloride is the result of the combined effect of the following processes: 1) Bentonite porewater evaporation near the heater, 2) Water condensation some distance away from heater, 3) Dilution due to inflow of granitic groundwater, 4) Advective displacement of hydration front, and 5) Solute diffusion which tends to dissipate solute gradients. All these processes lead to a pattern of decreasing chloride concentration with increasing radial distance (Fig. **29**). THC predictions for CGM-0 and CGM-2 are slightly larger than those of CGM-1 because initial Cl concentration of CGM-0 and CGM-2 (0.187 M) was derived from squeezing tests while initial Cl concentration for CGM-1 was derived from AET and is equal to 0.155 M. THC model overpredicts chloride concentrations near granite-bentonite interface (0.75 < r < 1.1 m) while underpredicts them for r < 0.75 m. While computed concentrations decrease monotonically with distance in a convex manner, measured values show a maximum at r = 0.6 m.

Inferred aqueous extract Ca concentrations from section 29 are largest near the heater and decrease with increasing radial distance. Near the bentonite-granite interface they increase again due to a sampling artefact caused by colloidal particles [71]. Similar to Cl, inferred Ca data show a maximum at r = 0.6 m. Ca inferred with CGM-0 and CGM-2 are similar and generally larger than those inferred with CGM-1. Such a discrepancy in Ca data is related to the fact that CGM-1 assumes that bentonite contains gypsum which may dissolve/precipitate at local equilibrium. THC predictions with CGM-0 and CGM-1 are similar and deviate from the cloud of inferred data. The best fit to inferred Ca data is achieved with CGM-2 (Fig. **30**).

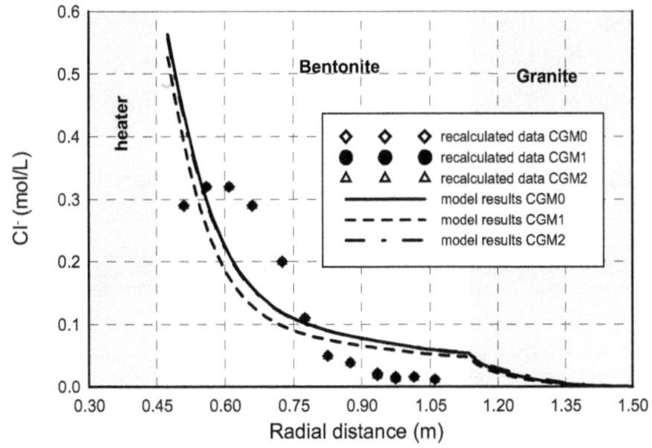

Figure 29: Comparison of predicted Cl^- concentrations with models CGM-0, CGM-1 and CGM-2 (lines) with inferred aqueous extract data (symbols) at section 29 after 1930 days of heating [51].

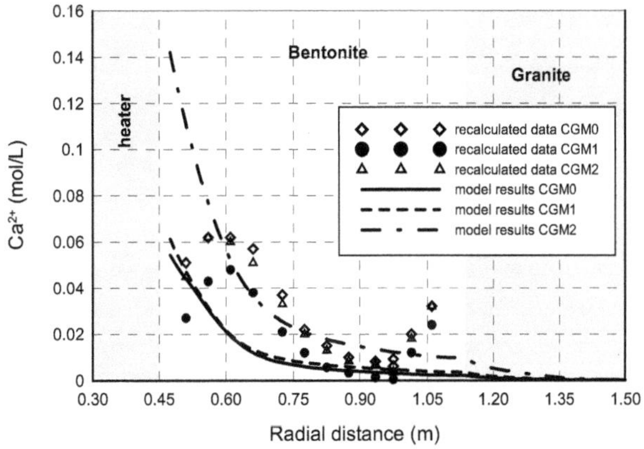

Figure 30: Comparison of predicted Ca^{2+} concentrations with models CGM-0, CGM-1 and CGM-2 (lines) with inferred aqueous extract data (symbols) at section 29 after 1930 days of heating [51].

The THC Model predictions capture also the trends of inferred aqueous extracts Mg^{2+}, Na^+ and K^+ concentrations. However, there are some discrepancies near the heater which are especially significant for SO_4^{2-} and HCO_3^-. The THC model could be improved by taking into account different types of water in bentonite, bentonite swelling, protonation/deprotonation by surface complexation, and CO_2 degassing and dissolution reactions.

6. CONCLUSIONS

Complex natural processes can be modeled by coupling flow, heat and solute transport processes and chemical reactions. CORE^{2D} V4 is a comprehensive code to model these processes. It can handle both saturated and unsaturated flow, heat transport and reactive transport and accounts for a wide range of geochemical reactions such as aqueous complexation, redox, acid-base, sorption (including cation exchange and surface complexation), mineral dissolution/ precipitation, gas dissolution/ ex-solution and microbial processes. Precipitation-dissolution and aqueous complexation can be modeled either at equilibrium or using kinetic laws. The finite element method is used for the spatial discretization. Hydraulic properties may change dynamically due to geochemical reactions. Strong interactions among flow and chemistry may occur in concrete degradation and bentonite-concrete interactions in a HLW repository in clay. CORE^{2D} V4 has been extensively verified with analytical solutions and against other numerical codes and applied to many laboratory experiments and field cases.

ACKNOWLEDGEMENTS

CORE development and its applications have been funded by ENRESA, the European Commission through the following projects of the Nuclear Fission Safety Programme: CERBERUS (F14W-CT95-0008), FEBEX (FI4W-CT95-0006 & FIKW-CT-2000-0016), BENIPA (FIKW-CT-2000-00015), FUNMIG (FP6-516514), NF-PPRO (FI6W-CT-2003-02389), and PAMINA (FP6-036404), the Spanish Ministry of Science and Technology through projects HID98-0282, REN2003-8882 and CGL2006-09080), the Galician Research Program through projects PGIDT04PX IC11801PM and PGIDT00PX 111802, and the University of La Coruña. The contributions of professors and students to the development of CORE during the last 15 years as well as the inputs of CORE users are greatly acknowledged. We specially acknowledge the contributions of J. Molinero, R. Juanes, R. Juncosa and J. Delgado to CORE-LE^{2D}, CORE^{2D}V2 and other versions of CORE.

REFERENCES

[1] D. J. Kirkner, A. A. Jennings, T. L. Theis, "Multisolute mass transport with chemical interaction kinetics", *Journal of Hydrology*, vol. 76, no. 1-2, pp. 107-117, 1985.

[2] J. J. Kaluarachchi and J. C. Parker, "Modeling multicomponent organic chemical transport in three-fluid-phase porous media", *Journal of Contaminant Hydrology*, vol. 5, no. 4, pp. 349-374, 1990.

[3] C. I. Steefel and P. Van Cappellen, "A new kinetic approach to modeling water-rock interaction: The role of nucleation, precursors, and Ostwald ripening", *Geochimica et Cosmochimica Acta*, vol. 54, no. 10, pp. 2657-2677, 1990.

[4] H. J. Lensing, M. Vogt, and B. Herrling, "Modeling of biologically mediated redox processes in the subsurface", *Journal of Hydrology*, vol. 159, no. 1-4, pp. 125-143, 1994.

[5] K. M. Salvage and G. T. Yeh, "Development and application of a numerical model of kinetic and equilibrium microbiological and geochemical reactions (BIOKEMOD)", *Journal of Hydrology*, vol. 209, no. 1-4, pp. 27-52, 1998.

[6] C. I. Steefel and P. C. Lichtner, "Multicomponent reactive transport in discrete fractures: II: Infiltration of hyperalkaline groundwater at Maqarin, Jordan, a natural analogue site", *Journal of Hydrology*, vol. 209, no. 1-4, pp. 200-224, 1998.

[7] C. J. Tebes-Stevens, A. Valocchi, J. M. VanBriesen, and B. E. Rittmann, "Multicomponent transport with coupled geochemical and microbiological reactions: model description and example simulations", *Journal of Hydrology*, vol. 209, no. 1-4, pp. 8-26, 1998.

[8] S. B.Yabusaki, C. I. Steefel, and B. D. Wood, "Multidimensional, multicomponent, subsurface reactive transport in nonuniform velocity fields: code verification using an advective reactive streamtube approach", *Journal of Contaminant Hydrology*, vol. 30, no. 3-4, pp. 299-331, 1998.

[9] A. Chilakapati, S. Yabusaki, J. Szecsody, and W. MacEvoy, "Groundwater flow, multicomponent transport and biogeochemistry: development and application of a coupled process model", *Journal of Contaminant Hydrology*, vol. 43, no. 3-4, pp. 303-325, 2000.

[10] M. W. Saaltink, J. Carrera, and C. Ayora, "A comparison of two approaches for reactive transport modeling", *Journal of Geochemical Exploration*, 69-70, 97-101, 2000.

[11] G. T. Yeh, *Computational subsurface hydrology, reactions, transport and fate*. Kluwer Academic Publishers, The Netherlands, 2000.

[12] T. R. Ginn, E. M. Murphy, A. Chilakapati, and U. Seeboonruang, "Stochastic-convective transport with nonlinear reaction and mixing: application to intermediate-scale experiments in aerobic biodegradation in saturated porous media", *Journal of Contaminant Hydrology*, vol. 48, no. 1-2, pp. 121-149, 2001.

[13] P. Regnier, J. P. O'Kane, C. I. Steefel, and J. P. Vanderborght, "Modeling complex multi-component reactive-transport systems: towards a simulation environment based on the concept of a Knowledge Base", *Applied Mathematical Modeling*, vol. 26, no. 9, pp. 913-927, 2002.

[14] M. W. Saaltink, C. Ayora, P. J. Stuyfzand, and H. Timmer, "Analysis of a deep well recharge experiment by calibrating a reactive transport model with field data", *Journal of Contaminant Hydrology*, vol. 65, no. 1-2, pp. 1-18, 2003.

[15] K. Pruess, J. Garcia, and T. Kovscek, C. Oldenbry, J.Rutqvist, C. Steefel, T.Xu. "Code intercomparison builds confidence in numerical simulation models for geologic disposal of CO_2", *Energy*, vol. 29, no. 9-10, pp. 1431-1444, 2004.

[16] K. Maher, C. I. Steefel, D. J. DePaolo, and B. E. Viani, "The mineral dissolution rate conundrum: Insights from reactive transport modeling of U isotopes and pore fluid chemistry in marine sediments", *Geochimica et Cosmochimica Acta*, vol. 70, no. 2, pp. 337-363, 2006.

[17] C. Yang, *Conceptual and numerical coupled thermal-hydro-bio-geochemical models for three-dimensional porous and fractured media*. Ph.D. Thesis, Univ. of La Coruña, La Coruña, Spain, 2006.

[18] C. Yang, J. Samper, J. Molinero, and M. Bonilla, "Modeling geochemical and microbial consumption of dissolved oxygen after backfilling a high level radiactive waste repository", *Journal of Contaminant Hydrology*, vol. 93, pp. 130-148, 2007.

[19] C. Yang, J. Samper, and L. Montenegro, "A coupled non-isothermal reactive transport model for long-term geochemical evolution of a HLW repository in clay", *Environmental Geology*, vol. 53, pp. 1627-1638, 2008.

[20] J. Samper, R. Juncosa, J. Delgado, and L. Montenegro, *CORE-LE*^{2D}*: A code for water flow and reactive solute transport*. Users Manual. University of A Coruña, Spain, 1998.

[21] T. Xu, J. Samper, C. Ayora, M. Manzano, and E. Custodio, "Modeling of non-isothermal multi-component reactive transport in field scale porous media flow systems", *Journal of Hydrology*, vol. 214, pp. 144-164, 1999.

[22] J. Samper, R. Juncosa, J. Delgado, and L. Montenegro, *CORE*^{2D} *A code for non-isothermal water flow and reactive solute transport*. Users manual version 2. ENRESA Technical Publication 6/2000, 2000.

[23] L. Montenegro, G. Zhang, J. Samper, and J. Delgado, *Documento de verificación del código CORE*^{2D}. Technical Report. Civil Engineering School, University of A Coruña, Spain, 1999.

[24] Z. Dai and J. Samper, *INVERSE-CORE2D: A code for inverse problem of water flow and reactive solute transport*, Users Manual, Version 0, University of La Coruña. pp. 240, 1999.

[25] Z. Dai and J. Samper, "Inverse problem of multicomponent reactive chemical transport in porous media: Formulation and applications": *Water Resource Research* 40, p. W07407, doi: 10.1029/2004-WR003248, 2004.

[26] Z. Dai and J. Samper, "Inverse modeling of water flow and multicomponent reactive transport in coastal aquifer systems", *Journal of Hydrology*, vol. 327, pp. 447-461, 2006.

[27] G. Zhang, *Nonisothermal hydrobiogeo-chemical models in porous media*. Ph.D. Thesis, Univ. of La Coruña, Spain, 2001.

[28] J. Samper, G. Zhang, and L. Montenegro, "Coupled microbial and geochemical reactive transport models in porous media: Formulation and application to synthetic and *in situ* experiments", *Journal of Iberian Geology*, vol. 32, no. 2, pp. 215-231, 2006.

[29] G. Zhang, J. Samper, and L. Montenegro, "Coupled thermo-hydro-bio-geochemical reactive transport model of CERBERUS heating and radiation experiment in Boom clay", *Applied Geochemistry*, vol. 23, pp. 932-949, 2008.

[30] J. Samper, C. Yang, and L. Montenegro, *CORE*^{2D} *version 4: A code for non-isothermal water flow and reactive solute transport*. Users Manual. University of A Coruña, Spain, 2003.

[31] C. Voss, *SUTRA, A finite element simulation model for saturated-unsaturated, fluid-density-dependent ground-water flow with energy transport or chemically-reactive single-species solute transport*, U.S.G.S., 1984.

[32] I. G. Richardson, "The nature of C-S-H in hardened cements", *Cement and Concrete Research*, vol. 29, pp. 1131-1147, 1999.

[33] C. I. Steefel and P. C. Lichtner, "Diffusion and reaction in rock matrix bordering a hyperalkaline fluid-filled fracture", *Geochimica et Cosmochimica Acta*, vol. 58, no. 17, pp. 3595-3612, 1994.

[34] J. Cuevas, M. V. Villar, M. Martin, J. C. Cobena, and S. Leguey, "Thermo-hydraulic gradients on bentonite: distribution of soluble salts, microstructure and modification of the hydraulic and mechanical behaviour", *Applied Clay Science*, vol. 22, no. 1-2, pp. 25-38, 2002.

[35] E. J. Garboczi and D. P. Bentz, "Computer simulation of the diffusivity of cement-based materials", *Journal of Materieal Sciences*, vol. 27, pp. 2083-2092, 1992.

[36] C. Yang and J. Samper, "A subgrid-scale stabilized finite element method for multicomponent reactive transport through porous media", *Transport in Porous Media*, vol. 78, no.1, pp. 101-126, 2009.

[37] J. Samper, T. Xu, and C.Yang, "A sequential partly iterative approach for multicomponent reactive transport with CORE^{2D}", *Computational Geosciences*, vol. 13, pp. 301-316, 2009.

[38] T. J. Wolery, *EQ3NR, a computer program for geochemical aqueous speciation-solubility calculations* (version 7.0). Lawrence Livermore National Laboratory, 1992.

[39] J. Samper, C. Yang, A. Naves. "A fully 3-D anisotropic model of DI-B *in situ* diffusion experiment in the Opalinus clay formation", *Physics and Chemistry of the Earth*, vol. 31, pp. 531-540, 2006.

[40] C. Yang, J. Samper, and J. Molinero, "Inverse microbial and geochemical reactive transport models in porous media", *Physics and Chemistry of the Earth*, vol. 33, pp. 1026-1034, 2008.

[41] J. Samper and C. Yang, "An approximate analytical solution for multicomponent cation exchange reactive transport in groundwater", *Transport in Porous Media*, vol. 69, pp. 67-88, 2007.

[42] C.I. Steefel, *1DREACT, One dimensional reaction-transport model.* User manual and programmer's guide. Pacific Northwest Laboratories, Batelle, Washington, 1993.

[43] C. W. Liu and T. N. Narasimhan, "Redox-controlled multiple species reactive chemical transport. 2. Verification and Application", *Water Resources Research*, vol. 25, pp. 883-910, 1989.

[44] P. Nienhuis, C. A. T. Appelo, and A. Willemsen, *Program PHREEQM: Modified from PHREE-QM for use in mixing cell flow tube*, 1991.

[45] P. Engesgaard, *Model for biological clogging in 3D, Brief user's manual and guide.* Department of Hydrodynamics and Water Resources. Danish Technology University, 2000.

[46] L. Zheng, J. Samper, L. Montenegro, and J. C. Mayor, "Multiphase flow and multicomponent reactive transport model of the ventilation experiment in Opalinus clay", *Physics and Chemistry of the Earth* Vol. 33, S186-S195, 2008.

[47] J. Samper, S. Dewonck, L. Zheng, Q. Yang, and A. Naves, "Normalized sensitivities and parameter identifiability of *in situ* diffusion experiments on Callovo-Oxfordian clay at Bure site", Physics and Chemistry of the Earth, vol. 33, pp. 1000-1008, 2008.

[48] J. Molinero and J. Samper, "Groundwater flow and solute transport in fracture zones: an improved model for a largescale field experiment at Äspö (Sweden)", *Journal of Hydraulic Research*, vol. 42(Extra Issue), pp. 157-172, 2004.

[49] J. Molinero and J. Samper, "Modeling of reactive solute transport in fracture zones of granitic bedrocks", *Journal of Contaminant Hydrology*, vol. 82, pp. 293-318, 2006.

[50] J. Molinero, J. Samper, C. Yang, and G. Zhang, "Biogeochemical reactive transport model of the Redox zone experiment of the Äspö hard rock laboratory (Sweden)", *Nuclear Technology*, vol. 48, pp. 151-165, 2004.

[51] J. Samper, L. Zheng, L. Montenegro, A. M. Fernández, and P. Rivas, "Coupled thermo-hydro-chemical models of compacted bentonite after FEBEX *in situ* test", *Applied Geochemistry* vol. 23, pp. 1186-1201, 2008.

[52] L. Zheng, J. Samper, and L. Montenegro, "Inverse hydrochemical models of aqueous extracts tests", *Physics and Chemistry of the Earth*, vol. 33, pp. 1009-1018, 2008.

[53] J. Samper, C. Lu, and L. Montenegro, "Reactive transport model of interactions of corrosion products and bentonite. *Physics and Chemistry of the Earth*, vol. 33, S306-S316, 2008.

[54] Z. Dai, J. Samper, and R. Ritzi, "Identifying geochemical processes by inverse modeling of multicomponent reactive transport in Aquia aquifer", *Geosphere*, vol. 4, pp. 210-219, 2006.

[55] W. Pfingsten and M. Shiotsuki, Modeling a cement degradation experiment by a hydraulic transport and chemical equilibrium coupled code. in I.G. McKinley,C. McCombie (Editors), *21st International Symposium on the Scientific Basis for Nuclear Waste Management.* Mat. Res. Soc. Symp. Proc., Davos, Switzerland, 1997.

[56] J. M. Galíndez, J. Molinero, J. Samper, and C. Yang, Simulating concrete degradation processes by reactive transport models. *Journal de Physique* IV, 136, 177-188. NUCPERF 2006, Corrosion and long term performance of concrete in NPP and waste facilities. Workshop 27/03/2006, Cadarache, France, 27-30 March 2006.

[57] C. Yang and J. Samper, " Numerical evaluation of multicomponent cation exchange reactive transport in physically and geochemically heterogeneous porous media", *Computational Geosciences*, vol. 13, no. 3, pp. 391-404, 2009.

[58] K. Romanak and C. Yang, *Field test of CO_2 sensor for carbon sequestration applications. Technical report.* Bureau of Economic Geology, University of Texas at Austin, pp. 14, 2009.

[59] J. Samper, S. Moreira, D. Alvares, L. Montenegro, C. Lu, C. López, M. Bonilla, H. Ma, Y. Li, B. Pisani, F. Arechaga, A. Gil, J.-A. Menéndez, T. Lucas, R. V.-García. Model predictions of water chemistry for the future pit lake in As Pontes, A Coruña (Spain). Mine Water and the Environment. *Proceedings of the 10th International Mine Water Association Congress*, Karlovy Vary, Czech Republic, pp. 2-5 June 2008.

[60] J. Imberger and J.C. Patterson, A dynamic reservoir simulation model - DYRESM:5. in H. Fischer (Ed.), *Transport Models for Inland and Coastal Waters*, Academic Press, New York, New York, pp. 310-361, 1981.

[61] CWR, *Centre for Water Research* (CWR). http://www.cwr.uwa.edu.au/services/models.php. The University of Western Australia, Australia, 2006.

[62] ENRESA, Full-scale engineered barriers experiment for a deep geological repository in crystalline host rock (FEBEX Project). *European Commission*. EUR 19147 EN, 2000.

[63] A. M. Fernández, B. Baeyens, M. Bradbury, and P. Rivas, "Analysis of the pore water chemical composition of a Spanish compacted bentonite used in an engineered barrier", *Physics and Chemistry of the Earth*, vol. 29, pp. 105-118, 2004.

[64] ENRESA, *FEBEX: Updated final report*. ENRESA Technical Publication PT 05-0/2006, 2006.

[65] ENRESA, *FEBEX: Post-mortem bentonite analysis*. ENRESA Technical Publication PT 05-1/2006, 2006.

[66] ENRESA, *FEBEX: Final report on thermo-hydromechanical modeling*. ENRESA Technical Publication PT 05-2/2006, 2006.

[67] ENRESA, *FEBEX: Final THG modeling report*. ENRESA Technical Publication PT 05-3/2006, 2006.

[68] J. Samper, A. Vázquez, and L. Montenegro, Inverse hydrochemical modeling of aqueous extracts experiments for the estimation of FEBEX bentonite pore water chemistry. in Alonso, E.E., Ledesma, A. (Eds.), *Advances in Understanding Engineered Clay Barriers*. A.A. Balkema Publishers, Leiden, The Netherlands, pp. 553-563, 2005.

[69] N. L. Plummer, D. L. Parkhurst, and T. M. L. Wigley, "Critical review of the kinetics of calcite dissolution and precipitation", in Jenne, E.A. (Ed.), Chemical Modeling in Aqueous Systems, *Amerocam Chemical Society Sympposium*, vol. 93, pp. 537-573, 1979.

[70] C. Ayora, C. Taberner, and J. Samper, "Modelización de transporte reactivo: Aplicación a la dedolomitización", *Estudios Geológicos*, vol. 50, pp. 397-410 (in Spanish), 1994.

[71] A. M. Fernández, P. Rivas, Task 141: post-mortem bentonite analysis. Geochemical behaviour. *CIEMAT Internal Note* 70-IMA-L-0-107 v0, 2003.

[72] C.I. Steefel, D.J. DePaolo, and P. Lichtner, "Reactive transport modeling: An essential tool and a new research approach for the Earth sciences", *Earth and Planetary Science Letters*, vol. 240, pp. 539-558, 2005.

CHAPTER 8

Reactive Transport Modeling in Variably Saturated Media with MIN3P: Basic Model Formulation and Model Enhancements

K. U. Mayer[1*], R. T. Amos[2], S. Molins[3] and F. Gérard[4]

[1]Department of Earth and Ocean Sciences, University of British Columbia, Vancouver, BC, Canada; [2]Department of Earth and Environmental Sciences, University of Waterloo, Waterloo, ON, Canada; [3]Lawrence Berkeley National Laboratory, Earth Sciences Division, Berkeley, CA, USA and [4]INRA-IRD-SupAgro, UMR 1222 Eco&Sols, Montpellier, France

Abstract: MIN3P was developed as a general purpose multicomponent reactive transport code for variably saturated media. The basic version of the code includes Richard's equation for the solution of variably-saturated flow, and solves mass balance equations for advective-diffusive solute transport and diffusive gas transport. Biogeochemical reactions are described by a partial equilibrium approach, using equilibrium-based law-of-mass-action relationships for fast reactions, and a generalized kinetic framework for reactions that are relatively slow in comparison to the transport time scale. MIN3P has been used to support multiple field and laboratory investigations involving the fate of inorganic and organic substances and has served as a platform for additional code development: MIN3P-Bubble, an enhanced version to simulate gas generation and exsolution in the saturated zone, as well as gas entrapment and release due to water table fluctuations; MIN3P-Dusty, a version of the code that includes gas advection and multicomponent gas diffusion based on the Dusty Gas Model (DGM); and MIN3P-Soil, a version that includes plant-soil interactions. The capabilities of the basic code and the follow-up developments are demonstrated by simulating the oxidation of pyrite in mine waste, associated metal release, and subsequent attenuation processes; the interactions between the formation of "excess air" and biogeochemical reactions in the vadose zone and below the water table; the evolution of vadose zone gas composition and transport processes at a petroleum hydrocarbon spill site undergoing natural attenuation; and the effect of plant-soil interactions on mineral weathering and secondary mineral formation in soils and surficial sediments.

Keywords: Vadose zone gas transport, gas entrapment and release, gas exsolution, dusty gas model, plant-soil interaction.

1. INTRODUCTION

Multicomponent reactive transport models have become essential tools to support the investigation of the fate of chemicals in both pristine and contaminated hydrogeologic systems, to facilitate the testing of hypotheses and conceptual models, and for the quantitative analysis of complex and non-linear interactions of migrating chemicals in laboratory experiments and field investigations [1].

The vadose zone is an important component of many hydrogeologic systems, providing both a connection, as well as a buffer zone between the ground surface and the saturated groundwater zone. In many cases, organic contaminants that are accidently released at the ground surface or within the vadose zone are naturally attenuated in this region by aerobic biodegradation processes enhanced by the relative ease of oxygen ingress from the atmosphere. On the other hand, contaminants may also be released from the unsaturated zone, for example in mine waste deposits where sulfide minerals undergo oxidation when exposed to atmospheric oxygen, causing the release of acidity and metals [2].

As a result, several numerical models with capabilities to simulate multicomponent reactive transport in the vadose zone have been developed in recent years [2-9]. These codes distinguish themselves from reactive

*Address correspondence to K.U. Mayer: Department of Earth and Ocean Sciences, University of British Columbia, Vancouver, BC, Canada; Tel: 1-604-822-1539; Email: umayer@eos.ubc.ca

Fan Zhang, Gour-Tsyh (George) Yeh, Jack C. Parker and Xiaonan Shi (Eds)

transport codes that focus on the saturated zone by the inclusion of a formulation for unsaturated flow or multiphase flow, the ability to simulate gas migration by diffusion and possibly advection, and the partitioning of gases between the gas phase and the vadose zone pore water.

Despite the versatility of these models, field observations and laboratory studies continue to reveal that additional processes play a role in controlling the biogeochemical evolution in the vadose zone and underlying aquifers. For example, Amos *et al.* [10] have demonstrated that biogeochemical degradation reactions taking place below the water table at a crude oil spill site drive gas generation, exsolution, and the formation of gas bubbles enriched in CH_4 and CO_2. In addition, it was hypothesized that the carbon balance between the vadose zone and the saturated zone is likely affected by the entrapment and release of gases due to water table fluctuations. These processes are not typically considered in reactive transport formulations and provide the motivation for additional model development with the goal of generating a more complete suite of modeling capabilities.

MIN3P [2] is one of the codes with abilities to simulate multi-component reactive transport in the vadose zone and has served as a suitable platform for the incorporation of several model enhancements [11-17]. The objectives of this chapter are: 1) to provide a summary of the formulation of the original MIN3P code, 2) to provide an outline on three selected follow-up developments that were motivated by the need to extend existing modeling capabilities. The model enhancements presented include MIN3P-Bubble (gas exsolution and gas bubble formation in unconfined aquifers [13]). MIN3P-Soil (plant soil interactions [11, 16]); and MIN3P-Dusty (multicomponent gas diffusion and advection [15]), and 3) providing example applications that demonstrate the model capabilities of the original and enhanced versions of the code.

2. BASIC MODEL FORMULATION AND CAPABILITIES

2.1. Model Formulation

The model formulation of the basic MIN3P code includes Richard's equation for unsaturated flow, advective-dispersive solute transport, and gas transport by diffusion. Transport processes are coupled with a reaction network that includes aqueous complexation, redox processes, non-competitive sorption, ion exchange and non-electrostatic surface complexation, phase partitioning of gases in the vadose zone, and mineral dissolution-precipitation reactions (Fig. **1**).

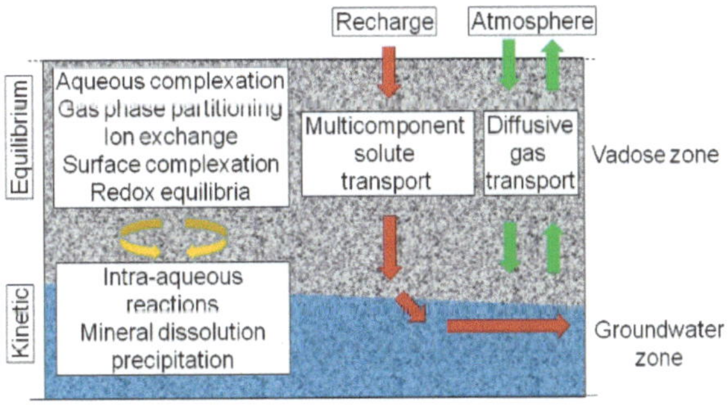

Figure 1: Capabilities of basic MIN3P code.

2.1.1. Flow Equations

Variably saturated flow is described by Richard's equation [18, 19] under the assumption of a passive air phase and neglecting hysteresis:

$$S_a S_s \frac{\partial h}{\partial t} + \phi \frac{\partial S_a}{\partial t} - \nabla \cdot \left[k_{ra} \mathbf{K} \nabla h \right] - Q_a = 0, \tag{1}$$

where S_a [m^3 H_2O m^{-3} void] defines the saturation of the aqueous phase, S_s [m^{-1}] is the specific storage coefficient, h [m] is hydraulic head, t [s] is time, ϕ [m^{-3} void m^{-3} porous medium] is porosity, k_{ra} [-] is the relative permeability, **K** [m s^{-1}] is the hydraulic conductivity tensor, and Q_a is a source sink term [m^3 H_2O m^{-3} porous medium s^{-1}]. The relative permeability and aqueous phase saturation are calculated using the soil hydraulic functions given by van Genuchten [20].

2.1.2. Reactive Transport Equations

The equations for reactive transport of N_c dissolved components in the aqueous phase can be written as [2]:

$$\frac{\partial}{\partial t}\left[S_a \phi T_j^a\right]+\frac{\partial}{\partial t}\left[S_g \phi T_j^g\right]+\frac{\partial}{\partial t}(T_j^s) + \nabla\cdot\left[\mathbf{q}_a T_j^a\right]-\nabla\cdot\left[S_a \mathbf{D}_a \nabla T_j^a\right]-\nabla\cdot\left[S_g \mathbf{D}_g \nabla T_j^g\right]-Q_j^{a,a}-Q_j^{a,m}=0 \qquad j=1,N_c \tag{2}$$

where T_j^a [mol L^{-1} H_2O] is the total aqueous component concentration, T_j^g [mol L^{-1} gas] and T_j^s [mol dm^{-3} bulk porous medium] are the total component concentrations of the gaseous and sorbed species respectively, S_g is the saturation of the gas phase, $\mathbf{q}_a$ is the Darcy flux vector, which can be obtained from the solution of equation 1, $\mathbf{D}_a$ is the hydrodynamic dispersion tensor applicable to all species in the system, and $\mathbf{D}_g$ is the diffusion tensor applicable to all gases. $Q_j^{a,a}$ and $Q_j^{a,m}$ are source-sink terms [mol dm^{-3} porous medium s^{-1}] due to intra-aqueous and dissolution-precipitation reactions.

In addition to maintaining the mass balance for aqueous, gaseous and sorbed species (provided by equation 2), it is necessary to provide a mass balance equation for the mineral phases:

$$\frac{d\phi_i}{dt}=V_i^m R_i^m \qquad i=1,N_m \tag{3}$$

where φ_i [-] is the volume fraction of the mineral in question, V_i^m [dm^3 mineral mol^{-1}] is the molar volume and R_i^m [mol dm^{-3} porous medium s^{-1}] is the rate of mineral dissolution-precipitation; N_m is the number of minerals considered.

2.1.3. Biogeochemical Reactions

MIN3P includes a suite of biogeochemical reactions (Fig. **1**) that can be described by the following set of stoichiometric relationships:

$$A_i^{a,eq} \leftrightarrow \sum_{j=1}^{N_c} \nu_{ij}^{a,eq} A_j^c \qquad i=1,N_{a,eq} \tag{4}$$

$$A_i^g \leftrightarrow \sum_{j=1}^{N_g} \nu_{ij}^g A_j^c \qquad i=1,N_g \tag{5}$$

$$0 \leftrightarrow \sum_{j=1}^{N_c} \nu_{ij}^{a,kin} A_j^c \qquad i=1,N_{a,kin} \tag{6}$$

$$A_i^m \leftrightarrow \sum_{l=1}^{N_i^{m,p}} \sum_{j=1}^{N_c} \nu_{ijl}^m A_j^c \qquad i=1,N_m \tag{7}$$

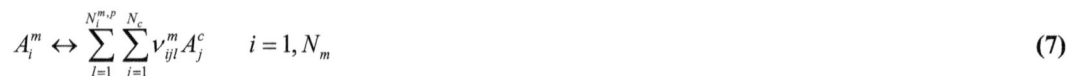

Here, $A_i^{a,eq}$, A_i^g, and A_i^m are the names of the aqueous complexes, gases, and minerals, respectively, and $\nu_{ij}^{a,eq}$, ν_{ij}^g, $\nu_{ij}^{a,kin}$, ν_{ijl}^m define the stoichiometric coefficients of the components A_j^c, which constitute the N_c primary unknowns present in the aqueous phase. $N_{a,eq}$, $N_{a,kin}$, and N_g define the number of aqueous complexes, kinetic intra-aqueous reactions, and gases; N_i^{mp} define the number of parallel pathways affecting the kinetic dissolution or precipitation of mineral A_i^m. Kinetic intra-aqueous reactions (equation 6)

define mass transfer between components, *e.g.* the kinetic oxidation of Dissolved Organic Carbon (DOC) to CO_3^{2-}. No additional species are created in this process, explaining the absence of a species on the left hand side of the reaction equation. Ion exchange and surface complexation reactions are also considered, using stoichiometric relationships similar to equations 4 and 5.

This reaction network includes both equilibrium (equations 4, 5, ion exchange and surface complexation reactions) and kinetically controlled reactions (equations 6, 7). All equilibrium relationships enter the reactive transport equations through component mass balances defined by total concentration terms, while kinetic reactions are considered through source-sink terms (equation 2, Fig. **1**). Kinetic relationships have been chosen for reactions that tend to be slow in relation to the time scale of transport, and include mineral dissolution-precipitation and microbially mediated reactions. However, these relationships can also be used to describe equilibrium conditions through the inclusion of an affinity term that accounts for the thermodynamic constraints of the reaction [2]. A kinetic formulation for surface complexation reactions is presently not available in the code.

Total concentration terms: The total component concentration for component A_j^c in the aqueous phase is defined as [21, 22]:

$$T_j^a = C_j^c + \sum_{i=1}^{N_{a,eq}} \nu_{ij}^{a,eq} C_i^{a,eq} \qquad j = 1, N_c, \tag{8}$$

where C_j^c are the concentrations of the components as species in solution [mol L^{-1} H_2O], and $C_i^{a,eq}$ are the concentrations of secondary species in the aqueous phase [mol L^{-1} H_2O]. Similar relationships can be defined for the gas phase and the sorbed phase. For example, the total concentration term for component A_j^c in the gas phase is given by:

$$T_j^g = \sum_{i=1}^{N_g} \nu_{ij}^g C_i^g, \qquad j = 1, N_c \tag{9}$$

where C_i^g is the concentration of the gas A_i^g [mol L^{-1} gas]. The concentrations of secondary species (*e.g.* $C_i^{a,eq}$) are determined based on law of mass action relationships subject to the appropriate activity corrections [2].

Source-sink terms: Kinetic reactions lead to the mass transfer between components within the aqueous phase or between minerals and the aqueous phase. For intra-aqueous kinetic reactions, the source-sink terms are defined as:

$$Q_j^{a,a} = -\phi \sum_{i=1}^{N_{a,kin}} \nu_{ij}^{a,kin} R_i^a \qquad j = 1, N_c \tag{10}$$

where R_i^a is the rate of the ith intra-aqueous kinetic reaction. In a similar fashion, the production or consumption of aqueous components due to dissolution-precipitation reactions is given by:

$$Q_j^{a,m} = -\phi \sum_{i=1}^{N_m} \sum_{l=1}^{N_i^{m,p}} \nu_{ijl}^m R_{il}^m \qquad j = 1, N_c \tag{11}$$

where R_{il}^m is the rate of dissolution or precipitation of the mineral A_i^m along the lth reaction pathway. MIN3P contains a flexible framework designed to include a variety of processes and dependencies when defining kinetic rate expressions [2].

2.1.4. Numerical Implementation

The reactive transport equations are solved in a fully coupled fashion using the Global Implicit Method (GIM) in combination with the Direct Substitution Approach (DSA) [21, 22]. The governing equations for variably saturated flow and reactive transport (equations 1 and 2) are both linearized using a modified Newton's method. Spatial discretization is based on the finite volume technique and includes options for upstream weighting, centered weighting or a flux limiter formulation for the advective terms. Time weighting is performed using a fully implicit scheme. The solution of the matrix equations is performed using sparse iterative methods [23]. All reactions considered in a simulation can be defined through a database and external input files.

2.2. Previous Model Applications

MIN3P has previously been used to investigate a variety of vadose zone scenarios including the weathering of mine waste deposits [2, 24-27], and land application of food-processing waste [28]. Although the MIN3P development was performed with vadose zone applications in mind, the code has also been used to investigate reactive transport problems in the saturated zone including the performance assessment of permeable reactive barriers for groundwater remediation [29-32], *in situ* chemical oxidation of chlorinated solvents [17], and the geochemical stability of crystalline rock formations [33, 34].

2.3. Example Application: Oxidation of Sulfide Minerals in Mine Tailings

Here we present a hypothetical scenario to demonstrate some of the key model capabilities for vadose zone reactive transport problems. The example focuses on the oxidation of sulfide minerals in a 5 m thick unsaturated mine waste deposit (Fig. **2**).

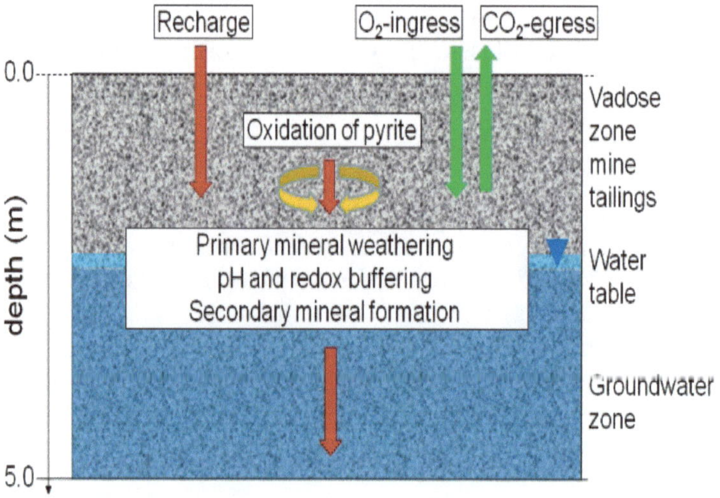

Figure 2: Conceptual model of sulfide mineral oxidation in mine tailings.

In addition to pyrite, the waste material contains the primary mineral phases calcite, siderite, gibbsite, gypsum, K-feldspar, and muscovite. Primary minerals are defined as the reactive minerals that are initially present in the mine tailings. Three secondary minerals (ferrihydrite, K-jarosite and amorphous silica) are also included. Secondary minerals are not initially present, but are allowed to precipitate, if the solution becomes supersaturated with respect to these phases. The mineral dissolution-precipitation reactions considered are summarized in Table **1**.

The simulation requires 12 aqueous components (Al^{3+}, Ca^{2+}, K^+, CO_3^{2-}, Cl^-, H_4SiO_4, H^+, $O_2(aq)$, SO_4^{2-}, HS^-, Fe^{2+}, Fe^{3+}). The iron and sulfur redox couples are assumed to be at equilibrium. Diffusive oxygen ingress through the vadose zone is simulated, and CO_2 produced by carbonate mineral dissolution can migrate towards the ground surface through the gas phase. Relevant aqueous complexes (21 in total) are considered to adequately describe mineral solubilities. The simulation is conducted for a 1D vertical domain with a spatial discretization of 0.05m, a maximum time step of 1 year, and a final simulation time of 10 years.

Table 1: Primary and secondary mineral reactions considered in acid mine drainage example.

	Primary minerals	log K_{sp}
Pyrite	$FeS_2 + \frac{7}{2}O_2(aq) + H_2O \rightarrow Fe^{2+} + 2SO_4^{2-} + 2H^+$	215.3
Calcite	$CaCO_3 \leftrightarrow Ca^{2+} + CO_3^{2-}$	-8.48
Siderite	$FeCO_3 \leftrightarrow Fe^{2+} + CO_3^{2-}$	-10.45
Gibbsite	$Al(OH)_3(am) + 3H^+ \leftrightarrow Al^{3+} + 3H_2O$	8.11
Gypsum	$CaSO_4 \cdot 2H_2O \leftrightarrow Ca^{2+} + SO_4^{2-} + 2H_2O$	-4.58
K-feldspar	$KAlSi_3O_8 + 4H^+ + 4H_2O \rightarrow K^+ + Al^{3+} + 3H_4SiO_4$	0.08
Biotite	$KAl_3Si_3O_{10}(OH)_2 + 10H^+ \rightarrow K^+ + 3Al^{3+} + 3H_4SiO_4$	12.99
	Secondary minerals	**log K_{sp}**
Ferrihydrite	$Fe(OH)_3 + 3H^+ \leftrightarrow Fe^{3+} + 3H_2O$	4.89
Jarosite	$KFe_3(SO_4)_2(OH)_6 + 6H^+ \rightarrow K^+ + 3Fe^{3+} + 2SO_4^{2-} + 6H_2O$	-9.21
Silica (am)	$SiO_2(am) + 2H_2O \leftrightarrow H_4SiO_4$	-2.71

A porosity of 0.5 is assumed and the vertical saturated hydraulic conductivity is $1x10^{-6}$ m s^{-1}. Van Genuchten soil hydraulic function parameters are $S_{ra} = 0.05$ [-] (residual saturation), $\alpha = 3.5$ [m^{-1}], and n = 1.4 [-]. Free phase diffusion coefficients are 2.4 $x10^{-9}$ m s^{-2} for all dissolved species and $2.1x10^{-5}$ m s^{-2} for all gases; the Millington formulation is used to estimate tortuosity and effective diffusion coefficients both for the aqueous and gas phase. A longitudinal dispersivity of 0.005 m is used.

The flow system is at steady state with a recharge rate of 9.5 x 10^{-9} m s^{-1}, corresponding to 300 mm yr^{-1}, and a constant head boundary of h = 2.5m at the bottom of the domain resulting in a water table located at an intermediate depth within the domain. For transport, a mixed boundary condition is applied at the ground surface with a third type (specified mass flux) boundary for dissolved species and a first type (specified concentration) boundary condition for O_2 and CO_2 gases. The composition of recharge water is relatively dilute, but reflects fertilizer application in support of revegetation efforts typically employed at tailings, and is in equilibrium with the atmosphere with respect to O_2 and CO_2. The chemical compositions of recharge and pore-water initially present in the tailings are summarized in Table **2**:

Table 2: Boundary and initial conditions for acid mine drainage example, all concentrations in units of [mol L^{-1}], unless otherwise noted.

Parameter	Recharge	Initial
Al^{3+}	1.3 x 10^{-8}	2.6 x 10^{-8}
Ca^{2+}	1.9 x 10^{-3}	1.4 x 10^{-2}
K^+	9.0 x 10^{-3}	9.0 x 10^{-3}
CO_3^{2-}	[a]3.2 x 10^{-4}	2.5 x 10^{-3}

Cl^-	1.1×10^{-4}	1.1×10^{-3}
H_4SiO_4	2.0×10^{-4}	1.9×10^{-3}
H^+ (as pH)	5.0	7.0
$O_2(aq)$	$^{b)}0.21$	$^{c)}-2.5$
SO_4^{2-}	7.5×10^{-3}	2.0×10^{-2}
HS^-	-	3.3×10^{-12}
Fe^{2+}	1.3×10^{-11}	1.5×10^{-4}
Fe^{3+}	5.4×10^{-5}	7.2×10^{-12}

a) as pCO_2 [atm], b) as pO_2 [atm], c) as pe.

The composition of the pore water initially present in the tailings reflects equilibrium with calcite, siderite, gibbsite, gypsum, and amorphous silica. The initial volume fraction and effective rate coefficients for the primary minerals are summarized in Table **3**.

Table 3: Primary and secondary mineral reactions considered in acid mine drainage example.

Mineral	Volume fraction [-]	k_{eff} [mol dm^{-3} s^{-1}]
Pyrite	1.37×10^{-2}	1×10^{-9}
Calcite	1.77×10^{-3}	5×10^{-8}
Siderite	3.49×10^{-4}	1×10^{-8}
Gibbsite	8.30×10^{-4}	1×10^{-8}
Gypsum	6.06×10^{-3}	a)
K-feldspar	2.68×10^{-2}	5×10^{-11}
Biotite	7.31×10^{-2}	1×10^{-11}

a) at quasi-equilibrium.

The initial mineralogical composition is typical for relatively low sulfide mine waste with low carbonate mineral content, and a high fraction of silicate minerals. With the exception of pyrite, K-feldspar, and biotite, mineral phases are only present at trace amounts of much less than 1%. The mineralogical composition and effective rate constants are similar in nature to measured and calibrated parameters from the Nickel Rim mine site, Sudbury, ON [2]. The effective rate coefficients for silicate minerals provide a strong kinetic control, reflecting the slow dissolution kinetics of these phases. Gypsum and all secondary phases are simulated as quasi-equilibrium reactions. This approach requires that the rate of reaction is fast in relation to the transport time scale. Under these conditions, knowledge of the reaction rate coefficients is not required. Large rate coefficients on the order of 1×10^{-7} mol L^{-1} s^{-1} are used to approximate equilibrium behavior.

Simulations lead to a water saturation ranging from fully saturated conditions to a saturation of $S_a = 0.74$ near the ground surface (not shown). Unsaturated conditions in the shallow region of the tailings facilitate oxygen ingress (Fig. **3a**). Pyrite oxidation is limited to the uppermost region of the tailings material (Fig. **3b**), constrained by the availability of oxygen (Fig. **3a**). The oxidative dissolution of pyrite results in acid production and triggers the dissolution of calcite, and the precipitation and dissolution of a series of mineral phases that buffer pH including siderite, gibbsite, and jarosite (Fig. **3b**, **3c**). The simulations reproduce the pH-buffering sequence reported in previous studies [2, 35], with the exception of ferrihydrite, which does not precipitate due to the abundance of potassium, favoring jarosite formation. Potassium is added with recharge due to fertilizer application and is further supplemented from biotite and K-feldspar dissolution (not shown). Biotite and feldspar weathering correlates with the region that is dominated by low pH conditions, as indicated by the zone of amorphous silica formation (Fig. **3b**). Only in this region are these phases highly undersaturated providing the driving force for dissolution.

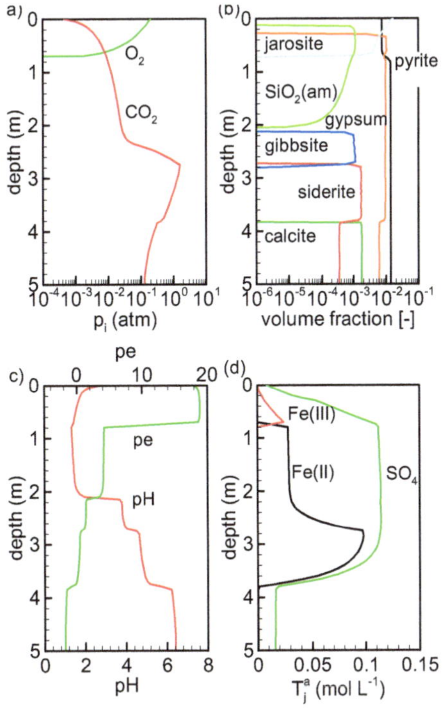

Figure 3: Oxidation of sulphide minerals in mine tailings. Simulation results as a function of depth, all profiles shown at T = 10 years (a) O_2 and CO_2 partial pressures, (b) mineral volume fractions, (c) pH and redox potential, (d) Total concentrations of Fe(II), Fe(III) and SO_4.

The dissolution of carbonate minerals results in the production of CO_2, which is released to the ground surface by diffusion through the gas phase - if dissolution reactions occur above the water table. If dissolution occurs below the water table, closed system conditions are approached and CO_2 is carried downwards with the recharging groundwater (Fig. **3a**). Fe and SO_4 concentrations are controlled by the oxidation of pyrite, and mineral equilibria with jarosite and gypsum (Fig. **3b**, **d**). The assumption of redox equilibrium leads to a pronounced redox zonation where O_2 disappears with Fe(III) dominating above, and Fe(II) controlling Fe concentrations below (Fig. **3c,d**). Although the present example is limited to one spatial dimension for the ease of presentation, the code is suitable for two and three-dimensional reactive transport problems [17, 24, 27, 29].

3. MODEL ENHANCEMENTS

The need to better describe gas migration processes in the vadose zone, the fate of gases in the region of the capillary fringe, and the feedback between plants and geochemical processes has motivated several model enhancements (Fig. **4**).

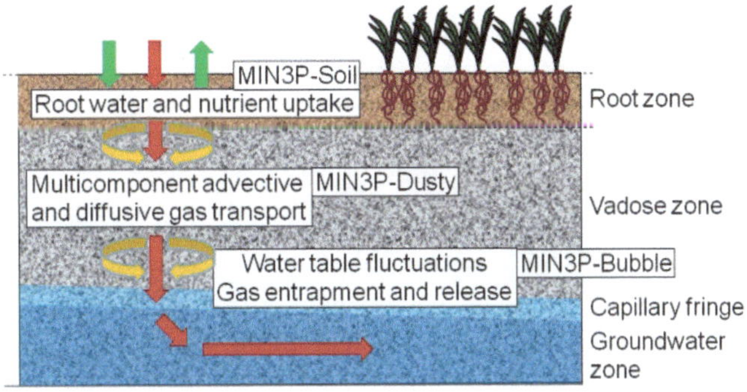

Figure 4: Summary of MIN3P model enhancements for reactive transport modeling in the soil zone (MIN3P-Soil), vadose zone (MIN3P-Dusty) and capillary fringe region (MIN3P-Bubble).

Specifically, modules for MIN3P were developed that are capable of describing the exsolution of gases and gas bubble formation in the saturated zone, permeability reduction due to the presence of gas bubbles, and gas exchange between the saturated zone and the vadose zone caused by water table fluctuations (MIN3P-Bubble [13]). In the vadose zone, the code was enhanced to simulate multi-component advective and diffusive gas migration, including inert gases, driven by the generation and consumption of reactive gases (MIN3P-Dusty [15]), and in the soil zone the code was enhanced to simulate root water and solute uptake (MIN3P-Soil [11, 16]).

The remainder of this chapter will focus on introducing key aspects of these model enhancements and will provide a set of demonstration examples that highlight their capabilities and the utility of these developments.

3.1. MIN3P-Bubble: Gas Entrapment and Release in the Groundwater Zone

3.1.1. Motivation

Dissolved gases in groundwater can be geochemically important constituents in contaminated aquifers and are also used extensively as indicators of various physical and geochemical phenomena. Because gases are relatively insoluble, gas bubbles can often form through biogenic processes such as methanogenesis, denitrification, or through physical mechanisms such as entrapment near the water table due to water table fluctuations. In either case the formation of gas bubbles will lead to geochemical and physical changes in the aquifer. For example, the entrapment of oxygen-rich gas bubbles near the water table can lead to significantly enhanced transport of oxygen into an otherwise anaerobic aquifer [36]. Furthermore, the presence of gas bubbles can lead to a reduction in hydraulic conductivity of an aquifer due to blocking of the pore spaces [37-39]. The formation of gas bubbles can also lead to ebullition, the vertical transport of gas bubbles driven by buoyancy forces, resulting in rapid transport of gases through the saturated sediment. In wetlands this process can lead to release of methane to the atmosphere as gas bubbles rapidly bypass oxidation zones in the sediment [40, 41].

The formation of gas bubbles in groundwater is often inferred through monitoring of stable gases. In methanogenic aquifers the formation of gas bubbles is accompanied by depletion in dissolved atmospheric gases such as Ar and N_2 [10, 42, 43]. Similarly in groundwater where gas bubbles form through denitrification, Ar concentrations become depleted [44]. Near the water table, the entrapment of gas bubbles can lead to the formation of excess air. This phenomenon results from dissolution of gas bubbles due to over-pressuring of the bubbles as the water table rises. The degree of excess air formation must be considered when interpreting dissolved noble gas concentrations for the reconstruction of groundwater infiltration temperature and infiltration elevations [45-47], and determining groundwater age using dissolved gas tracer methods [48, 49]. Analytical methods have been developed to quantify the degree of biogenic gas bubble formation [10, 43] and to quantify excess air formation [50]. Although these analytical tools can provide important information, they generally lack the sophistication to account for the complexities of coupled solute transport and reaction.

The basic version of MIN3P has been enhanced to quantify gas bubble formation and contraction through biogenic gas production or consumption, quantify gas bubble formation due to entrapment near the water table and subsequent equilibration of the gas bubbles with the surrounding groundwater, and quantify permeability changes due to changes in gas phase saturation as a result of gas bubble formation and contraction [13].

3.1.2. Model Formulation

Gas Pressure Induced Changes in Gas Phase Saturation: In the saturated zone gases may be produced or consumed by a number of chemical or biogenic processes resulting in changes in the partial pressure of individual dissolved gases, and the total dissolved gas pressure. If the total gas pressure exceeds a threshold pressure a gas bubble may form [51] and gases will partition between the aqueous and gas phases based on Henry's Law. Once a bubble is formed, the bubble may grow or shrink depending on whether gases are produced or consumed. In MIN3P the partial pressures and concentrations of gaseous species are given by:

$$P_l = \left(K_l^g\right)^{-1} \prod_{j=1}^{N_C} \left(C_j^C \gamma_j^C\right)^{v_{lj}^g}, \tag{12}$$

$$C_l^g = \frac{P_l}{RT}, \tag{13}$$

where P_l (atm) and K_l^g (mol atm^{-1}) are the partial pressure and Henry's Law constant for gas species l, respectively, R is the gas constant [0.08206 L atm mol^{-1} K^{-1}] and T [K] is the temperature. For simplicity we neglect capillary forces such that a gaseous phase will form below the water table under the condition:

$$\sum P_l > P_H \quad P_H > 0, \tag{14}$$

where P_H (atm) is the hydrodynamic pressure equal to the sum of the atmospheric pressure and water pressure. Under equilibrium conditions the following condition will apply:

$$\sum_{l=1}^{N_g} P_l - P_H = 0 \qquad if\ S_{gt} > 0, \tag{15}$$

where S_{gt} is the trapped gas phase saturation. The model does not explicitly account for individual gas bubbles but rather the total pore space occupied by the gas bubbles, *i.e.* the trapped gas phase saturation. For any solution to the mass conservation equation (Equation 2), the total mass of each gas component in the aqueous and gaseous phases is fixed

$$T_k^T = S_a T_k^a + S_{gt} T_k^g. \tag{16}$$

where the subscript k refers to a specific component associated with each gas. The partial pressure of each gas can then be constrained by the concentration of the gas in the aqueous phase (Equation 12) and also by the gas saturation S_{gt}. Cirpka and Kitanidis [52] provide a formulation that relates the partial pressure P_l to S_{gt},

$$P_{l(k)} = \frac{T_k^T}{\left(1 - S_{gt}\right)\Psi_k + \frac{S_{gt}}{RT}}. \tag{17}$$

where Ψ_k describes the equilibrium of gaseous species to aqueous components in a multicomponent framework (see [13] for full development of this expression). A solution for S_{gt} can be obtained by minimizing Equation 15 with $P_{l(k)}$ given by Equation 17.

Bubble Entrapment: Gas bubbles may also be entrapped near the water table during imbibition [36, 53] or conversely be released to the unsaturated zone if drainage occurs. Here the entrapment of gas bubbles is described by the formulation for hysteretic saturation-capillary pressure relationships in a two-phase system given by Kaluarachchi and Parker [54],

$$S_{egt} = \left\{ \frac{1 - S_{ea}^{\min}}{1 + R_L(1 - S_{ea}^{\min})} - \frac{1 - S_{aa}}{1 + R_L(1 - S_{aa})} \right\}, \tag{18}$$

where S_{egt} is the effective gas phase saturation, S_{ea}^{min} is the minimum effective aqueous saturation, and S_{aa} is the apparent aqueous phase saturation given by S_{egt} plus S_{ea}, which is the effective aqueous saturation. R_L is the empirically derived Land's parameter given by,

$$R_L = \frac{1}{S_{egt}^{\max}} - 1, \tag{19}$$

where S_{egt}^{max} is the maximum effective trapped gas saturation, which is the only additional parameter needed to determine the trapped gas saturation, beyond the parameters already needed to determine non-hysteretic aqueous phase saturations, and must be determined empirically for particular porous media. Effective saturations are related to actual saturations by [53],

$$S_{ea} = \frac{S_a - S_{ra}}{1 - S_{ra}}, \tag{20}$$

and

$$S_{egt} = \frac{S_{gt}}{1 - S_{ra}}, \tag{21}$$

where S_{ra} is the residual saturation. Once a particular control volume becomes saturated ($S_{aa} = 1$), S_{gt} is no longer controlled by the saturation/pressure head relation but is instead a function of total dissolved gas pressure as described in the preceding section.

Relative Permeability Relationship: In unsaturated porous media relative permeability, k_{ra}, can be expressed as a function of aqueous phase saturation;

$$k_{ra} = S_{ea}^{\,l}\left(1 - \left[1 - S_{ea}^{\,1/m}\right]^m\right)^2, \tag{22}$$

where l and m are soil hydraulic function parameters [55]. Fry *et al.* [56] showed that this relationship is valid in quasi-saturated sands (*i.e.* where a trapped gas phase exists) ranging from fine to coarse grained, with gas saturations ranging from ~ 10 to 60%. The MIN3P bubble model extends the use of Equation 22 to account for permeability loss as a result of the formation of a trapped gas phase.

Numerical Implementation: The formation of gas bubbles results in coupling of geochemical reactions and the physical flow system due to the growth and contraction of the bubbles displacing water, and changes in permeability due to changes in gas phase saturation. The MIN3P bubble model employs a three step iterative approach to solve this non-linear coupling. The three steps include the flow solution (Equation 1), the reactive transport solution (Equation 2) and the bubble solution (Equations 15 and 17). At each timestep the model solves each of these equations in series then iterates until the primary unknowns converge to a user-defined tolerance. The relative permeability is updated at the end of each timestep.

3.1.3. Example Application: Excess Air

Previously, Amos and Mayer [13] conducted detailed simulations of natural attenuation processes at the petroleum hydrocarbon spill site near Bemidji, MN focusing on the fate and transport of methane from the methanogenic degradation of hydrocarbons. Here we conduct simplified 1-D simulations of excess air formation to demonstrate the capabilities of the MIN3P bubble module and also to illustrate the coupling of physical and geochemical processes involving excess air (Fig. **5**).

The first set of simulations consists of a 10 m deep sediment column with water in equilibrium with the atmosphere (WEA) infiltrating into the top of the column. The column is initially drained. As water infiltrates from the top through the column, the column becomes progressively saturated from the bottom up; resulting in the accumulation of entrapped air. Seven components are included in the simulations: O_2, Ar, N_2, HCO_3^-, H^+, Ne and Xe, along with 6 gas species O_2, CO_2, N_2, Ar, Ne and Xe. Two simulations are conducted, the first with no geochemical reactions. The second includes solid phase organic matter in the sediments that reacts with O_2;

$$CH_2O + O_2 \rightarrow HCO_3^- + H^+, \tag{23}$$

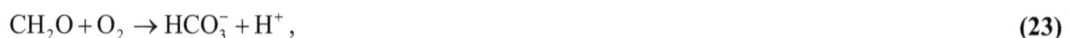

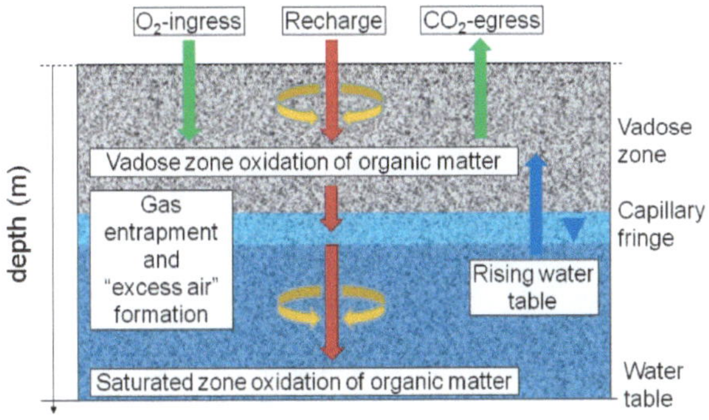

Figure 5: Conceptual model of gas entrapment and excess air formation under rising water table conditions, O_2 consumption and CO_2 production may occur in the vadose zone or the saturated groundwater zone.

resulting in the depletion of O_2 in the sediment column. The gas phase and aqueous phase are assumed to be in equilibrium with Henry's constants for O_2, N_2, and Ar from Weiss [57] and for Ne and Xe from the CRC Handbook [58]. Other relevant physical parameters are given in Table **4**.

Table 4: Hydrological parameters for excess air simulations.

van Genuchten – alpha (m^{-1})	3.5
van Genuchten – n (-)	4
residual saturation (-)	0.05
longitudinal dispersivity (m)	0.1
transverse dispersivity (m)	0.001
aqueous diffusion coeff. ($m^2\ s^{-1}$)	2×10^{-09}
gas phase diffusion coeff. ($m^2\ s^{-1}$)	1×10^{-05}
temperature (K)	283

The results of the simulations are shown in Fig. **6**. Initially the Ne, Ar and O_2 partial pressure are in equilibrium with the atomsphere as the column is fully unsaturated.

At a later time the water table (defined where the pressure head is equal to zero) is approximately at 5 m depth. Above this point the sediment is unsaturated, *i.e.* there is a connected gas phase and the gas concentrations remain in equilibrium with air. Below this point the water saturations range from 0.9 to 1 as a result of a trapped gas phase (*i.e.* gas bubble entrapment). For the non-reactive column the Ne, Ar and O_2 concentrations exceed WEA as the hydrostatic pressure on the trapped gas forces dissolution of the gas bubbles. Note that the aqueous concentrations of a particular gas are a function of the solubility of the gas and also the solubility and mole fraction of each of the gases present.

For the simulations including reactive organic matter, the O_2 concentration drops to near zero above the water table. Bicarbonate is produced in this reaction, resulting in CO_2 generation, and overall the total pressure of the gas phase remains relatively constant. Nevertheless, the amount of excess air, shown for Ne and Ar in Fig. **6**, is increased compared to the non-reactive simulation. In this case the increase in excess air is a result of the difference in solubility of O_2 and CO_2. This simulation highlights the complex coupling between geochemical processes, *i.e.* organic matter oxidation, and physical processes, *i.e.* gas bubble entrapment and gas dissolution.

Another set of simulations is run to mimic bubble entrapment in a contaminated aquifer. Here a 1 m initially unsaturated column with atmospheric gas concentrations is filled from the bottom, corresponding to a rising water table. For the initial simulation the infiltrating water is in equilibrium with the atmosphere and no reactions occur. For the second simulation the infiltrating water is anaerobic and contains dissolved organic matter. As the water level rises, atmospheric gas bubbles are entrapped resulting in mixing of oxygen with the anaerobic groundwater and aerobic degradation of the dissolved organic matter as per equation 23. All physical parameters are identical as those used in the previous set of simulations.

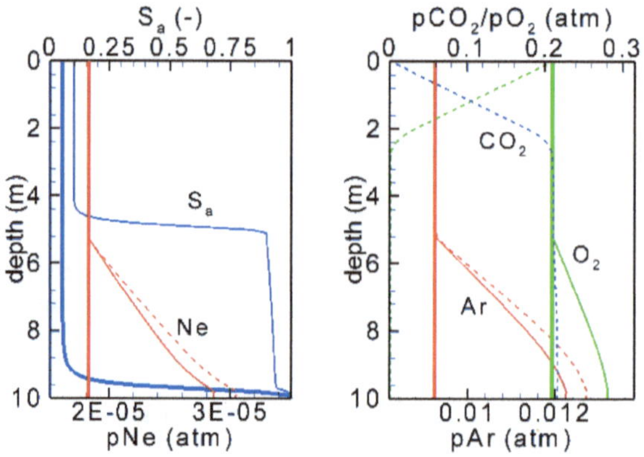

Figure 6: Gas concentrations and water saturation for excess air simulations. Thick solid lines represent initial conditions; thin solid lines present results with ~ 5 m hydraulic head for non-reactive simulations; and dashed lines depict results with ~ 5 m hydraulic head for reactive simulations.

As with the previous simulations, initial gas concentrations are in equilibrium with the atmosphere (Fig. 7). For the non-reactive simulation, the rising water table created a trapped gas phase and excess air develops as the water table continues to rise and increases the hydrostatic pressure. For the reactive simulations, as the water table rises, incoming contaminated water mixes with oxygen in the entrapped gas phase, and oxygen is consumed. It is assumed that CO_2 is not produced in significant quantities, resulting in a decrease of the total amount of entrapped gas, leading to a reduction in the saturation of the gas phase and consequently an increase in the partial pressure of the remaining gases in the entrapped bubbles. The concentrations of the gases in the water must also increase to maintain equilibrium. In this case the excess neon is approximately 40% at the peak, which is substantially higher than in the non-reactive case, and is primarily due to the effects of the biochemical removal of O_2. A lack of CO_2 production is possible if the reaction takes place under circum-neutral to alkaline pH conditions, if CO_2 is removed by carbonate mineral precipitation, or if the organic matter oxidation reaction (equation 23) is replaced by the oxidation of Fe(II), Mn(II), or reduced sulfur species.

These simulations demonstrate the utility of MIN3P-Bubble to evaluate the importance of biogeochemical processes on excess air formation.

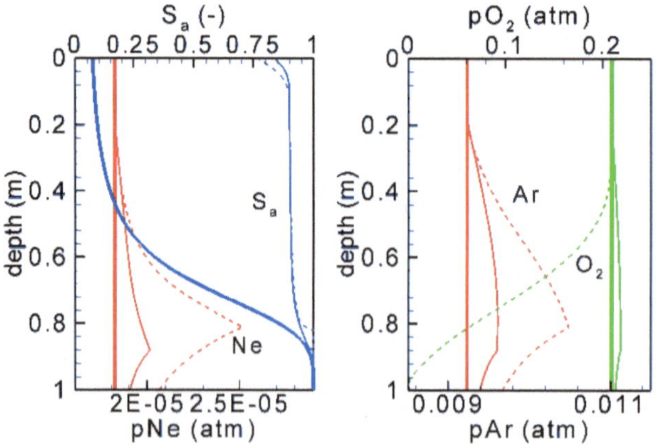

Figure 7: Gas concentrations and water saturation for contaminated aquifer simulations. Thick solid lines represent initial conditions; thin solid lines present results with 0.78 m hydraulic head for non-reactive simulations; dashed lines depict results with 0.78 m hydraulic head for reactive simulations.

3.2. MIN3P-Dusty: Multicomponent Gas Transport in the Vadose Zone

3.2.1. Motivation

Contaminants present in the vadose zone commonly exist as part of a multicomponent mixture of gases and tend to react with atmospheric gases to produce chemical species with different chemical and physical properties. The result is an inherently coupled system where both transport and reaction processes determine the gas phase composition and the transport regime. This dynamic coupling has been observed experimentally and at field sites where consumption or production of gases by microbially mediated reactions is significant enough to alter gas transport. Examples include gas production and release in partially saturated landfill covers [59, 60], methanogenesis and methane oxidation in aquifers contaminated by organic compounds [10], and oxidation of sulfide minerals in mine waste deposits [61].

Most existing reactive transport models, including the basic version of MIN3P, simply describe reaction-induced gas transport by means of Fick's law [2, 3, 7, 62-65]. However, this approach does neither account for multicomponent diffusion, nor for reaction-driven advection. In multi-component mixtures diffusion of each gas component is not only a function of its own concentrations but also a function of the concentration of all other gas components as described by the Dusty Gas Model [66, 67]. In addition, in reactive systems, gas advection can be a relevant gas transport process because even small pressure gradients generated by geochemical processes (< 1 Pa m^{-1}) can drive significant viscous fluxes [66]. Gas advection has previously been considered in models of injection and extraction of air or vapor [68, 69], volatilization of organic compounds [70-72], underpressurization in basements of buildings and the adjacent sediments [73, 74], barometric pumping [67], displacement due to infiltrating recharge water [75], and temperature changes [62].

In order to provide a rigorous framework for multicomponent gas diffusion and advection in a reactive transport model, the basic version of MIN3P is enhanced with the inclusion of an advective term and a diffusive term based on the DGM [66].

3.2.2. Model Formulation

The mass balance equation for component A_j^c (Equation 2) is modified to include both the flux contributions from multicomponent diffusion ($\nabla \cdot \mathbf{N}_j^{T,g}$) and advection ($\nabla \cdot [\mathbf{q}_g T_j^g]$):

$$\frac{\partial}{\partial t}\left[S_a \phi T_j^a\right] + \frac{\partial}{\partial t}\left[S_g \phi T_j^g\right] + \frac{\partial}{\partial t}(T_j^s) + \nabla \cdot \left[\mathrm{q}_a T_j^a\right] - \nabla \cdot \left[S_a \mathbf{D}_a \nabla T_j^a\right] + \nabla \cdot \left[\mathrm{q}_g T_j^g\right] - \nabla \cdot \mathbf{N}_j^{T,g} - \mathbf{Q}_j^{a,a} - \mathbf{Q}_j^{a,m} = 0 \qquad j = 1, \mathbf{N}_c \tag{24}$$

where $\mathbf{q}_g$ [m s^{-1}] is the Darcy flux vector in the gaseous phase, and $\mathbf{N}_j^{T,g}$ [mol m dm^{-3} porous medium s^{-1}] is the total diffusive flux vector for component A_j^c in the gaseous phase.

The advective flux in the gaseous phase is directly substituted in the mass balance equation (24) and calculated using Darcy's law as a function of the total pressure in the gas phase (p_g [Pa])

$$\mathbf{q}_g = -\frac{k_{rg}\mathbf{k}}{\mu_g}\left(\nabla p_g + \rho_g g \nabla z\right) \tag{25}$$

where $\mathbf{k}$ [m^2] is the permeability tensor, k_{rg} [-] is the relative gas permeability [76], ρ_g [kg m^{-3}] and μ_g [kg m^{-1} s^{-1}] are the gaseous phase density and viscosity, respectively. Total pressure in the gas phase is related to the sum of the partial pressures of all gas species (p_i^g [Pa]), which in turn are related to the molar concentrations of gas species (C_i^g [mol L^{-1} gas]) by means of the ideal gas law [15]. Therefore, since concentrations of the gas species are obtained from the solution of equation 24, gas phase velocities adjust to compositional changes in the gas phase. This substitution approach ensures that a direct link between gas transport processes and chemical reactions is provided: any change in gas concentrations caused by chemical reactions has a direct effect on the gradient that drives gas transport.

Diffusive fluxes in the gas phase ($\mathbf{N}_i^g$ [mol m dm^{-3} porous medium s^{-1}]) are calculated using the Stefan-Maxwell equations within the framework of the Dusty Gas Model [77]:

$$\sum_{\substack{k=1\\k\neq i}}^{N_g} \frac{\mathbf{N}_i^g \chi_k^g - \mathbf{N}_k^g \chi_i^g}{S_g \phi \tau_g D_{ik}^g} + \frac{\mathbf{N}_i^g}{D_i^{K,g}} = -\nabla C_i^g - \frac{M_i^g C_i^g g}{RT} \nabla z \qquad i = 1,\ldots,N_g \tag{26}$$

where χ_i^g [-] is the molar fraction of gas species A_i^g, and M_i^g [kg mol^{-1}] is the molecular weight of gas species A_i^g. The right hand side of equation 26 describes the driving force for diffusive transport of gas species A_i^g [66, 78]. The driving force is balanced by the friction between gas species A_i^g and all other gas species and the sediment particles as expressed by the left hand side of equation 26. The first term on the left hand side accounts for molecule-molecule interactions, with D_{ik}^g [m^2 s^{-1}] being the free phase binary diffusion coefficient between gas species A_i^g and A_k^g, and τ_g the gas tortuosity coefficient. The second term accounts for molecule-sediment interactions, with $D_i^{K,g}$ [m^2 s^{-1}] being the Knudsen diffusion coefficient for species A_i^g. The Knudsen diffusion coefficient can be calculated as a function of the Klinkenberg parameter [15].

When $D_{ik}^g >> D_i^{K,g}$, the first term in equation 26 becomes negligible, and the system is said to be in the Knudsen flow regime. In this regime, gas molecules do not interact with each other but only with the sediment particles. On the other hand, when gas molecules interact only with other gas molecules ($D_{ik}^g << D_i^{K,g}$), the system is in the molecular regime When D_{ik}^g and $D_i^{K,g}$ have similar magnitudes, the system is in the so-called transition regime, and both free phase and Knudsen diffusion processes contribute to the diffusive fluxes.

The total component concentrations (T_j^g) and diffusive fluxes ($\mathbf{N}_j^{T,g}$) can be calculated as a sum of the species concentrations (C_i^g) and diffusive fluxes ($\mathbf{N}_i^g$) weighted by the appropriate stoichiometric coefficients (ν_{ij}^g) [2, 15].

3.2.3. Example Application: Fate of Organic Contaminants in the Vadose Zone

A conceptual simulation of the attenuation of volatile organic contaminants is presented here to illustrate the capabilities of the enhanced code (Fig. **8**).

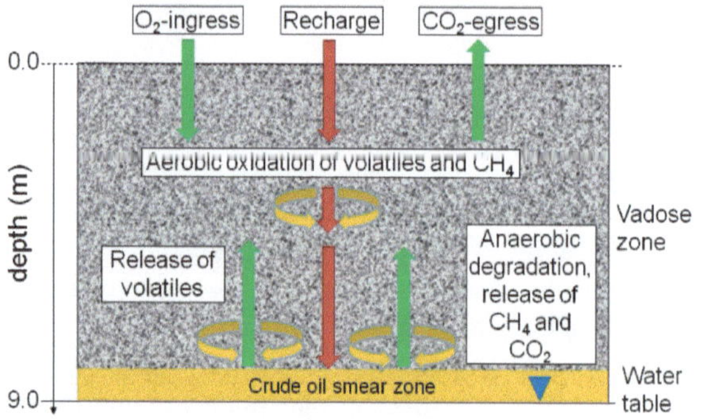

Figure 8: Conceptual model of major vadose zone reactive transport processes at a crude oil spill site.

This simulation is based on an earlier study of the oil spill site near Bemidji, MN [79] but is intended to be a generic case representing the conditions encountered at many sites affected by LNAPL contamination that are undergoing natural attenuation processes.

The simulation comprises a 9-m-deep one-dimensional column of a sandy vadose zone spanning from 20 cm above the water table to the ground surface. Porosity is 0.38 and permeability 5x10^{-12} m^2, with aqueous saturations ranging from 0.86 at the lower boundary to 0.24 at the surface. A high-saturation (0.54), low-porosity (0.25) area at a depth of 4 m hinders transport in the gas phase. At the lower boundary of the domain, a 30 cm thick region of residual oil phase is present. The different oil fractions are described by three relevant groups: 15% of a volatile fraction that is the source of VOC contamination at early stages of plume evolution, represented by C_4H_{10}; 25% of a less volatile, yet soluble fraction that provides a long term source driving biodegradation reactions, represented by C_7H_{14}; and a residual fraction that is assumed non-reactive and not soluble on the time scale of interest.

Both C_4H_{10} and C_7H_{14} degrade under aerobic and anaerobic conditions. Anaerobic degradation occurs initially by reductive dissolution of Fe(III)- and Mn(IV)-minerals that are present initially with a volume fraction of 2.5×10^{-4} and 5×10^{-5}, respectively. When these phases are depleted, anaerobic degradation proceeds by methanogenesis. To account for methanogenesis taking place in the smear zone below the domain of the current simulation, a CH_4 flux of 0.13 mol m^{-2} d^{-1} is applied at the lower boundary when methanogenesis in the domain is taking place. The main reaction products of anaerobic degradation processes are CH_4, CO_2, Fe(II), and Mn(II). Under aerobic conditions, organic contaminants and CH_4 can in turn be oxidized to form CO_2. The geochemical system used in the simulation includes 11 aqueous components, 7 gas equilibrium reactions (O_2(g), CH_4(g), CO_2(g), Ar(g), N_2(g), and C_7H_{14}(g), C_4H_{10}(g)), 15 aqueous complexation reactions, and 3 mineral reactions ($CaCO_3$, FeOOH, and MnO_2). The simulation is run for a period of 28 years.

Simulation results show distinct gas phase compositions in the source zone as the spill ages progressively (Fig. **9**). At early stages, the unsaturated zone gas plume is characterized by elevated VOC concentrations due to fast volatilization of C_4H_{10}. These concentrations decrease relatively rapidly over the 5 first years of the simulation until all C_4H_{10} in the oil is depleted. In contrast, CH_4 concentrations are relatively low initially, but increase significantly over time, as FeOOH and MnO_2 become depleted. Methanogenesis becomes the dominant pathway for oil degradation after 15 years leading to CH_4 concentrations of 20% near the oil body.

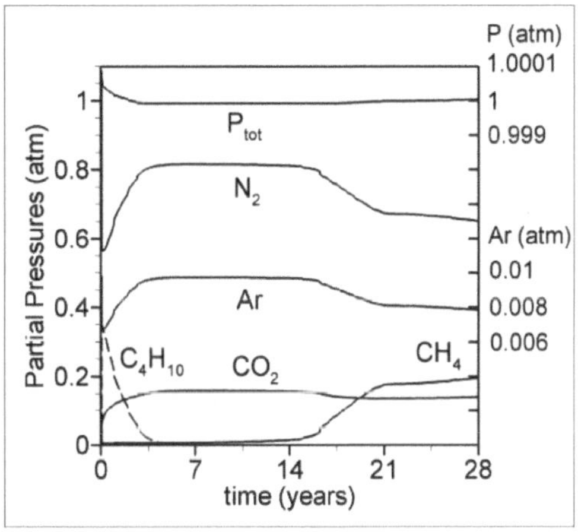

Figure 9: Evolution of gas composition overlying the smear zone.

Simultaneously to compositional changes, pressure in the gas phase evolves (Fig. **9**). At early stages, the large volume of volatiles causes an increase in the gas pressure. As C_4H_{10} concentrations decrease with time, pressure also decreases. However, when CH_4 concentrations increase due to methanogenesis, pressure builds up again marginally. Associated with pressure increases, a decrease in partial pressures of non-reactive gases such as N_2 and Ar is predicted, such that below-atmospheric concentrations of N_2 and Ar are found in the source zone. Pressure gradients generated in reactive processes highlight the role of advection in these environments: as contaminant concentrations increase significantly (C_4H_{10} and CH_4), they are transported away from the reaction zone not only by diffusion but also by advection.

The organic contaminants are transported away from the source zone and mix with atmospheric gases entering the vadose from the ground surface (Fig. **10**). Aerobic oxidation attenuates the concentrations of the contaminants emitted towards the atmosphere. At early stages, C_4H_{10} are reduced due to oxidation with O_2 between 2.5 and 4 m of depth. As a result, pressure decreases with respect to conditions deeper in the vadose zone, but also with respect to atmospheric levels. The region of lowered pressure drives advection toward the aerobic reaction zone from the source zone and the ground surface as indicated by the pressure gradients (Fig. **10**). This results in slight increases of N_2 concentrations relative to atmospheric values in the reaction zone.

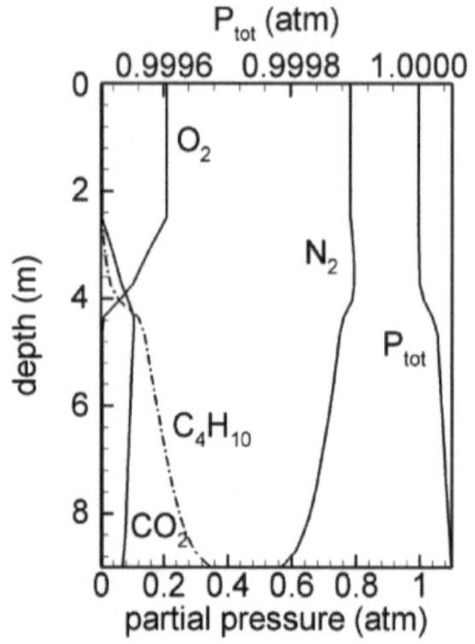

Figure 10: Vertical profile of gas composition in vadose zone at 0.1 years.

At later stages, a similar situation develops with CH_4 playing the role of C_4H_{10} (Fig.**11**). However, methane concentrations near the source zone are not as high and therefore the flow reversal pattern develops more clearly.

As proposed by Amos *et al.* [10] on the basis of field observations and modeling, concentrations of non-reactive gases such as N_2 can be used as proxy for reaction processes that produce or consume gases. As mentioned above, a similar pattern develops for N_2 at early stages but this has not been confirmed in the field.

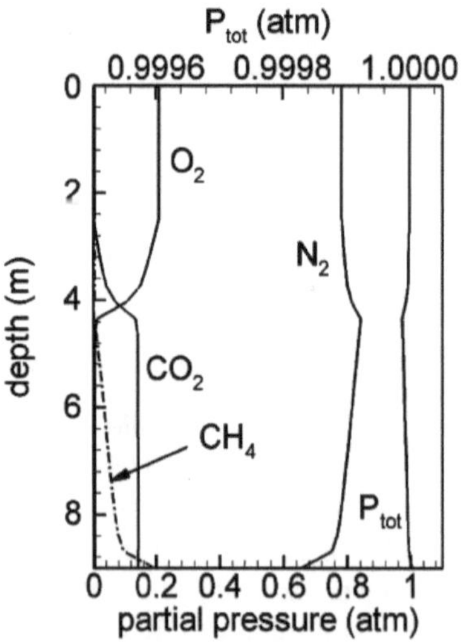

Figure 11: Vertical profile of gas composition in vadose zone at 28 years.

This demonstration example highlights the capabilities of the current model enhancement as it involves multicomponent mixtures of gases, including atmospheric gases and contaminants, which are subject to both advective and diffusive transport. The model allowed to confirm hypothesis and quantify observations made in the field by Amos *et al.* [10]. The enhanced code provides a versatile tool that can be applied to other unsaturated systems with different geochemical reaction networks, where gas generation and consumption play a role.

3.3. MIN3P-Soil: Plant-Soil Interactions

3.3.1. Motivation

A version of the MIN3P code is developed to simulate water and solute dynamics in the root zone of soils and the underlying vadose zone. To this end, several processes have been implemented including physical evaporation, plant transpiration, solute uptake by plants, and preferential flow (equilibrium scheme). The implementation of plant transpiration and preferential flow was required to accurately simulate soil moisture variations in a forest soil [11]. Plant uptake of solutes was added to support the modeling investigation of the Si-cycle [16]. The long-term goal of the development of MIN3P-Soil is to generate a code that links processes in the shallow soil zone with vadose zone reactive transport.

3.3.2. Model Formulation

To account for physical evaporation and transpiration, separate sink terms have been implemented in equation 1 to yield:

$$S_a S_s \frac{\partial h}{\partial t} + \phi \frac{\partial S_a}{\partial t} - \nabla \cdot \left[k_{ra} \mathbf{K} \nabla h \right] - Q_a - Q_e - Q_u = 0, \tag{27}$$

where Q_e [m^3 H_2O m^{-3} porous medium s^{-1}] is the evaporative flux and Q_u [m^3 H_2O m^{-3} porous medium s^{-1}] is the transpiration or uptake flux.

Physical Evaporation: To account for physical evaporation, the modified code includes a formulation that distributes the evaporative flux through a sink term to the relevant soil horizons. Actual evaporation is constrained by potential evaporation and soil moisture content. The transient (*e.g.* daily) distribution of potential evaporation (noted E_u, in m s^{-1}) can be specified through an external input file. Regions where physical evaporation can occur, such as the top soil horizon, are also user-defined. Actual physical evaporation is calculated based on the assumption that evaporation occurs as a one-dimensional process in vertical direction along a "soil column" l that contains j control volumes including control volume i. If water saturation in a control volume exceeds field capacity, the evaporative flux attributed to control volume i ($Q_{e,i}$ [m^3 H_2O s^{-1}]) is calculated as:

$$Q_{e,i} = E_u A_l \frac{V_i}{\sum_{j \in l} V_j} \tag{28}$$

where A_l is the area of soil column l exposed to atmospheric conditions at the ground surface (m^2), and V_j are the control volumes (m^3) located in soil column l.

If the local water saturation ranges between field capacity (S_a^f) and the saturation under air-dried conditions (S_a^{dry}), the local evaporation flux is reduced to:

$$Q_{e,i} = E_u \, A_l \left(\frac{S_{a,i} - S_a^{dry}}{S_a^f - S_a^{dry}} \right) \frac{V_i}{\sum_{j \in l} V_j} \tag{29}$$

For the case when water saturation is at or below saturations representative of air-dried conditions, the local evaporation flux is set to zero.

Plant Transpiration: For this biological process, the potential plant transpiration rate (noted T, in m s^{-1}) is calculated from the evaporative budget:

$$PET = T + E_u + F_i I \tag{30}$$

where *PET* is the potential evapotranspiration rate (m s^{-1}), *I* the rate of water interception by the plant canopy (m s^{-1}) and F_i (dimensionless) is a crop factor that represents the rate at which intercepted water evaporates. The values for *PET* and *I* are both provided through an external file (daily values can be specified). The actual transpiration rate is determined based on the soil moisture and is constrained by the potential plant transpiration rate. This allows accounting for the negative effect of water stress on plant transpiration. Three options have been implemented:

In the first case, the reduction function for water stress given by Battaglia and Sands [80] is averaged across the entire spatial domain:

$$\alpha = \frac{REW^2 \exp\left(p_1 REW\right)}{REW_0^2 \exp\left(p_1 REW_0\right) + REW^2 \exp(p_1 REW)} \tag{31}$$

where REW defines the reserve of extractable water, α is the ratio of the actual transpiration over the potential transpiration, and p_1 and REW_0 are calibrated dimensionless parameters with:

$$REW = \frac{S_a^{avg} - S_a^{\lim}}{S_a^f - S_a^{\lim}} \tag{32}$$

where S_a^{avg} is an average water saturation for the soil zones penetrated by roots and S_a^{lim} is the water saturation at the wilting point.

Accordingly, the uptake rate of water in control volume *i* contained in soil column *l* is calculated as:

$$Q_{u,i} = \alpha T A_l \left(\frac{V_i R_{k(i)} S_{a,i}}{\sum_{j \in l} V_j R_{k(j)} S_{a,j}} \right) \tag{33}$$

where $R_{k(i)}$ is the root length density assigned to control volume *i* (m root m^{-3} bulk).

In the second option, the reduction function α is calculated separately for each control volume. The reduction function becomes:

$$\alpha_{ij} = \frac{REW_{ij}^2 \exp(p_1 REW_{ij})}{REW_0^2 \exp(p_1 REW_0) + REW_{ij}^2 \exp(p_1 REW_{ij})} \tag{34}$$

with:

$$REW_i = \frac{S_{a,i} - S_a^{\lim}}{S_a^f - S_a^{\lim}} \tag{35}$$

and:

$$Q_{u,i} = \alpha_i T A_l \left(\frac{V_i R_{i(k)} S_{a,i}}{\sum_{j \in l} V_j R_{k(j)} S_{a,j}} \right) \tag{36}$$

This method is more rigorous from a porous media perspective, because it accounts for the spatial variability of the soil parameters. Conversely, it is a less rigorous method from a tree physiology point of view, because plant leaves are assumed to exhibit a range of stomatal closures as a function of the *REW* value.

For the third option, the reduction factor α is set to unity, implying that the effect of water stress is neglected.

Solute Uptake: This biological process is considered by means of an additional sink term in the mass balance equation 2, which becomes:

$$\frac{\partial}{\partial t}\left[S_a\phi T_j^a\right]+\frac{\partial}{\partial t}\left[S_g\phi T_j^g\right]+\frac{\partial}{\partial t}(T_j^s) \\ +\nabla\cdot\left[\mathbf{q}_a T_j^a\right]-\nabla\cdot\left[S_a\mathbf{D}_a\nabla T_j^a\right]-\nabla\cdot\left[S_g\mathbf{D}_g\nabla T_j^g\right]-Q_j^{a,a}-Q_j^{a,m}+Q_u\beta T_j^a=0 \quad j=1,N_c \tag{37}$$

where β is a unitless constant. When β equals unity, passive uptake is considered, implying that solute uptake occurs at the same rate as water uptake. When β is less or greater than unity, solute uptake is either rejective or active, respectively. Rejective uptake implies that solute is left behind during water uptake by plant roots, leading to build-up of solute in the root zone. Active uptake implies that solute is preferentially taken up by the plants, leading to solute depletion in the root zone and enhancing diffusive solute transport towards the plant roots.

Preferential Flow: The preferential flow scheme implemented into MIN3P-Soil is based on the work by Mohanty *et al.* [81], who proposed an equilibrium preferential flow scheme associated with an equivalent representation of preferential flow paths (*i.e.* the geometry of preferential flow paths and their relationships with soil matrix are not explicitly represented). According to this method, the following hydraulic conductivity function was implemented:

$$K^*(\psi)=K(\psi)+\kappa\left(\exp\left(\psi-\psi^{pf}\right)-1\right) \tag{38}$$

where $K(\psi)$ is the standard function of van Genuchten [20] expressed in terms of pressure head, κ is an additional hydraulic conductivity term referring to preferential flow paths (m s^{-1}), and ψ^{pf} is a threshold pressure head (m).

This method is attractive, because only two parameters are required to account for preferential flow. However, the simplicity of the formulation also implies limitations, which have been pointed out by Gérard *et al.* [11] and, more recently, by Köhne *et al.* [82].

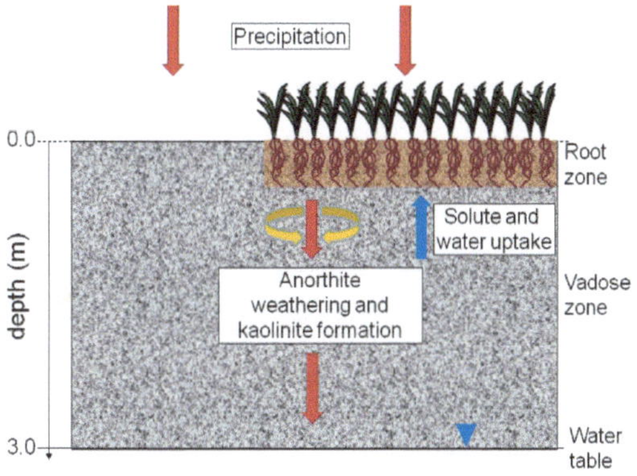

Figure 12: Conceptual model to illustrate the effect of plant water and solute uptake on reactive transport in a soil column. The optional root zone extends to a depth of 0.5 m.

3.3.3. Example Application: Plant Water and Solute Uptake

This example aims at illustrating the effect of water and solute uptake by plants on mineral weathering reactions. A 3 m deep homogeneous soil column with steady-state boundary conditions for flow and transport is considered (Fig. **12**). The boundary conditions for flow have been specified so that the 1-D medium remains unsaturated. The soil column contains a primary silicate, anorthite, and a secondary silicate, kaolinite. Dilute water (Al, Si and Ca concentrations of 10^{-8} mol L^{-1}) infiltrates the domain at a constant flow rate. For simplicity, the initial composition of the pore water is set to the same concentrations. Advection is the only solute transport process considered, while soil temperature is allowed to vary with time. Daily temperature variations such as those used in Gérard *et al.* [16] are considered in the simulations that follow.

Anorthite dissolution kinetics are assumed independent of pH, exhibit a first order behaviour with respect to chemical affinity, and vary with temperature (Arrhenius law). Kaolinite reaction kinetics are subjected to the same assumptions.

Simulations without plants and with plants are carried out by considering water uptake and various extents of solute uptake. Fig. **13** depicts the variations in the volume fraction of kaolinite vs. depth, as calculated by MIN3P-Soil after 1000 days with the no-plant-scenario and by considering different solute uptake fluxes. Fig. **14** presents the variations of the concentration of Si over time at 0.5 m depth; *i.e.* directly below the root zone. Results demonstrate that the influence of transpiration and plant solute uptake can affect reactive transport processes in an important way.

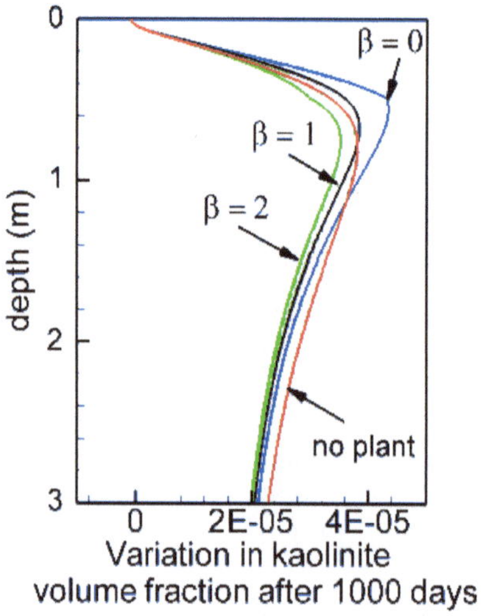

Figure 13: Influence of plant transpiration and solute uptake on the variations of kaolinite volume fraction as a function of soil depth. The red line represents the results calculated without considering plant processes. The black line stands for the results obtained by considering plant transpiration and passive solute uptake ($\beta = 1$). Blue lines stand for the results obtained with different extents of rejective uptake ($\beta < 1$). Green lines stand for the results obtained with different extents of active uptake ($\beta > 1$).

First, the addition of the transpiration process lowers the soil moisture, particularly in the root zone, and therefore decreases the flow rate in the unsaturated zone. As a result, water and solute residence time increase compared to the simulation performed without plants. Accordingly, the precipitation front of kaolinite occurs closer to the soil surface. Only a preferential or active uptake of Si with a β-value of 1.5 could compensate for this effect (see Fig. **13**).

Conversely, at a greater soil depth, less kaolinite is precipitated because anorthite, the source of Si and Al for kaolinite formation, also becomes saturated closer to the soil surface under the effect of plant transpiration and solute uptake. A preferential uptake of Si cannot compensate this effect as this process occurs in the top soil; the root zone.

Consistent with these results, larger Si concentrations are obtained at 0.5 m depth when plant processes are considered. Larger Si concentrations are observed during the entire simulation period with some exceptions encountered locally for large β-values (see Fig. **14**). Indeed, this example application clearly demonstrates how plant processes can alter solution chemistry over the short term and soil mineralogy over the long term. MIN3P-Soil and follow-up developments are designed to provide a tool for investigating these processes.

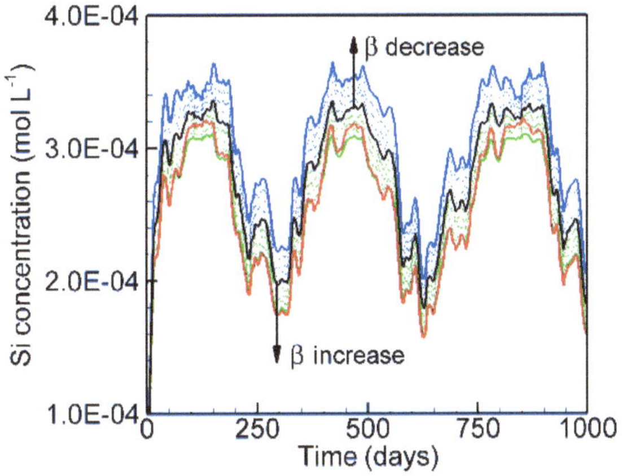

Figure 14: Influence of plant transpiration and solute uptake on Si-concentrations simulated at 0.5 m depth. The red line represents the results calculated without considering plant processes. The black line stands for the results obtained by considering plant transpiration and passive solute uptake ($\beta = 1$). Blue lines stand for the results obtained with different extents of rejective uptake ($\beta < 1$). Green lines stand for the results obtained with different extents of active uptake ($\beta > 1$).

4. CONCLUSIONS AND OUTLOOK

The reactive transport code MIN3P and its various follow-up developments provide a tool-box for a range of reactive transport problems with a focus on vadose zone applications. The code enhancements target specific applications that are not available in most standard reactive transport codes such as multicomponent gas diffusion and advection, the generation and fate of gas bubbles below the water table, gas bubble entrapment and release due to water table fluctuations, and plant-soil interactions. A variety of different hydrological, geochemical and biological processes can be considered and the code family has been and will continue to be applied to quantitatively investigate processes at field sites and in laboratory experiments. Current code development includes the implementation of the Pitzer ion interaction model for high ionic strength solutions and electrostatic adsorption models.

REFERENCES

[1] C. I. Steefel, D. J. DePaolo, and P.C. Lichtner, "Reactive transport modeling: An essential tool and a new research approach for the Earth sciences", *Earth and Planetary Science Letters*, vol. 240, pp. 539-558, 2005.

[2] K. U. Mayer, E. O. Frind, and D. W. Blowes, "Multicomponent reactive transport modeling in variably saturated porous media using a generalized formulation for kinetically controlled reactions", *Water Resources Research*, vol. 38, pp. 1174, doi: 10:1029/2001WR000862, 2002.

[3] J. Šimůnek and D. L. Suarez, "Two dimensional transport for variably saturated porous media with major ion chemistry", *Water Resources Research*, vol. 30, pp. 1115-1134, 1994.

[4] P. C. Lichtner, Continuum formulation of multi-component multiphase reactive transport. Ch. 1 in: Reactive Transport in Porous Media, Eds.: Lichtner, P. C., C. I. Steefel and E. H. Oelkers, *Reviews in Mineralogy*, Vol. 34, Mineralogical Society of America. Washington, DC, 1996.

[5] M. D. Wunderly, D. W. Blowes, E. O. Frind, and C. J. Ptacek, "Sulfide mineral oxidation and subsequent reactive transport of oxidation products in mine tailings impoundments: A numerical model", *Water Resources Research*, vol, 32, pp. 3173-3187, 1996.

[6] T. Xu and K. Pruess, "Modeling multiphase non-isothermal fluid flow and reactive geochemical transport in variably saturated fractured rock: 1. Methodology", *American Journal of Science*, vol. 301, pp. 16-33, 2001.

[7] M. W. Saaltink, F. Batlle, C. Ayora, J. Carrera, and S. Olivella, "RETRASO, a code for modeling reactive transport in saturated and unsaturated porous media", *Geologica Acta*, vol. 2, pp. 235-251, 2004.

[8] C. Linklater, D. Sinclair, and P. Brown, "Coupled chemistry and transport modeling of sulphidic waste rock dumps at the Aitik mine site, Sweden", *Applied Geochemistry*, vol. 20, pp. 275-293, 2005.

[9] P. Acero, C. Ayora, J. Carrera, M. W. Saaltink, and S. Olivella, "Multiphase flow and reactive transport in vadose tailings", *Applied Geochemistry*, vol. 24, pp. 1238-1250, 2009.

[10] R. T. Amos, K. U. Mayer, B. A. Bekins, G. N. Delin, and R. L. Williams, "Use of dissolved and vapor phase gases to investigate methanogenic degradation of petroleum hydrocarbon contamination in the subsurface", *Water Resources Research*, vol. 41, W02001, doi:10.1029/2004WR00 3433, 2005.

[11] F. Gérard, M. Tinsley, and K. U. Mayer, "Preferential flow revealed by hydrologic modeling based on predicted hydraulic properties and intensive water content monitoring", *Soil Sciences Society of America Journal*, vol. 68, pp. 1526-1538, 2004.

[12] L. Cheng, Dual porosity reactive transport modeling, Ph.D thesis, University of Sheffield, Sheffield, U.K., 2005.

[13] R. T. Amos and K. U. Mayer, "Investigating the role of gas bubble formation and entrapment in contaminated aquifers: Reactive transport modeling", *Journal of Contaminant Hydrology*, vol. 87, pp. 123-154, 2006a.

[14] R. T. Amos and K. U. Mayer, "Investigating ebullition in a sand column using dissolved gas analysis and reactive transport modeling", *Environmental Science & Technology*., vol. 40, pp. 5361-5367, 2006b.

[15] S. Molins and K. U. Mayer, "Coupling between geochemical reactions and multi-component gas diffusion and advection A reactive transport modeling study", *Water Resources Research*, vol. 43, W05435, doi:10.1029/2006WR005206, 2007.

[16] F. K. Gérard, U. Mayer, M. J. Hodson, and J. Ranger, "Modeling the biogeochemical cycle of silicon in soils: application to a temperate forest ecosystem", *Geochimica et Cosmochimica Acta*, vol. 72, pp. 741-758, 2008.

[17] T. H. Henderson, K. U. Mayer, B. L. Parker, and T. A. Al., "Three-dimensional density-dependent flow and multicomponent reactive transport modeling of chlorinated solvent oxidation by potassium permanganate", *Journal of Contaminant Hydrology*, vol. 106, pp. 183-199, 2009.

[18] S. P. Neuman, "Saturated-unsaturated seepage by finite elements", *Journal of Hydrology Division. American Society Civil Engineers*, vol. 99, pp. 2233-2250, 1973.

[19] S. Panday, P. S. Huyakorn, R. Therrien, and R. L. Nichols, "Improved three-dimensional finite-element techniques for field simulation of variably-saturated flow and transport", *Journal of Contaminant Hydrology*, vol. 12, pp. 3-33, 1993.

[20] M. T. Van Genuchten, "A closed form equation for predicting the hydraulic conductivity of unsaturated soils", *Soil Sciences Society of America Journal*. vol. 44, pp. 892-898, 1980.

[21] G. T. Yeh and V. S. Tripathi, "A critical evaluation of recent developments in hydrogeochemical transport models of reactive multi-chemical components", *Water Resources Research*, vol. 25, pp. 93-108, 1989.

[22] C. I. Steefel, A. C. Lasaga, "A coupled model for transport of multiple chemical species and kinetic precipitation/ dissolution reactions with application to reactive flow in single phase hydrothermal systems", *American Journal of Science*, vol. 294, pp. 529-592, 1994.

[23] J. E. vanderKwaak, P. A. Forsyth, K. T. B. MacQuarrie, and E. A. Sudicky, WatSolv - Sparse Matrix Iterative Solver, user's guide for version 2.16, Univ Waterloo, Waterloo, Ontario, Canada, 1997.

[24] A. D. Brookfield, W. Blowes, and K. U. Mayer, "Integration of field measurements and reactive transport modeling to evaluate contaminant transport at a sulfide mine tailings impoundment", *Journal of Contaminant Hydrology*, vol. 88, pp. 1-22, 2006.

[25] J. Molson, M. Aubertin, B. Bussiere, and M. Benzaazoua, "Geochemical transport modeling of drainage from experimental mine tailings cells covered by capillary barriers", *Applied Geochemistry*, vol. 23, pp. 1-24, 2008.

[26] M. Ouangrawa, J. Molson, M. Aubertin, B. Bussiere, and G. J. Zagury, "Reactive transport modeling of mine tailings columns with capillarity-induced high water saturation for preventing sulfide oxidation", *Applied Geochemistry*, vol. 24, pp. 1312-1323, 2009.

[27] K. U. Mayer, S. G. Benner, E. O. Frind, S. F. Thornton, and D. L. Lerner, "Reactive transport modeling of processes controlling the distribution and natural attenuation of phenolic compounds in a deep sandstone aquifer", *Journal of Contaminant Hydrology*, vol. 53, pp. 341-368, 2001a.

[28] G. R. Miller, Y. Rubin, K. U. Mayer, and P. H. Benito, "Modeling vadose zone processes during land application of food-processing waste water in California's Central Valley", *Journal of Environmental Quality*, vol. 37, pp. S-43-S-57, 2008.

[29] K. U. Mayer, D. W. Blowes, and E. O. Frind, "Reactive transport modeling of groundwater remediation by an *in situ* reactive barrier for the treatment of hexavalent chromium and trichloroethylene", *Water Resources Research*, vol. 37, pp. 3091-3103, 2001b.

[30] R. T. Amos, K. U. Mayer, D. W. Blowes, and C. J. Ptacek, "Reactive transport modeling of column experiments for the remediation of acid mine drainage", *Environmental Science & Technology*, vol. 38, pp. 3131-3138, 2004.

[31] K. U. Mayer, S. G. Benner, and D. W. Blowes, "Process-based reactive transport modeling of a permeable reactive barrier for the treatment of mine drainage", *Journal of Contaminant Hydrology*, vol. 85, pp. 195-211, 2006.

[32] S.-W. Jeen, K. U. Mayer, R. W. Gillham, and D. W. Blowes, "Reactive transport modeling of trichloroethene treatment with declining reactivity of iron", *Environmental Science & Technology*, vol. 41, pp. 1432-1438, 2007.

[33] S. M. Spiessl, K. T. B. MacQuarrie, and K. U. Mayer, "Identification of key parameters controlling dissolved oxygen migration and attenuation in fractured crystalline rocks", *Journal of Contaminant Hydrology*, vol. 95, pp. 141-153, 2008.

[34] K. T. B. MacQuarrie, K. U. Mayer, B. Jin, and S. M. Spiessl, "The importance of conceptual models in the reactive transport simulation of oxygen ingress in sparsely fractured crystalline rock", *Journal of Contaminant Hydrology*, vol. 112, pp. 64-76, 2010.

[35] A. L. Walter, E. O. Frind, D. W. Blowes, C. J. Ptacek, and J. W. Molson, "Modeling of multicomponent reactive transport in groundwater. 2. Metal mobility in aquifers impacted by acidic mine tailings discharge", *Water Resources Research*, vol. 30, pp. 3149-3158, 1994.

[36] M. D. Williams and M. Oostrom, "Oxygenation of anoxic water in a fluctuating water table system: an experimental and numerical study". *Journal of Hydrology*, vol. 230, pp. 70-85, 2000.

[37] M. C. Ryan, K. T. B. MacQuarrie, J. Harman, and J. Mclellan, "Field and modeling evidence for a "stagnant flow" zone in the upper meter of sandy phreatic aquifers", *Journal of Hydrology*, vol. 233, pp. 223-240, 2000.

[38] C. W. Beckwith and A. J. Baird, "Effect of biogenic gas bubbles on water flow through poorly decomposed blanket peat". *Water Resources Research*, vol. 37, pp. 551-558, 2001.

[39] W. D. Reynolds, D. A. Brown, S. P. Mathur, R. P. Overend, "Effect of *in situ* gas accumulation on the hydraulic conductivity of peat", *Soil Science*, vol. 153, pp. 397-408, 1992.

[40] S. C. Whalen, "Biochemistry of methane exchange between natural wetlands and the atmosphere", *Environmental Engineering Science*, vol. 22, pp. 73-94, 2005.

[41] B. P. Walter and M. Heimann, "A process-based, climate-sensitive model to derive methane emissions from natural wetlands: Application to five wetland sites, sensitivity to model parameters, and climate", *Global Biogeochemical Cycles*, vol. 14, pp. 745-765, 2000.

[42] K. Revesz, T. B. Coplen, M. J. Baedecker, P. Glynn, M. Hult, "Methane production and consumption monitored by stable H and C isotope ratios at a crude oil spill site, Bemidji, Minnesota", *Applied Geochemistry*, vol. 10, pp. 505-516, 1995.

[43] N. P. M. Fortuin and A. Willemsem, "Exsolution of nitrogen and argon by methanogenesis in Dutch ground water", *Journal of Hydrology*, vol. 301, pp. 1-13, 2005.

[44] G. Blicher-Mathiesen, G. W. McCarty, and L. P. Nielsen, "Denitrification and degassing in groundwater estimated from dissolved dinitrogen and argon", *Journal of Hydrology*, vol. 208, pp. 16-24, 1998.

[45] W. Aeschbach-Hertig, F. Peeters, U. Beyerle, and R. Kipfer, "Interpretation of dissolved atmospheric noble gases in natural waters", *Water Resources Research*, vol. 35, pp. 2779-2792, 1999.

[46] C. J. Ballentine and C. M. Hall, "Determining paleotemperature and other variables by using an error-weighted, nonlinear inversion of noble gas concentrations in water", *Geochimica Cosmochimica Acta*, vol. 63, pp. 2315-2336, 1999,

[47] A. H. Manning, and D. K. Solomon, "Using noble gases to investigate mountain-front recharge", *Journal of Hydrology*, vol. 275, pp. 194-207, 2003.

[48] P. Schlosser, M. Stute, C. Dörr, C. Sonntag, and K. O. Münnich, "Tritium/3He-dating of shallow groundwater", *Earth and Planetary Science Letters*, vol. 89, pp. 353- 362, 1988.

[49] E. Busenberg and N. L. Plummer, "Dating young groundwater with sulfur hexafluoride: Natural and anthropogenic sources of sulfur hexafluoride", *Water Resources Research*, vol. 36, pp. 3011- 3030, 2000.

[50] W. Aeschbach-Hertig, F. Peeters, U. Beyerle, and R. Kipfer, "Palaeotemperature reconstruction from noble gases in ground water taking into account equilibration with entrapped air", *Nature*, vol. 405, pp. 1040-1044, 2000.

[51] X. Li and Y. C. Yortsos, "Theory of multiple bubble growth in porous media by solute diffusion", *Chemical Engineering Science*, vol. 50, pp. 1247-1271, 1995.

[52] O. A. Cirpka and P. K. Kitanidis, "Transport of volatile compounds in porous media in the presence of a trapped gas phase", *Journal of Contaminant Hydrology*, vol. 49, pp. 263-285, 2001.

[53] J. C. Parker and R. J. Lenhard, "A model for hysteretic constitutive relations governing multiphase flow 1. Saturation-pressure relations". *Water Resources Research*, vol. 23, pp. 2187-2196, 1987.

[54] J. J. Kaluarachchi and J. C. Parker, "Multiple flow with a simplified model of oil entrapment", *Trans. Por. Med.*, vol. 7, pp. 1-14, 1992.

[55] J. H. M. Wösten and M. T. van Genuchten, "Using texture and other soil properties to predict the unsaturated soil hydraulic functions", *Soil Sciences Society of America Journal*, vol. 52, pp. 1762-1770, 1988.

[56] V. A. Fry, J. S. Selker, and S. M. Gorelick, "Experimental investigations for trapping oxygen gas in saturated porous media for *in situ* bioremediation", *Water Resources Research*, vol. 33, pp. 2687-2696, 1997.

[57] R. F. Weiss, "The solubility of nitrogen, oxygen and argon in water and seawater", *Deep-Sea Research*, vol. 17, pp. 721-735, 1970.

[58] CRC Press, CRC Handbook of Chemistry and PhysicsI, 84th ed., Boca Raton, Fla., 2004.

[59] A. De Visscher, D. Thomas, P. Boeckx, and O. Van Cleemput, "Methane Oxidation in Simulated Landfill Cover Soil Environments", *Environmental Science & Technology*, vol. 33, pp. 1854-1859, 1999.

[60] C. Scheutz and P. Kjeldsen, "Capacity for Biodegradation of CFCs and HCFCs in a Methane Oxidative Counter-Gradient Laboratory System Simulating Landfill Soil Covers", *Environmental Science & Technology*, vol. 37, pp. 5143-5149, 2003.

[61] A. I. M. Ritchie, Oxidation and Gas Transport in Piles of Sulfidic Material in Environmental Aspects of Mine Wastes, eds. J.L. Jambor, D.W. Blowes and A.I.M. Ritchie, Mineralogical Association of Canada, Short Course Series, vol. 31, Vancouver, BC, 2003.

[62] S. P. White, "Multiphase Nonisothermal Transport of Systems of Reacting Chemicals", *Water Resources Research*, vol. 31, pp. 1761-1772, 1995.

[63] T. Xu, S. P. White, K. Pruess, and G. H. Brimhall, "Modeling of Pyrite Oxidation in Saturated and Unsaturated Subsurface Flow Systems", *Trans Por. Med*, vol. 39, pp. 25-56, 2000.

[64] C. I. Steefel, Evaluation of the field-scale cation exchange capacity of Hanford sediments in Proceedings of the 11th International Symposium on Water-Rock Interaction, R.B. Wanty and R.R. Seal (eds.), Taylor and Francis Group, London, pp. 999-1002, 2004.

[65] P. C. Lichtner, S. Yabusaki, K. Pruess, and C. I. Steefel, "Role of Competitive Cation Exchange on Chromatographic Displacement of Cesium in the Vadose Zone beneath the Hanford S/SX Tank Farm", *Vadose Zone Journal*, vol. 3, pp. 203-219, 2004.

[66] D. C. Thorstenson and D. W. Pollock, "Gas-Transport in Unsaturated Zones – Multicomponent Systems and the Adequacy of Fick's Law", *Water Resources Research*, vol. 25, pp. 477-507, 1989.

[67] J. Massmann and D. F. Farrier, "Effects of Atmospheric Pressures on Gas Transport in the Vadose Zone", *Water Resources Research*, vol. 28, pp. 777-791, 1992.

[68] J. W. Massmann, "Applying groundwater flow models in vapor extraction system design", *Journal of Environmental Engineering*, vol. 115, pp. 129-149, 1989.

[69] R. W. Falta, K. Pruess, I. Javandel and P. A. Witherspoon, "Numerical Modeling of Steam Injection for the Removal of Nonaqueous Phase Liquids from the Subsurface. 1. Numerical Formulation. *Water Resources Research*, vol. 28, pp. 433-449, 1992.

[70] R. W. Falta, I. Javandel, K. Pruess, and P. Witherspoon, Density-driven flow of gas in the unsaturated zone due to evaporation of volatile compounds", *Water Resources Research*, vol. 25, pp. 2159-2169, 1989.

[71] C. Mendoza and E. Frind, "Advective-Dispersive Transport of Dense Organic Vapors in the Unsaturated Zone. 1. Model Development", *Water Resources Research*, vol. 26, pp. 379-387, 1990.

[72] P. Gaganis, P. Kjeldsen, and V. N. Burganos, "Modeling Natural Attenuation of Multicomponent Fuel Mixtures in the Vadose Zone: Use of Field Data and Evaluation of Biodegradation Effects", *Vadose Zone Journal*, vol. 3, pp. 1262-1275, 2004.

[73] I. Hers, R. Zapf-Gilje, D. Evans, and L. Li, "Comparison, Validation and Use of Models for Predicting Indoor Air Quality from Soil and Groundwater Contamination", *Soil and Sediment Contamination*, vol. 11, pp. 491-527, 2002.

[74] L. V. Abreu and P. C. Johnson, "Effect of Vapor Source-Building Separation and Building Construction on Soil Vapor Intrusion as Studied with a Three-Dimensional Numerical Model", *Environmental Science & Technology*, vol. 39, pp. 4550-4561, 2005.

[75] M. A. Celia and P. Binning, "A mass conservative numerical solution for two-phase flow in porous media with application to unsaturated flow", *Water Resources Research*, vol. 28, pp. 2819-2828, 1992.

[76] J. C. Parker, R. J. Lenhard, and T. Kuppusamy, "A Parametric Model For Constitutive Properties Governing Multiphase Flow In Porous Media", *Water Resources Research*, vol. 23, pp. 618-624, 1987.

[77] E. A. Mason and A.P. Malinauskas, Gas transport in porous media: The dusty-gas model, Chemical engineering Monographs 17. Elsevier Science Publishers, Amsterdam, 1983.

[78] B. E. Sleep, "Modeling transient organic vapor transport in porous media with the Dusty Gas model", *Advances in Water Resources*, vol. 22, pp. 247-56, 1998.

[79] S. Molins, K. U. Mayer, R. T. Amos, and B. A. Bekins, "Vadose zone attenuation of volatile organic compounds at a crude oil spill site - Interactions between multicomponent gas transport and biogeochemical reactions", *Journal of Contaminant Hydrology*., vol. 112, pp. 15-29, 2010.

[80] M. Battaglia and P. Sands, "Modeling site productivity of Eucalyptus Globulus in response to climatic and site factors". *Australian Journal of Plant Physiology.*, vol. 24, pp. 831-850, 1997.

[81] B. P. Mohanty, R. S. Bowman, J. M. H. Hendrick, and M. T. van Genuchten, "New piecewise-continuous hydraulic functions for modeling preferential flow in an intermittent flood-irrigated field", *Water Resources Research*, vol. 33, pp. 2049-2063, 1997.

[82] J. M. Köhne, S. Köhne, and Simunek J., "A review of model applications for structured soils: a) Water flow and tracer transport", *Journal of Contaminant Hydrology*, vol. 104, pp. 4-35, 2009.

CHAPTER 9

Overview of NUFT: A Versatile Numerical Model for Simulating Flow and Reactive Transport in Porous Media

Y. Hao*, Y. Sun, and J. J. Nitao

Lawrence Livermore National Laboratory, Livermore, CA, USA

Abstract: Sophisticated and robust numerical modeling is essential to developing a good understanding of complex physical and chemical phenomena in the subsurface. In this chapter we provide a general overview of NUFT (**N**onisothermal **U**nsaturated-saturated **F**low and **T**ransport) code, which is a highly flexible computer software package for modeling multiphase, multi-component heat and mass flow and reactive transport in unsaturated and saturated porous media. An integrated finite difference method is used for numerical discretization. Several mathematical models are implemented in order to address various flow and reactive transport processes in porous media. The governing equations for each sub-model are solved by implicit time-integration. In particular a globally implicit approach is employed to solve transport and reaction equations simultaneously. The code is designed based on object-oriented principles, and equipped with efficient solvers and massively parallel computation capability. We present two examples involving reactive transport modeling to demonstrate capabilities of the code.

Keywords: NUFT, multiphase and multicomponent flow, reactive transport, porous media, numerical model.

1. INTRODUCTION

Coupled flow and reactive transport phenomena often occur in a wide range of subsurface systems. It has long been appreciated that a numerical model would provide a fundamental understanding of various combined physical and chemical processes in geologic media as well as quantified performance and risk assessment for many engineering applications ranging from groundwater remediation to geologic CO_2 sequestration and storage. For this reason, numerical modeling of subsurface flow and reactive transport has attracted considerable attention and a variety of numerical techniques and computational tools have been proposed and developed, especially as computer power and processing capabilities have significantly increased in the past decades [1-12].

In this chapter we present an overview of NUFT (**N**onisothermal **U**nsaturated-saturated **F**low and **T**ransport) code, a highly flexible computer software package for modeling multiphase, multi-component heat and mass flow and reactive transport in unsaturated and saturated porous media [2-3]. This computer code has been developed over decades by the third author of this chapter, and represents a current state-of-the-art in subsurface flow and reactive transport modeling.

To account for a variety of flow and reactive transport processes in porous media several submodels are implemented in the NUFT software package, which are:

- UCSAT – unconfined and confined aquifer, saturated flow module;
- US1P – unsaturated single phase flow module (Richard's equation);
- USNT – fully coupled unsaturated multiple phases, multiple components flow module with isothermal and nonisothermal options;
- US1C – unsaturated single component transport module;
- TRANS – geochemical multiphase transport module;
- RESHEAT – electrical resistive heating module.

*Address correspondence to Y. Hao: Lawrence Livermore National Laboratory, Livermore, CA, USA; Tel: 1-925-422-9657; Email: hao1@llnl.gov.

It is important to keep in mind that each of above-mentioned modules can either work as a stand-alone model or collaborate with another sub-model through internal inter-module data exchange/transfer. For example, in order to perform flow and reactive transport simulations an explicit coupling between the flow and reactive transport models is achieved by solving them sequentially.

The NUFT framework is written in the C^{++} language, and designed based on an object-oriented approach. This helps transform the code into a highly efficient and flexible numerical modeling tool, which is easy to maintain and extend. The NUFT code is capable of running on major operating platforms such as Microsoft Windows, Linux, and various versions of UNIX systems. In addition to conventional serial computing NUFT has been designed for running in a parallel fashion, which allows for large-scale complex simulations on high-performance massive parallel processing systems.

The code has been widely used for numerical modeling of subsurface multiphase flow and reactive transport processes. Application examples include geological disposal of nuclear waste [13-16], CO_2 geologic sequestration and storage [17-21], groundwater monitoring and remediation [22-29], and subsurface hydrocarbon production [30], to name a few.

The rest of the chapter is organized as follows. We will first provide a general overview of mathematical formulations implemented in the NUFT code. The above-mentioned sub-modules will be addressed respectively. This is followed by an outline of numerical solution as well as software design and architecture. Finally we present two numerical examples associated with flow and reactive transport modeling to illustrate the capabilities of the code.

2. MATHEMATICAL FOMULATION

It is assumed that the porous medium is stationary and the solid matrix is nondeformable. The mathematical equations used to describe flow and transport processes in porous media are based on the principle of mass, momentum and energy conservation. Darcy's law is the well-known approximate momentum balance for fluid flow through a porous medium. Note that all the balance equations are presented at a macroscopic level, and obtained by averaging system variables over a representative elementary volume (REV) [31].

In view of the fact that multiphase multicomponent flow and reactive transport problems often draw great interest for many practical applications, and also pose more significant modeling challenges here we intend to more focus on mathematical implementations of USNT (multiphase flow) and TRANS (reactive transport) models in NUFT. However, for completeness we also provide brief review of saturated (UCSAT), variably saturated (US1P) flow, and single-component solute transport (US1C) modules.

2.1. Confined and Unconfined Aquifer, Saturated Flow Module (UCSAT MODUEL)

Commonly used for hydraulic groundwater modeling, the mathematical formulation for mass transport of a single fluid phase in porous media can be written in a general form as [32],

$$S_0 \frac{\partial h}{\partial t} = -\nabla \cdot \mathbf{q} + \Gamma \,, \tag{1}$$

where t, ∇, and Γ respectively represent time, spatial derivative operator, and source/sink terms. S_0 is the specific storativity that depends on compressibility of fluid phase and porous media, and the Darcy flux vector $\mathbf{q}$ is expressed as:

$$\mathbf{q} = -\mathbf{K} \cdot \nabla h \,, \tag{2}$$

in which h is piezometric head, and $\mathbf{K}$ denotes hydraulic conductivity tensor.

In NUFT a single-phase flow in either confined or unconfined aquifer can be handled by solving appropriate variant or extension of equation (1) along with proper specifications of boundary conditions. More detailed descriptions can be found in [31, 33].

2.2. Variably Saturated Flow (Richard's Equation) Model (US1P MODULE)

The governing equation for the flow of a single fluid-phase in the saturated-unsaturated zone, also referred to as Richard's Equation, is implemented in the US1P module,

$$\phi \frac{\partial S}{\partial t} = -\nabla \cdot \mathbf{q} + \Gamma , \tag{3}$$

$$\mathbf{q} = -\mathbf{K} \cdot \nabla(\Psi + z) \tag{4}$$

Here ϕ is the porosity, S is the fluid phase saturation, z denotes vertical coordinate, and ψ represents the pressure head which equals to $p/\rho g$ with p, ρ, g indicating pressure, fluid phase density, and acceleration due to gravity. For a flow in an unsaturated zone both hydraulic conductivity **K** and pressure head ψ are functions of phase saturation S, which are defined by relative-permeability- and capillary-pressure-saturation curves. Thus treated as the primary variable, the saturation is readily obtained by solving equation (3) with the substitution of equation (4). However, equation (3) can not be used directly to address fully saturated flow (S = 1), and must therefore be rewritten in another form [31]-[32],

$$\phi \frac{\partial \rho}{\partial t} = -\nabla \cdot (\rho \mathbf{q}) + \Gamma \tag{5}$$

Since the fluid density is known as a function of pressure $\rho = \rho(\Psi)$ due to fluid compressibility, the pressure head can be uniquely defined in equation (5), and consequently selected as the primary variable for the solution. For combined saturated-unsaturated flow modeling both equations (3) and (5) are considered, and an adaptive technique enabling primary variable switching between phase saturation and pressure head is developed in NUFT in order to effectively handle the phase transition between saturated and unsaturated flow conditions and ponded boundary conditions.

2.3. Multiphase and Multicomponent Flow and Transport Model (USNT MODULE)

The USNT module is designed to solve multiphase and multicomponent flow and heat and mass transport problems. Additionally it has the option to deal with kinetic-controlled chemical reactions that are fully coupled with flow and transport balance equations. Essentially the USNT model is formulated as a compositional model in which an arbitrary number of components and phases can be solved based on mass conservation for each of the components. According to the extended Gibbs phase rule [32, 34] for a porous medium with nondeformable solid matrix, the total number of degrees of freedom or primary variables is,

under non-isothermal conditions,

$$NF = NC - NP + 2 + (NP - 1) = NC + 1 \tag{6}$$

under isothermal conditions,

$$NF = NC \tag{7}$$

with NC denoting the number of primary (basis) components (species), NP the number of fluid phases. Note that the phase rule (6) or (7) is also applicable to a chemical reaction system. When equilibrium reactions are involved we have the number of primary components/species reduced to,

$$NC = NS - NR_{eq}$$

where NS is the total number of components (species) in the chemical system, and NR_{eq} represents the number of equilibrium reactions. More discussion follows when reactive transport (TRANS) module is reviewed.

Therefore, in order to define the thermo-dynamic state of a non-isothermal flow and (or) chemical system we must choose NC+1 independent primary variable, and as well identify NC+1 corresponding balance equations. The governing equations for compositional model consist of mass balance equations for each of NC components, and energy balance equation if non-isothermal conditions are considered. The mass balance equation for γ–component is expressed in a general form,

$$\frac{\partial}{\partial t}\left[\sum_{\alpha}\phi\rho_{\alpha}S_{\alpha}\omega_{\alpha}^{\gamma}+(1-\phi)\rho_{s}K_{d}\omega_{\alpha}^{\gamma}\right]=q_{adv}+q_{disp}+q_{diff}+\Gamma^{\gamma}+\Gamma_{react}^{\gamma} \tag{8}$$

where

$$q_{adv}=-\sum_{\alpha}\nabla\cdot\phi\rho_{\alpha}S_{\alpha}\omega_{\alpha}^{\gamma}\mathbf{V}_{\alpha}\qquad q_{disp}=-\sum_{\alpha}\nabla\cdot\phi\rho_{\alpha}S_{\alpha}\mathbf{J}_{h\alpha}^{\gamma}$$
$$q_{diff}=-\sum_{\alpha}\nabla\cdot\phi\rho_{\alpha}S_{\alpha}\mathbf{J}_{\alpha}^{\gamma}\qquad \Gamma^{\gamma}=-\sum_{\alpha}\lambda_{\alpha}^{\gamma}\phi\rho_{\alpha}S_{\alpha}\omega_{\alpha}^{\gamma}$$
$$\Gamma_{react}^{\gamma}=\sum_{r}\Gamma_{r}^{\gamma}\qquad \alpha=1\cdots NP\qquad \gamma=1\cdots NC$$
$$r=1\cdots NR_{kin}$$

The subscripts α and r, and superscript γ denote α–phase of fluid, r-kinetic reaction, and γ–component, respectively. NR_{kin} is the number of kinetic reactions. Here, and in what follows, $\mathbf{V}_{\alpha}$, ρ_{α} and S_{α} are flow velocity, mass density and saturation of an α–phase fluid, ω_{α}^{γ} is mass fraction of a γ–component in an α–phase. While the left hand side of equation (8) corresponds to mass accumulation rate of γ–component summed over all the phases, on the right hand side are the overall mass advective q_{adv}, hydrodynamic dispersive q_{disp} and diffusive q_{diff} fluxes along with source terms Γ^{γ} and Γ^{γ}_{react} due to adsorption-desorption processes and kinetic chemical reactions, respectively. Fick's Law is used to represent hydrodynamic dispersive and diffusive fluxes as:

$$\mathbf{J}_{h\alpha}^{\gamma}=-\mathbf{D}_{h\alpha}^{\gamma}\cdot\nabla\omega_{\alpha}^{\gamma}$$

and

$$\mathbf{J}_{\alpha}^{\gamma}=-D_{\alpha}^{\gamma}\nabla\omega_{\alpha}^{\gamma}$$

where $\mathbf{D}_{h\alpha}^{\gamma}$ and D_{α}^{γ} are hydrodynamic dispersion tensor and molecular diffusive coefficient, respectively.

The use of Darcy's law for each phase yields,

$$\phi S_{\alpha}\mathbf{V}_{\alpha}=-\frac{k_{\alpha}(S_{\alpha})}{\mu_{\alpha}}(\nabla p_{\alpha}+\rho_{\alpha}\mathbf{g}) \tag{9}$$

where k_{α}, μ_{α} and $\mathbf{g}$ denote the permeability function, phase viscosity, and gravitational acceleration vector, respectively. The capillary pressure relationships are given as:

$$P_{\alpha}=P_{g}-P_{ca}(S_{\alpha})\qquad \alpha\neq g$$

in which P_{ca} is the retention pressure function. The subscript g represents the least wetting phase, normally gas phase. It is observed that both permeability and capillary pressure are dependent on phase saturation. Such dependence is able to be quantified by using empirical correlations and models such as Brooks and Corey [35], and van Genuchten [36] models. For example, the relationships between permeability, saturation, and capillary pressure can be derived based on the van Genuchten formulation as, for capillary pressure,

$$P_{ca} = \frac{1}{\alpha}\left(S_{e,l}^{-\frac{1}{m}} - 1\right)^{\frac{1}{n}}$$

for relative permeability of a liquid phase,

$$k_{r,l} = S_{e,l}^{\frac{1}{2}}\left[1 - \left(1 - S_{e,l}^{\frac{1}{m}}\right)^{m}\right]^{2}$$

where $k_{r,l}$ is relative permeability for liquid phase, α is a curve-fitting parameter in unit of inverse pressure, n is a dimensionless curve-fitting parameter, and m = 1-1/n. $S_{e,l}$ is effective liquid phase saturation, and defined by:

$$S_{e,l} = \frac{S_l - S_{l,r}}{1 - S_{l,r}}$$

with S_l and $S_{l,r}$ as the liquid phase saturation and residual saturation, respectively.

The adsorption-desorption processes are assumed to be controlled by a linear equilibrium isotherm and a first-order decay constant [31], and approximated by $(1-\phi)\rho_s K_d \omega_\alpha^\gamma$ and Γ^γ terms of equation (8). ρ_s represents bulk density of soil, K_d denotes partitioning coefficient in adsorption, λ_α^γ is a first-order decay or degradation rate constant of the adsorbed component. When the component is radioactive λ_α^γ is referred to as coefficient of radioactive decay, which equals 1/T with T as the half-life of the radioactive component.

Γ^γ_r is the γ–component production/loss term associated with chemical reaction r, and will be further discussed later.

With local thermodynamic equilibrium assumed partitioning of components between phases is determined by:

$$n_\alpha^\gamma = K_{\alpha,\beta}^\gamma n_\beta^\gamma \tag{10}$$

where $K^\gamma_{\alpha,\beta}$ is the partitioning coefficient for γ–component between phase α and phase β. n^γ_α is the mole fraction of γ–component in α–phase, and related to the mass fraction ω_α^γ by:

$$n_\alpha^\gamma = \left(\omega_\alpha^\gamma / M^\gamma\right)\left(\sum_\kappa \omega_\alpha^\kappa / M^\kappa\right)$$

with M^γ as the molecular weight of γ–component. For a gas-liquid phase system, the relationship (10) is usually referred to as Henry's law, which is written as,

$$P_g^\gamma = H_{g,l}^\gamma n_l^\gamma$$

with P_g^γ as the partial pressure of γ-component in gas phase, n_l^γ the mole fraction of γ-component in liquid phase, and $H^\gamma_{g,l}$ the Henry's coefficient.

Equation (8) are also supplemented by the following constraints,

$$\sum_{\alpha} S_{\alpha} = 1 \quad \text{and} \quad \sum_{\gamma} \omega_{\alpha}^{\gamma} = 1 \tag{11}$$

For a non-isothermal system, local thermal equilibrium conditions are assumed, and thus we have the energy balance over all the fluid phases and solid matrix in a porous medium expressed in a general form,

$$\begin{aligned} &\frac{\partial}{\partial t}\left[\sum_{\alpha} \phi \rho_{\alpha} S_{\alpha} u_{\alpha} + (1-\varphi)\rho_{s} C_{p}\left(T - T_{ref}\right)\right] = \\ &\qquad -\sum_{\gamma}\sum_{\alpha} \nabla \cdot \varphi h_{\alpha}^{\gamma} \rho_{\alpha} S_{\alpha}\left(\omega_{\alpha}^{\gamma} \mathbf{V}_{\alpha} + \mathbf{J}_{h\alpha}^{\gamma} + \mathbf{J}_{\alpha}^{\gamma}\right) + \nabla \cdot \left(\mathbf{K}_{T} \nabla T\right) + Q \end{aligned} \tag{12}$$

where T, u_{α}, C_p, T_{ref}, h_{α}^{γ}, $\mathbf{K}_T$ and Q are temperature, internal energy of α–phase fluid, specific heat of solid, reference temperature, partial enthalpy of γ–component in α–phase, thermal conductivity, and heat source term, respectively. Note that $\mathbf{K}_T$ is the averaged thermal conductivity of both fluid and solid phases.

An important numerical feature of the USNT model is the capability of handling disappearance and appearance of any of the three fluid phases that can occur during evaporation/condensation or immiscible fluid displacement processes. As already described for US1P model the technique of primary variable switching is used to ensure the appropriate primary variable selection in response to phase change and transition.

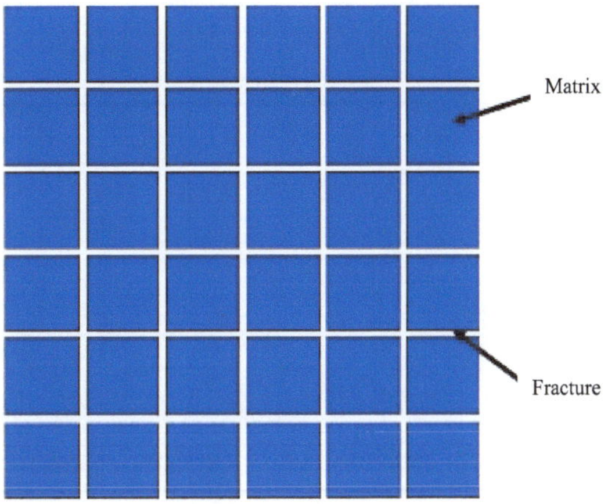

Figure 1: Conceptual schematic of dual-continuum (matrix-fracture) system.

In order to effectively approximate flow in fractured porous media two numerical approaches are considered in NUFT, namely, Effective Continuum Model (ECM) and dual porosity/permeability model (DKM). The ECM model uses equivalent porous media with averaged rock properties to represent fractured system, and is simple but less accurate. In DKM model the fracture and matrix systems are treated as two separate overlapping continua and multiphase flow and heat transfer are conceptually addressed in both matrix and fracture continua. Each continuum has a complete set of its own mass and energy balance equations such as (8) and (12). The balance equations for two continua are coupled through inter continuum mass and heat transfer. This approach is useful and efficient particularly when a subsurface system (like a heterogeneous reservoir) has too many fractures or too complex fracture networks to be explicitly represented. Fig. **1** illustrates conceptual schematic of a dual-continuum (matrix and fracture) system. This dual-continuum matrix-fracture model has been used in thermal hydrological modeling for geological disposal of nuclear waste [13, 14], and underground coal gasification [37].

For modeling chemical reactions we consider kinetic-controlled reactions among chemical species, which can be written in a general stoichiometric form,

$$\sum_{\gamma} \nu_r^{\gamma} A^{\gamma} \underset{k_r}{\overset{k_f}{\rightleftharpoons}} 0 \tag{13}$$

for r-reaction, where A^{γ} indicates generic chemical species γ, ν_r^{γ} is stoichiometric coefficient of γ–species in r-reaction with positive value for reactants and negative value for products, and k_f and k_r denote forward and reverse chemical reaction parameters. Generally the rate law is applied to describe the reaction kinetics as follows,

$$R_r = \left[k_{fc} \prod_{\nu_r^{\gamma}>0}^{\gamma} \left(c^{\gamma}\right)^{a^{\gamma}} - k_{rc} \prod_{\nu_r^{\gamma}<0}^{\gamma} \left(c^{\gamma}\right)^{a^{\gamma}} \right] \tag{14}$$

where c^{γ} represents concentration of γ-species, R_r is the overall reaction rate for r-chemical reaction, k_{fc} and k_{rc} are forward and reverse reaction rate constants, and a^{γ} indicates exponent or reaction order with respect to γ-species. To account for temperature dependence Arrhenius law is adopted and the reaction rate (14) is rewritten as:

$$R_r = \left[k_{fc} \prod_{\nu_r^{\gamma}>0}^{\gamma} \left(c^{\gamma}\right)^{a^{\gamma}} - k_{rc} \prod_{\nu_r^{\gamma}<0}^{\gamma} \left(c^{\gamma}\right)^{a^{\gamma}} \right] \cdot \exp\left(-\frac{E_a}{RT}\right) \tag{15}$$

E_a is referred to as the activation energy, and R is the gas constant.

For many bioremediation applications Monod model [29, 31] is used to characterize reaction kinetics of contaminant biode-gradation processes in porous media. The Monod equation implemented in NUFT takes the following format,

$$R_r = k_{max} \cdot \prod_{\gamma} \left(c^{\gamma}\right)^{a^{\gamma}} \cdot \prod_{\gamma} \frac{c^{\gamma}}{s^{\gamma} + c^{\gamma}} \cdot \prod_{\gamma} \frac{b^{\gamma}}{b^{\gamma} + c^{\gamma}} \cdot \exp\left(-\frac{E_a}{RT}\right) \tag{16}$$

where k_{max} is maximum reaction rate, and s^{γ} and b^{γ} are saturation and inhibition constants for γ–species. It should be noted that the reaction we consider can be either a homogenous (within a single phase) or heterogeneous (between two phases) reaction. In view of this the mass production/loss rate of γ –component associated with r-reaction in equation (8) can be generally expressed as:

$$\Gamma_r^{\gamma} = \begin{cases} -M^{\gamma} \rho_{\alpha} S_{\alpha} \varphi \nu_r^{\gamma} R_r \\ \quad \text{for homogeneous } \alpha\text{-phase reaction} \\ -M^{\gamma} A_{\alpha\text{-}\beta} \nu_r^{\gamma} R_r \\ \quad \text{for heterogeneous } \alpha\text{-}\beta \text{ phase reaction} \end{cases} \tag{17}$$

$A_{\alpha-\beta}$ indicates the specific surface area between α and β phases. R_r is in unit of mol/kg-s or mol/m^2-s for homogeneous or heterogeneous reactions.

In summary equations (8) - (12), and (17) together form a complete set of mass and energy balance equations for multiphase flow and reactive transport problems, which can be solved with specification of appropriate boundary and initial conditions.

It's worth noting that the flow and reactive transport system strongly depends on fluid and rock properties in porous media, and the proper selection of these properties is of particular importance in subsurface flow modeling. For this reason in addition to generic user-input options a variety of well-known functions/correlations are implemented in NUFT to calculate fluid and rock properties such as capillary pressure, relative permeability, rock compressibility, fluid PVT properties, solubility, compressibility, solid-fluid adsorption, geochemical kinetics, *etc.*

2.4. Single Component Solute Transport Model (US1C MODULE)

The US1C module is constructed to simulate a single-component transport in either an aqueous or gaseous phase. The use of this module requires that the Darcy flow conditions be either known or obtained from coupling with a flow module. The mass balance equation for γ-component transport in α-phase takes a similar form as (8) except that no reaction is considered,

$$\frac{\partial}{\partial t}\left[\varphi\rho_\alpha S_\alpha\omega_\alpha^\gamma + (1-\varphi)\rho_s K_d\omega_\alpha^\gamma\right] = \\ -\nabla\cdot\phi\rho_\alpha S_\alpha\left(\omega_\alpha^\gamma \mathbf{V}_\alpha + \mathbf{J}_{h\alpha}^\gamma + \mathbf{J}_\alpha^\gamma\right) - \lambda_\alpha^\gamma\varphi\rho_\alpha S_\alpha\omega_\alpha^\gamma$$

Note that the flow variables of $\mathbf{V}_\alpha$, ρ_α, and S_α are either transferred from flow model solutions at each time step (*e.g.* UCSAT, US1P, or USNT sub-module) or pre-defined by user input. In most applications it is assumed that the solution is dilute, say, $\omega^\gamma_\alpha \ll 1$, and therefore the phase density is kept constant in the equation.

2.5. Reactive Transport with Equilibrium and Kinetic Chemistry Model (TRANS MODULE)

In this section we review mathematical formulations of transport processes combined with both equilibrium and kinetic reactions.

As we know a chemical system is made up of species and equilibrium and kinetic chemical reactions, and can be described in a general form [38].

$$\frac{d\mathbf{C}}{dt} = \mathbf{v}\cdot\mathbf{R} \qquad (18)$$

C and **R** are concentration and reaction rate vectors in the form of,

$$\mathbf{C} = \begin{Bmatrix} C_1 \\ \vdots \\ C_\gamma \\ \vdots \\ C_{NS} \end{Bmatrix} \quad \gamma = 1\cdots NS \quad \mathbf{R} = \begin{Bmatrix} r_1 \\ \vdots \\ r_r \\ \vdots \\ r_{NR} \end{Bmatrix} \quad r = 1\cdots NR$$

in which C_γ represents the concentration of γ–species in unit of mole/m^3, and r_r (mole/m^3-s) is the reaction rate of r-reaction, and NS and NR denote the total numbers of species and reactions, respectively. **v** is a matrix containing stoichiometric coefficients for the chemical reactions,

$$\mathbf{v} = \begin{Bmatrix} \nu_1^1 & \cdots & \cdots & \cdots & \nu_{NR}^1 \\ \vdots & \ddots & & \cdot^{\cdot^{\cdot}} & \vdots \\ \vdots & & \nu_r^\gamma & & \vdots \\ \vdots & \cdot^{\cdot^{\cdot}} & & \ddots & \vdots \\ \nu_1^{NS} & \cdots & \cdots & \cdots & \nu_{NR}^{NS} \end{Bmatrix} \quad \begin{matrix} \gamma = 1\cdots NS \\ r = 1\cdots NR \end{matrix}$$

with ν_r^γ as the stoichiometric coefficient for γ–species in r-reaction. For the sake of convenience we rewrite equation (18) in terms of mass concentration vector, **c** ,

$$\mathbf{M}^{-1} \cdot \frac{dc}{dt} = \mathbf{v} \cdot \mathbf{R}$$

$$\mathbf{c} = \begin{Bmatrix} c_1 \\ \vdots \\ c_\gamma \\ \vdots \\ c_{NS} \end{Bmatrix} \quad \mathbf{M}^{-1} = \begin{Bmatrix} \frac{1}{M^1} & 0 & \cdots & \cdots & 0 \\ 0 & \ddots & 0 & 0 & \vdots \\ \vdots & 0 & \frac{1}{M^\gamma} & 0 & \vdots \\ \vdots & 0 & 0 & \ddots & 0 \\ 0 & \cdots & \cdots & 0 & \frac{1}{M^{NS}} \end{Bmatrix} \quad \textbf{(19)}$$

$$\gamma = 1 \cdots NS$$

M is a diagonal matrix holding species molecular weights. Note that equation (18) or (19) is a general representation of chemical reaction system. In view of the fact that **R** takes account of both equilibrium and kinetic reactions we can further divide reactions into equilibrium and kinetic types, and define equilibrium and kinetic reaction rate vectors, $\mathbf{R}_{eq}$ and $\mathbf{R}_{kin}$, as:

$$\mathbf{R}_{eq} = \begin{Bmatrix} r_1 \\ \vdots \\ r_\gamma \\ \vdots \\ r_{NR_{eq}} \end{Bmatrix} \qquad \mathbf{R}_{kin} = \begin{Bmatrix} r_1 \\ \vdots \\ r_r \\ \vdots \\ r_{NR_{kin}} \end{Bmatrix}$$

$$r = 1 \cdots NR_{eq} \qquad r = 1 \cdots NR_{kin}$$

NR_{eq} and NR_{kin} are the total numbers of equilibrium and kinetic reactions, with $NR = NR_{eq} + NR_{kin}$. Due to apparent interdependence among species variables in equilibrium reactions, a reduced number of species can be chosen as primary or basis species to determine the chemical system under consideration of (18) or (19). Consequently, all the remaining species become dependent (or non-basis) species that can be decided from their associated equilibrium reactions. Here, we define NC as the number of primary (or basis) species, and ND as the number of dependent (or non-basis) species. To uniquely determine a chemical equilibrium system it is required that the equilibrium reactions be linearly independent and $ND = NS - NC = NR_{eq}$. Under this constraint, we reformulate equation (19) in terms of primary and dependent species, and as well as equilibrium and kinetic reaction rates,

$$\mathbf{M}_b^{-1} \cdot \frac{dc_b}{dt} = \mathbf{v}_{eq}^b \cdot \mathbf{R}_{eq} + \mathbf{v}_{kin}^b \cdot \mathbf{R}_{kin}$$

$$\mathbf{c}_b = \begin{Bmatrix} c_1 \\ \vdots \\ c_\gamma \\ \vdots \\ c_{NC} \end{Bmatrix}, \quad \mathbf{M}_b^{-1} = \begin{Bmatrix} \frac{1}{M^1} & 0 & \cdots & \cdots & 0 \\ 0 & \ddots & 0 & 0 & \vdots \\ \vdots & 0 & \frac{1}{M^\gamma} & 0 & \vdots \\ \vdots & 0 & 0 & \ddots & 0 \\ 0 & \cdots & \cdots & 0 & \frac{1}{M^{NC}} \end{Bmatrix} \quad \textbf{(20)}$$

$$\mathbf{M}_d^{-1} \cdot \frac{d\mathbf{c}_d}{dt} = \mathbf{v}_{eq}^{d} \cdot \mathbf{R}_{eq} + \mathbf{v}_{kin}^{d} \cdot \mathbf{R}_{kin}$$

$$\mathbf{c}_d = \begin{Bmatrix} c_1 \\ \vdots \\ c_\gamma \\ \vdots \\ c_{ND} \end{Bmatrix}, \quad \mathbf{M}_d^{-1} = \begin{Bmatrix} \frac{1}{M^1} & 0 & \cdots & \cdots & 0 \\ 0 & \ddots & 0 & 0 & \vdots \\ \vdots & 0 & \frac{1}{M^\gamma} & 0 & \vdots \\ \vdots & 0 & 0 & \ddots & 0 \\ 0 & \cdots & \cdots & 0 & \frac{1}{M^{ND}} \end{Bmatrix} \quad \textbf{(21)}$$

where the super/sub-scripts b and d denote primary (basis) and dependent (non-basis) species, and eq and kin indicate equilibrium and kinetic reactions. $\mathbf{v}$ represents stoichiometric coefficient matrices, specifically $\mathbf{v}^b_{eq}$ is a NC × NR_{eq} matrix with entry v^i_j (i=1...NC and j=1...NR_{eq}), $\mathbf{v}^b_{kin}$ is a NC × NR_{kin} matrix with entry v^i_j (i=1...NC and j=1...NR_{kin}), $\mathbf{v}^d_{eq}$ is a ND × NR_{eq} (or NR_{eq} × NR_{eq}) matrix with entry v^i_j (i=1...ND or NR_{eq} and j=1...NR_{eq}), and $\mathbf{v}^d_{kin}$ is a ND × NR_{kin} (or NR_{eq} × NR_{kin}) matrix with entry v^i_j (i=1...ND or NR_{eq} and j=1...NR_{kin}). The vector of $\mathbf{R}_{eq}$ can be obtained by solving equation (21) as,

$$\mathbf{R}_{eq} = \mathbf{v}_{eq}^{d\,-1} \cdot \mathbf{M}_d^{-1} \cdot \frac{d\mathbf{c}_d}{dt} - \mathbf{v}_{eq}^{d\,-1} \cdot \mathbf{v}_{kin}^{d} \cdot \mathbf{R}_{kin} \quad \textbf{(22)}$$

The substitution of equation (22) into (20) eliminates $\mathbf{R}_{eq}$ and yields,

$$\frac{d\mathbf{c}_b}{dt} - \mathbf{M}_b \cdot \mathbf{v}_{eq}^{b} \cdot \mathbf{v}_{eq}^{d\,-1} \cdot \mathbf{M}_d^{-1} \cdot \frac{d\mathbf{c}_d}{dt} = \mathbf{M}_b \cdot \left(\mathbf{v}_{kin}^{b} - \mathbf{v}_{eq}^{b} \cdot \mathbf{v}_{eq}^{d\,-1} \cdot \mathbf{v}_{kin}^{d}\right) \cdot \mathbf{R}_{kin} \quad \textbf{(23)}$$

Rearranging equation (23), we have:

$$\frac{d}{dt}(\mathbf{c}_{b,total}) \equiv \frac{d}{dt}\left(\mathbf{c}_b - \mathbf{M}_b \cdot \mathbf{v}_{eq}^{b} \cdot \mathbf{v}_{eq}^{d\,-1} \cdot \mathbf{M}_d^{-1} \cdot \mathbf{c}_d\right) = \mathbf{M}_b \cdot \left(\mathbf{v}_{kin}^{b} - \mathbf{v}_{eq}^{b} \cdot \mathbf{v}_{eq}^{d\,-1} \cdot \mathbf{v}_{kin}^{d}\right) \cdot \mathbf{R}_{kin} \quad \textbf{(24)}$$

Here, $\mathbf{c}_{b,total}$ denotes the vector holding total mass concentrations of primary (basis) species.

The dependent relationships between mass concentration $\mathbf{c}_d$ of dependent (non-basis) species and $\mathbf{c}_b$ of primary (basis) species can be established by mass action equations with respect to NR_{eq} equilibrium reactions that are discussed below.

An equilibrium chemical system is typically defined by a set of chemical species and a set of independent equilibrium reactions as well,

$$\sum_\gamma \nu_r^\gamma A^\gamma = 0 \qquad \gamma = 1 \cdots NS, \quad r = 1 \cdots NR_{eq} \quad \textbf{(25)}$$

The dependence among species in each reaction is derived by its corresponding mass action equation in the form of

$$K_r = \prod_i (a_i)^{\nu_r^i} \cdot \prod_g (f_g)^{\nu_r^g} \cdot \prod_m (a_m)^{\nu_r^m} \quad \text{or} \quad \log K_r = \sum_i \nu_r^i \log a_i + \sum_g \nu_r^g \log f_g + \sum_m \nu_r^m \log a_m \quad r = 1 \cdots NR_{eq}$$

in which K_r is the equilibrium constant for r-reaction, a_i and a_m represent the activities of aqueous and mineral species, f_g is the fugacity of gas species, and ν_r^i, ν_r^g, and ν_r^m are respective stoichometric coefficients for aqueous, gaseous and mineral species in r-reaction.

For a pure mineral phase the activity a_m is always unity. The aqueous activity a_i and gas fugacity f_g are related to molality m_i and partial pressure P_g or mole fraction n_g by activity and fugacity coefficients γ_i and γ_g,

$$a_i = \gamma_i \cdot m_i \qquad f_g = \gamma_g \cdot P_g = \gamma_g \cdot n_g \cdot P$$

P is the total pressure of the gas phase. The activity coefficient γ_i of an ion in an aqueous solution can be estimated by the application of the extended Debye-Hückel model,

$$\log \gamma_i = -\frac{A z_i^2 \sqrt{I}}{1 + a_i B \sqrt{I}}$$

where z_i is the electrical charge, a_i is the ion size parameter, A and B are temperature-dependent parameters, and the ionic strength I is obtained by

$$I = \frac{1}{2}\sum_i m_i z_i^2$$

As discussed before, each dependent (non-basis) species corresponds to one equilibrium reaction. Hence, it is convenient to rewrite (25) in a so-called canonical form [32] in which the dependent (non-basis) species is disassociated into a set of primary (basis) species, namely,

$$A^\lambda = \sum_i \nu_r^i B^i + \sum_j \nu_r^j B^j + \sum_k \nu_r^k B^k \quad r = 1 \cdots NR_{eq} \qquad \textbf{(26)}$$

where A^λ represents dependent (non-basis) species, and B^i, B^j, and B^k denote aqueous, gaseous and mineral basis species, respectively. Thus, for an aqueous dependent species we have,

$$m_\lambda = \frac{\prod_i (a_i)^{\nu_r^i} \cdot \prod_j (f_j)^{\nu_r^j} \cdot \prod_k (a_k)^{\nu_r^k}}{\gamma_\lambda \cdot K_r} \quad \text{or} \quad \log m_\lambda = \sum_i \nu_r^i \log a_i + \sum_j \nu_r^j \log f_j + \sum_k \nu_r^k \log a_k - \log K_r - \log \gamma_\lambda$$

$$r = 1 \cdots NR_{eq} \qquad \textbf{(27)}$$

Note that K_r is the equilibrium constant for r-reaction (26).

Similarly, for a gaseous dependent species, we have,

$$P_\lambda = \frac{\prod_i (a_i)^{\nu_r^i} \cdot \prod_j (f_j)^{\nu_r^j} \cdot \prod_k (a_k)^{\nu_r^k}}{\gamma_\lambda \cdot K_r} \quad \text{or} \quad \log P_\lambda = \sum_i \nu_r^i \log a_i + \sum_j \nu_r^j \log f_j$$

$$+ \sum_k \nu_r^k \log a_k - \log K_r - \log \gamma_\lambda \qquad \textbf{(28)}$$

$$r = 1 \cdots NR_{eq}$$

After converting molality (for aqueous species) in (27) or partial pressure (for gaseous species) in (28) into mass concentration, we have mass concentration of dependent (non-basis) species expressed in terms of mass concentration of the primary (basis) species, namely,

$$\mathbf{c}_{d} = f(\mathbf{c}_{b}) \tag{29}$$

Therefore, the mass balance equation (24) for the chemical system, supplemented by equation (29), can be specifically rearranged as:

$$\begin{aligned}
&\frac{d}{dt}(\mathbf{c}_{b,total}) \equiv \frac{d}{dt}\left[\mathbf{c}_{b} - \mathbf{E}(\mathbf{c}_{b})\right] = \mathbf{G} \cdot \mathbf{R}_{kin} \\
&\mathbf{E}(\mathbf{c}_{b}) = \mathbf{M}_{b} \cdot \mathbf{v}_{eq}^{b} \cdot \mathbf{v}_{eq}^{d\ -1} \cdot \mathbf{M}_{d}^{-1} \cdot \mathbf{c}_{d} \\
&\mathbf{G} = \mathbf{M}_{b} \cdot \left(\mathbf{v}_{kin}^{b} - \mathbf{v}_{eq}^{b} \cdot \mathbf{v}_{eq}^{d\ -1} \cdot \mathbf{v}_{kin}^{d}\right) \\
&b = 1 \cdots NC
\end{aligned} \tag{30}$$

where **E** is a vector representing mass distribution of basis species among dependent species, and **G** is a NC× NR_{kin} coefficient matrix arising from kinetic reactions with matrix entry $G_{i,j,}$ i=1 … NC, and j = 1 … $NR_{kin.}$ It is noted that the balance equation (30) takes account both equilibrium and kinetic reactions. When only equilibrium reactions are considered equation (30) is reduced to,

$$\frac{d}{dt}(\mathbf{c}_{b,total}) \equiv \frac{d}{dt}\left[\mathbf{c}_{b} - \mathbf{E}(\mathbf{c}_{b})\right] = 0 \quad b = 1 \cdots NC \quad \text{or} \quad \mathbf{c}_{b,total} \equiv \mathbf{c}_{b} - \mathbf{E}(\mathbf{c}_{b}) = \text{Total Concentration} \tag{31}$$

With known total concentrations of primary species we can obtain primary species concentrations by solving a set of nonlinear equations represented by (31), and consequently determine the concentrations of the remaining dependent species. This process is referred to as speciation that is used to define the chemical state of an equilibrium chemical system. For a kinetic chemical system equation (30) is transformed into,

$$\frac{d}{dt}(\mathbf{c}_{b,total}) \equiv \frac{d}{dt}(\mathbf{c}_{b}) = \mathbf{G} \cdot \mathbf{R}_{kin} \equiv \mathbf{M}_{b} \cdot \mathbf{v}_{kin}^{b} \cdot \mathbf{R}_{kin} \qquad b = 1 \cdots NS \tag{32}$$

To address fully coupled reactive transport problems we incorporate the mass balance (30) due to chemical reactions into multiphase flow and transport formulations, and have:

$$\begin{aligned}
&\frac{\partial}{\partial t}\sum_{\alpha} \varphi S_{\alpha}\rho_{\alpha}\omega_{\alpha}^{b} = -\sum_{\alpha} \nabla \cdot \varphi\rho_{\alpha}S_{\alpha}\left(\omega_{\alpha}^{b}\mathbf{V}_{\alpha} + \mathbf{J}_{h\alpha}^{b} + \mathbf{J}_{\alpha}^{b}\right) + \sum_{r} G_{b,r}R_{kin,r} \\
&\mathbf{J}_{h\alpha}^{b} = -\mathbf{D}_{h\alpha}^{b} \cdot \nabla\omega_{\alpha}^{b} \qquad \mathbf{J}_{\alpha}^{b} = -D_{\alpha}^{b}\nabla\omega_{\alpha}^{b} \qquad b = 1 \cdots NC
\end{aligned} \tag{33}$$

Here, b denotes the primary species and ω^{b}_{α} is the total mass fraction of b–basis species in α–phase, related to the total mass concentration in (30) by $c_{b,total} = \rho_{\alpha}\omega^{b}_{\alpha}$.

The last term of the right-hand side in equation (33) is the rate of mass change due to kinetic reactions with $R_{kin,r}$ denoting reaction rate of r-kinetic reaction.

The kinetic reactions can be either homogeneous or heterogeneous reactions. One of examples for heterogeneous kinetic reactions is mineral dissolution/precipitation, which is commonly encountered in subsurface reactive transport processes. A mineral dissolution/ precipitation reaction is typically expressed as the dissolution of a mineral A^{m} into a set of primary and secondary aqueous species A^{γ}, namely,

$$A^{m} \underset{k_{rev}}{\overset{k_{for}}{\rightleftharpoons}} \sum_{\gamma} \nu_{m}^{\gamma} A^{\gamma} \tag{34}$$

with ν^{γ}_{m} denoting stoichiometric coefficient of species A^{γ}. k_{for} and k_{rev} stand for forward (dissolution) and reverse (precipitation) reactions. The dissolution/precipitation rate R_m (mol/m^3-s) is calculated based on the rate law,

$$R_m = S_m \cdot k_m \cdot T_a \cdot \prod_{\gamma} \left(a^{\gamma}\right)^{p^{\gamma}} \left(1 - \frac{Q_m}{K_{eq}}\right) \tag{35}$$

where S_m is mineral specific surface area (m^2/m^3) dependent on mineral volume fraction θ_m, k_m is the rate constant, T_a is Arrhenius factor,

$$T_a = \exp\left(-\frac{E_a}{RT}\right) ,$$

to account for temperature dependence, a^{γ} and p^{γ} are the γ-species activity and exponent, K_{eq} denotes the equilibrium constant, and Q_m is the activity product.

$$Q_m = \prod_{\gamma} \left(a^{\gamma}\right)^{\nu_m^{\gamma}}$$

The saturation index SI_m is defined as $\log(Q_m/K_{eq})$ and used to measure a fluid's saturation with respect to a mineral A^m. While the mineral dissolves in fluid as SI_m becomes negative or $Q_m/K_{eq} < 1$, the positive saturation index or $Q_m/K_{eq} > 1$ indicates the fluid is under supersaturated conditions with mineral and the mineral precipitation tends to proceed. Considering the fact that dissolution and precipitation processes may have different reaction rate constants we further express the rate constant k_m in a more general way as:

$$k_m = \begin{cases} k_{for} H(\theta_m) & \text{for dissolution } (SI_m < 0) \\ k_{rev} & \text{for precipitation } (SI_m > 0) \end{cases}$$

Here k_{for} and k_{rev} are forward (dissolution) and reverse (precipitation) reaction rate constants, respectively. $H(\theta_m)$ is Heaviside step function and set to be 1 if $\theta_m > 0$ and 0 otherwise. The mineral volume fraction θ_m is updated by using the following relationship,

$$\frac{d\theta_m}{dt} = R_m V_m \tag{36}$$

with V_m as mineral molar volume. The resulting porosity change at each time step is computed by,

$$\Delta\phi = \phi - \phi_0 = \sum_m \left(\theta_{m,0} - \theta_m\right) \tag{37}$$

with ϕ_0 as the initial porosity and $\theta_{m,0}$ the initial volume fraction of mineral m. The dependence of permeability on porosity change is derived from a cubic law based upon Kozeny-Carman relationship.

3. NUMERICAL SOLUTION

An integral finite difference method, also referred to as finite volume method, is used to discretize the partial differential equations that govern mass and energy balance for each sub-model described earlier. Even though NUFT has the option to use explicit method for time integration, the implicit procedure is mainly chosen for numerically solving each sub-model because it is numerically stable, and allows larger simulation time-stepping, particularly when the numerical solution varies slowly. The resulting non-linear equations at each time step are solved by the Newton-Raphson method. Each nonlinear iteration step requires the solution of a set of linear algebraic equations. Hence multiple linear solver and decomposition options are provided in NUFT, which include Gauss elimination, preconditioned conjugate method, block Gauss-Seidel preconditioning, incomplete block LU decomposition, and so on. Additionally some special numerical techniques/algorithms are adopted and implemented in NUFT to help enhance its numerical accuracy, robustness and stability. Examples of these techniques include primary variable switching in UCSAT, US1P, and USNT models, basis-species switching in TRANS model, artificial diffusion technique and flux-correction methods to reduce numerical dispersion. Another important feature of NUFT in terms of computational capability is its transparent parallelization. It has been shown that massive parallel computing can effectively support large-scale simulations and produce significant speed-ups. Except for adding a few of parallel-related control parameters running NUFT in parallel uses the same input file as that for serial computing. All the parallel operations and processes such as inter-processer data transfer, synchronization and mesh partitioning routines are totally transparent to users. The parallel implementation in the code is intended for a distributed memory parallel system, and the inter-processor communication and data exchange are achieved with the Message Passing Interface (MPI) standard protocols [39]. The numerical routines from the Portable, Extensible Toolkit for Scientific Computation (PETSc) [40] are adopted to solve the large sets of linear equations resulted from Newton-Raphson iterations.

Following the general description of numerical solutions in NUFT we extend our discussion to more specifically focus on the coupled solution of flow and reactive transport model. Because of its significant complexity and strong nonlinearity, and large number of unknown variables reactive transport modeling has drawn considerable attention, and many numerical techniques and modeling tools have been proposed and developed [4, 5, 8, 9]. The key issue in numerical simulations of reactive transport is how to couple transport and reaction models in an accurate and efficient way. Normally most of available numerical models are based upon two major approaches, (1) global implicit (2) operator splitting approaches. The operator-splitting approach separates the transport and chemical models, and solves them in a sequential fashion. The representative numerical implementations for this kind of method are the Sequential Iteration Approach (SIA), and the Sequential Non-Iteration Approach (SNIA). The former includes iterations for convergence between the two solution steps, which thus requires more CPU time, but has more accurate solutions than the latter. Most of computational codes [4, 5, 8] are based on either SIA or SNIA methods. In a global implicit method transport and chemical reaction equations are fully coupled and solved simultaneously. Despite of the fact that there is higher numerical cost associated with solving a large highly nonlinear system the global-implicit formulation is an intuitive and effective way to couple all the physical and chemical processes without any concern about inconsistency, particularly as today computer power increases significantly and parallel computing enjoys great popularity. The global implicit procedure is also numerically stable even with larger simulation time-stepping. In TRANS module of NUFT the equilibrium and kinetic chemical equations (30) and the transport equations (33) are combined and solved in a fully implicit manner.

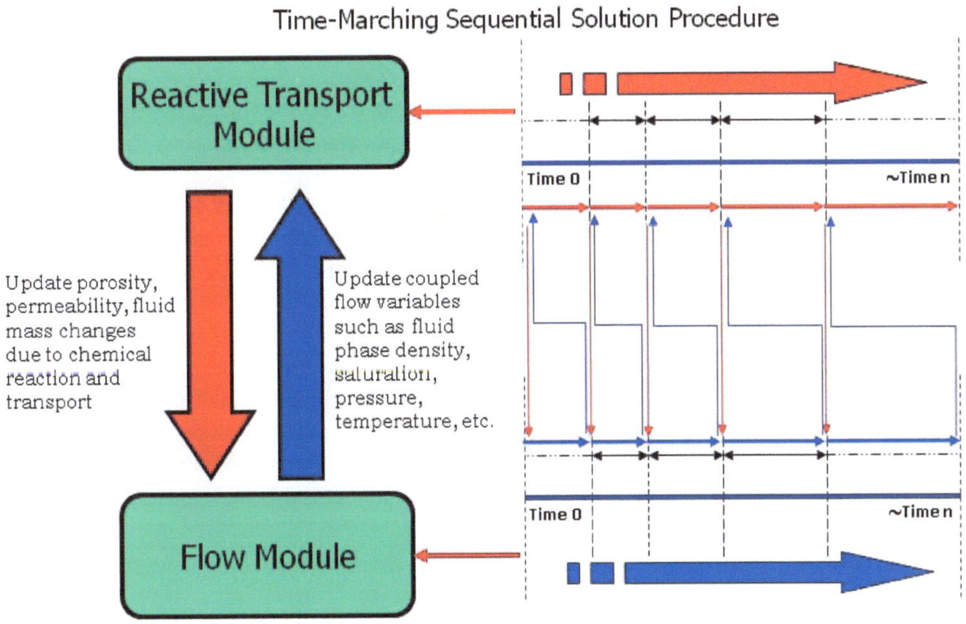

Figure 2: Coupling procedure between flow and reactive transport modules.

In most of flow and reactive transport simulations a reactive transport model (*e.g.* TRANS, US1C module) is required to be combined with a flow model (*e.g.* UCSAT, US1P, USNT modules). The coupling between flow and reactive transport modules is achieved by a time-marching sequential solution procedure that helps avoid otherwise significantly large Jacobian matrix resulting from fully coupled flow and transport solutions. Given a time step the mass and energy balance equations in the flow model are solved first, and the characterized flow variables such as Darcy flow fluxes, fluid densities, or phase saturations under multiphase conditions are calculated and transferred to reactive transport module. Subsequently the reactive transport simulation is performed within the same flow time step. Due to the fact that the time scale of chemical reactions is often smaller than that of flow processes multiple time steps are usually required for reactive transport calculations in order to reach the given flow time step. At the end of the calculation, the porosity and permeability changes caused by chemical reactions such as mineral dissolution and precipitation are updated in the flow module for the next time step. If the fluid phase density is highly concentration-dependent, the fluid mass changes arising from chemical reaction and transport can also be coupled to the flow module. The coupling procedure is illustrated in Fig. **2**.

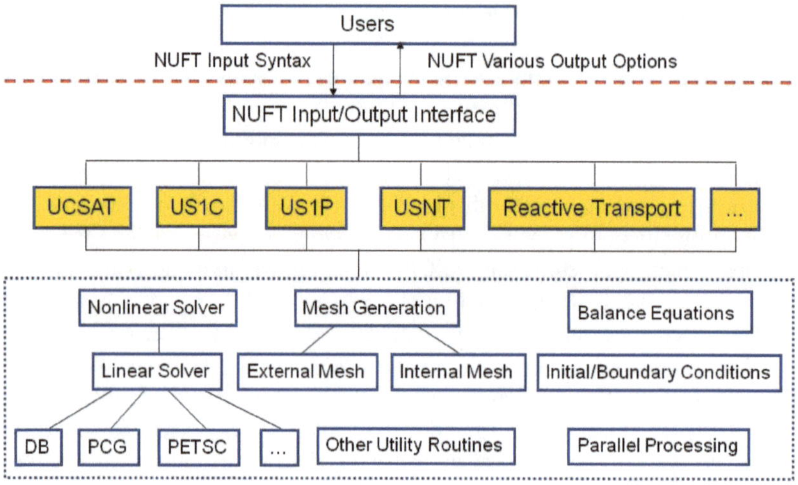

Figure 3: The architecture of the NUFT code.

4. SOFTWARE ORGANIZATION

The NUFT code is written in the C^{++} language and implemented within an object-oriented framework. As introduced before, there are several different sub-modules in NUFT. Following the modular and object-oriented approach the code is designed with a multi-layer architecture as shown in Fig. **3**. It is also noted that typically the balance equations for all the sub-models share the same partial differential equation form with the terms falling into several categories as temporal accumulation, advection, diffusion, and source/sink terms. Based on this consideration all the sub-models are implemented in a way such that they internally not only follow the same numerical discretization steps, but also utilize a common set of utility modules to support various numerical routines for both serial and parallel computations such as built-in mesh generation, initial and boundary condition specification, module-to-module interactions, and so on. Through a generic input interface the user is allowed to select the sub-modules and numerical method that are most appropriate to his/her modeling task, or combine the existing modules to perform more complex modeling like coupled flow and reactive transport simulations, without code recompilation. The modular approach greatly facilitates extensive code re-use, and helps isolate numerical/coding errors arising from one module not to affect others, and thus lowers the code maintenance cost, and also enables rapid and convenient future submodel design and development without impacting the existing ones.

The sub-models implemented in NUFT have already been extensively verified by comparison with analytical solutions of flow and transport problems, and with numerical solutions by other computer codes, and also validated against results from lab experiments and field tests [26, 27, 41].

5. EXAMPLES OF APPLICATION

In this section we shall now present two numerical examples for modeling subsurface flow and reactive transport processes with the application of the NUFT code. The first example is the use of the USNT module for simulating subsurface biodegradation processes. In the second example we perform multi-phase flow and reactive transport modeling of CO_2 injection into a carbonate reservoir.

5.1. Coupled Flow and Reactive Transport with Geochemical Kinetics

One-dimensional Reactive Transport Process

The generalized kinetic Monod formulation (16) can be specified to express the first-order sequential reactions using a small value of saturation constant, s, and a large value of inhibition constant, b, relatively to the concentration magnitude. As a result the inhibition terms approach one and the kinetics is only controlled by the power law format. In order to test kinetic reactions in the USNT module, a one-dimensional reactive transport problem with first-order sequential reactions is considered. Uniform and constant velocity and dispersivity are assumed. The boundary conditions are referred to P268 in [33].

Transport processes are calibrated first using the analytical solution of P268 in [33] assuming zero reaction rate. Both analytical and numerical solutions were computed for a one-dimensional column of length 150 m, which is discretized by using 75 evenly spaced nodal points. Fig. **4** shows the comparison between the analytical solution and NUFT simulations at 1, 2, and 3 years along the one-dimensional column. A slight difference at the outlet is due to the boundary effect in the numerical model.

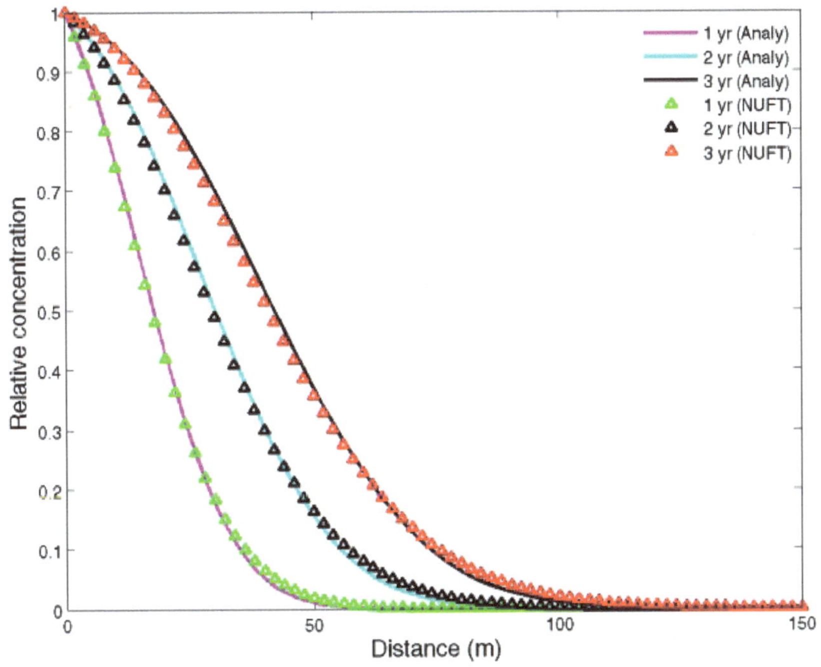

Figure 4: Concentration profiles of a single species in a one-dimensional column simulated using the analytical and numerical solutions. Flow velocity V = 0.0031 m/d and dispersivity α =10.0 m.

The transport of three synthetical species A, B, and C with three-step sequential first-order reactions is considered to calibrate the kinetic reactions in the USNT module,

$$A \xrightarrow{k_1} B \xrightarrow{k_2} C \xrightarrow{k_3} D\,.$$

The final reaction product, D, is excluded in the simulations. The reactive transport is governed by,

$$\frac{\partial C_A}{\partial t} - D\frac{\partial^2 C_A}{\partial x^2} + V\frac{\partial C_A}{\partial x} = -k_1 C_A$$

$$\frac{\partial C_B}{\partial t} - D\frac{\partial^2 C_B}{\partial x^2} + V\frac{\partial C_B}{\partial x} = k_1 C_A - k_2 C_B$$

$$\frac{\partial C_C}{\partial t} - D\frac{\partial^2 C_C}{\partial x^2} + V\frac{\partial C_C}{\partial x} = k_2 C_B - k_3 C_C$$

where C_A, C_B, and C_C denote the concentrations of the species, k_1, k_2, and k_3 are the first-order reaction rates, and D and V represent the dispersion coefficient and flow velocity, respectively. The above equations are subjected to the following initial and boundary conditions.

$$\text{I.C. } C_A(x,0) = C_B(x,0) = C_C(x,0) = 0 \quad \forall x > 0$$

$$\text{B.C. } C_A(0,t) = 1,\ C_B(0,t) = C_C(0,t) = 0 \quad \forall t \geq 0$$

When a quasi-steady state is reached we compare the NUFT simulations results with the analytical solutions of the steady-state concentrations, which are provided by Petersen and Sun [42]. Fig. **5** indicates good agreement of the spatial concentrations simulated using the analytical solution and the NUFT code.

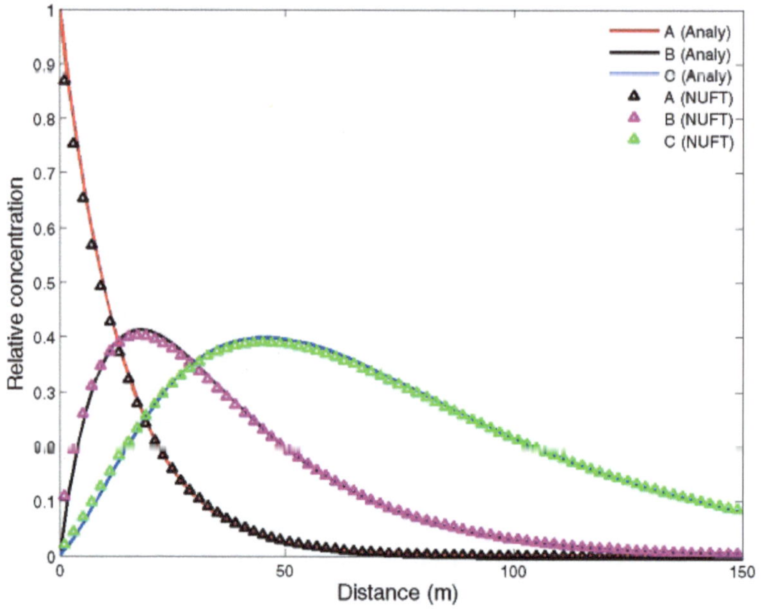

Figure 5: Concentration profiles of three sequential first-order decaying species in a one-dimensional column simulated using the analytical solution and NUFT code. V = 0.0031 m/d, α =5.0 m, k_1 = 0.0031, k_2 = 0.5 k_1, and k_3 = 0.25 k_1 1/d. Note that molecular weight is used in NUFT to convert the first-order reaction rate (1/d) to k_{max} (mol/kg s).

Two-dimensional Reactive Transport Process

The generalized kinetics formula can be used to express the power-law kinetics and competitive Monod reactions. To demonstrate the NUFT capability of modeling kinetic reactions, a sequential dechlorination process is considered where Trichloro-ethylene (TCE) reacts to produce dichloroethylene (DCE), and the daughter species further reacts to produce Vinyl Chloride (VC), and finally VC reacts to produce ethylene (ETH). The biomass involved in the reactions is assumed to be constant and uniformly distributed. TCE is the only original contaminant and released from two point sources at a discharge rate of 10^{-7} kg/s in a two-dimensional heterogeneous domain of 204 m long and 80 m wide as shown in Fig. **6**. Constant pressure conditions are set to be 1.4×10^5 Pa and 1.0×10^5 Pa respectively on the left and right boundaries. The corresponding flow velocity magnitude under the pressure gradient is also shown in Fig. **6**.

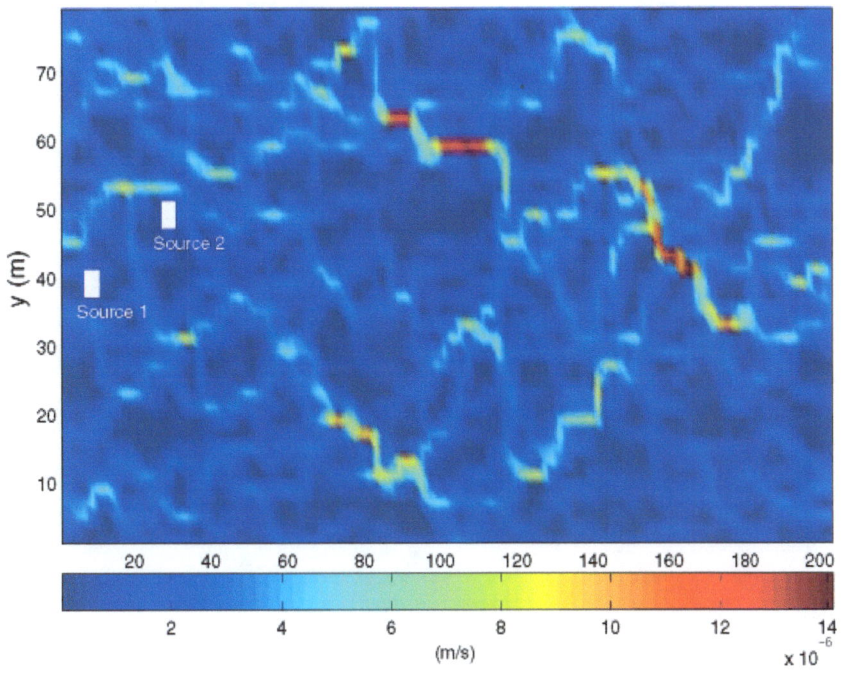

Figure 6: Velocity magnitude of flow field and contaminant sources.

Fig. **7** shows the concentration distribution of TCE and its daughter species at two years with reaction rates $k_1 = 4.5446\times10^{-3}$ (1/d), $k_2 = 0.5\ k_1$, and $k_3 = k_4 = 0.25\ k_1$. The figure indicates that daughter species plumes, located downstream of its parent species, spread far from the original contaminant sources. Therefore the peak concentration of a daughter species may not necessarily reflect the origins of contaminant sources. If the first reaction from TCE to DCE is assumed to be inhibited by DCE concentration (taking b_1=10^{-5}), then the transformation rate from TCE to DCE is simply reduced. As seen in Fig. **8** in which the concentration profiles of TCE and its daughter species are plotted the TCE plume is bigger than that without considering the inhibition. Correspondingly, DCE, VC, and ETH plumes are smaller with lower concentrations.

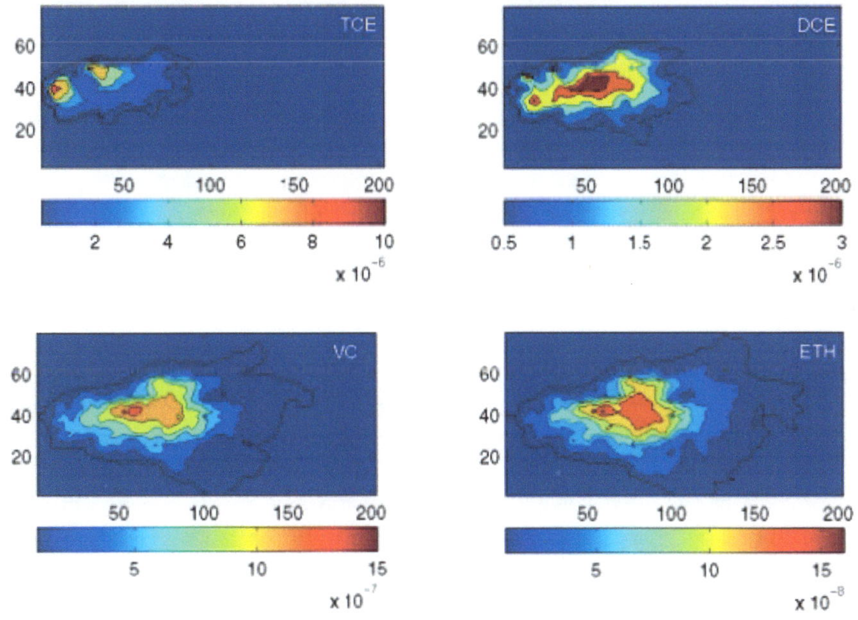

Figure 7: Mole fractions of TCE and its daughter species at two years without considering inhibition effect.

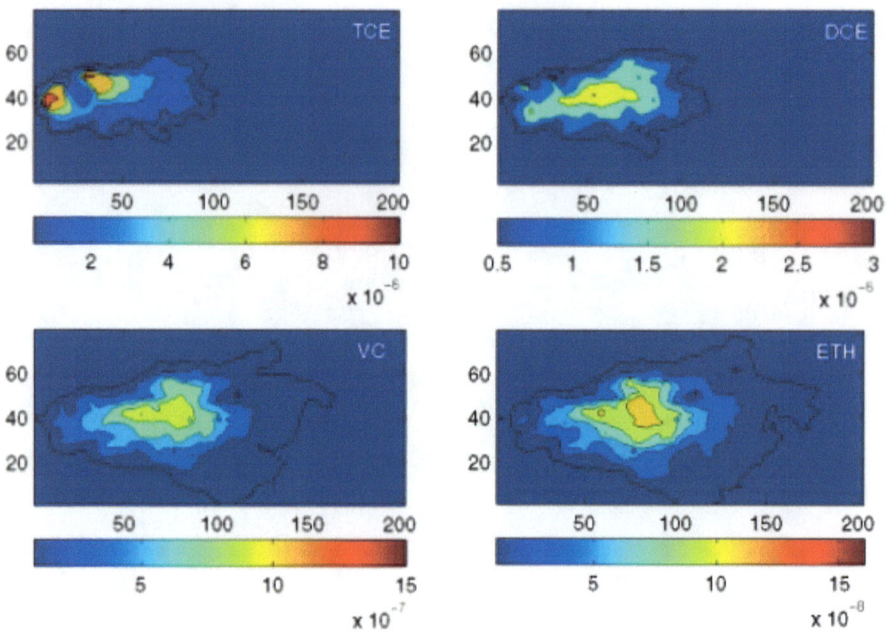

Figure 8: Mole fractions of TCE and its daughter species at two years with considering inhibition effect. Note that the inhibition constant takes the same unit as the concentration and the value needs to be adjusted accordingly to the concentration magnitude of the inhibiting species.

5.2. Simulation of CO_2 Transport in a Carbonate Reservoir

Geologic CO_2 sequestration and storage has a large potential to reduce net CO_2 released into atmosphere and, therefore, mitigate man-made global warming. The basic idea of such geologic disposal processes is that CO_2 captured from power plants is injected as a supercritical fluid into deep subsurface reservoirs such as deep saline aquifers, depleted or depleting oil and gas reservoirs, and coal beds [43]. Normally, CO_2 can be stored in deep geologic formations by three main mechanisms, hydrodynamic or permeability, solubility, and mineral trappings [44-46]. A number of field tests and projects that have been or are being carried out provide useful field data and experience for practical applications. Examples of these field projects include Sleipner in Norway, In Salah in Algeria, and Weyburn field. The flow and reactive transport modeling of CO_2 injection and migration in deep reservoirs is another critical component to the design and operation of geologic CO_2 disposal. In particular it can help provide fundamental understanding on flow dynamics, and chemical and mechanical system perturbations in response to CO_2 injection, and also assess the long-term performance and effectiveness of potential storage reservoirs. For this reason many modeling efforts have been devoted to this area.

This example serves as the purpose to illustrate the capabilities of the NUFT code for reservoir-scale reactive flow modeling of CO_2 injection, transport and storage. The simulations are emphasized on exploring the effects of reservoir heterogeneity, CO_2 injection rates, and aqueous geochemistry and mineralogy on supercritical CO_2 transport and storage in carbonate formations. A two-dimensional simulation domain, as seen in Fig. **9**, is selected to represent a 25m-thick and 4000m-long reservoir at a depth of 3000m. The model domain is set as 50m in width. The temperature through the reservoir is assumed to be 95 °C. While the top and bottom boundaries are kept impermeable the hydrostatic pressure and constant geo-chemical conditions are assigned along lateral boundaries. A horizontal injection well with cross-section area as 1 m^2 is placed at the center and near the bottom of the reservoir. The supercritical CO_2 is injected at constant rates of 10^3 and 10^4 ton/yr for 10 and 2 years, respectively. In this study there is also a 10-year post-injection period for each injection scenario. In order to account for heterogeneity the reservoir is vertically divided into six geologic layers with permeability ranging from 1 mD to 30 mD and porosity from 1.5% to 15%. It is also assumed that the mineralogy of geologic formations is mainly composed by calcite (15%), dolomite (75%), and anhydrate (10%). The geologic layer structure is detailed in Table **1**.

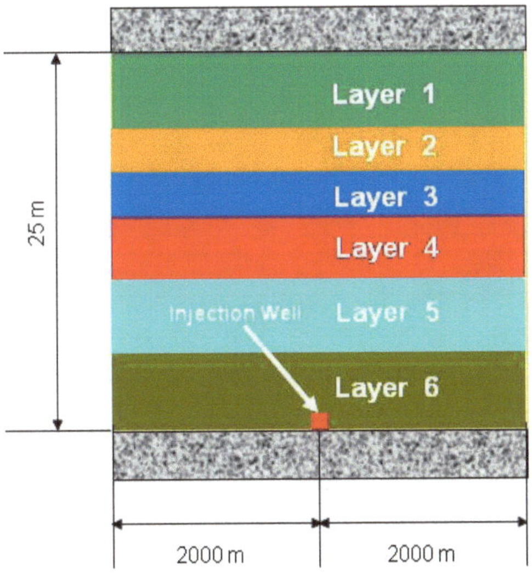

Figure 9: Configuration of the problem.

Table 1: Geologic model used in the simulation.

Layers	Porosity (%)	Permeability (mD)
1	5	10
2	1.5	3
3	5	10
4	1.5	1
5	15	30
6	5	10

The hypothetical initial brine chemistry is assumed and obtained by using the EQ3/6 geochemical modeling tool [47]. The thermo-dynamic data are generated from the SUCRPT database [48]. The geochemical reactions and initial brine composition used in the calculations are the same as those from previous work [20], and also listed in Tables **2** and **3**, respectively. All the reactions for brine are assumed to be equilibrium except mineral dissolution and precipitation reactions, which are treated as kinetically controlled, and described by the rate law (35).

Table 2: Equilibrium and kinetic reactions considered in the simulations and corresponding equilibrium reaction constants at temperature of 95 °C and pressure of 300 bars.

Reaction	logK
Dolomite + $2H^+ = Ca^{2+} + Mg^{2+} + 2\ HCO_3^-$	0.4777
Calcite + $H^+ = Ca^{2+} + HCO_3^-$	0.9697
Anhydrite = $Ca^{2+} + SO_4^{2-}$	-5.0872
Magnesite + $H^+ = Mg^{2+} + HCO_3^-$	0.7980
$NaCl(aq) = Na^+ + Cl^-$	0.5337
$CaSO_4(aq) = Ca^{2+} + SO_4^{2-}$	-2.4304
$CaCl^+ = Ca^{2+} + Cl^-$	-0.1385
$CO_2(aq) + H_2O = H^+ + HCO_3^-$	-6.2439
$KSO_4^- = K^+ + SO_4^{2-}$	-1.1494

Table 2: cont….

$KCl(aq) = K^+ + Cl^-$	1.8390
$CaCl_2(aq) = Ca^{2+} + 2Cl^-$	0.5246
$CaHCO_3^+ = Ca^{2+} + HCO_3^-$	-1.3440
$MgCl^+ = Mg^{2+} + Cl^-$	-0.1082
$NaHSiO_3^- + H^+ = Na^+ + SiO_2(aq) + H_2O$	7.6693
$CaCO_3(aq) + H^+ = Ca^{2+} + HCO_3^-$	5.9441
$MgHCO_3^+ = Mg^{2+} + HCO_3^-$	-1.3513
$HSiO_3^- + H^+ = SiO_2(aq) + H_2O$	8.8690
$OH^- + H^+ = H_2O$	12.2384
$CO_3^{2-} + H^+ = HCO_3^-$	9.9340
$HSO_4^- = H^+ + SO_4^{2-}$	-2.8355
$NaOH(aq) + H^+ = Na^+ + H_2O$	12.4061
$MgCO_3(aq) + H^+ = Mg^{2+} + HCO_3^-$	6.5392
$KHSO_4(aq) = K^+ + H^+ + SO_4^{2-}$	-0.2031
$CO_2(g) + H_2O = H^+ + HCO_3^-$	-8.3463

Table 3: Initial brine composition.

Species	**Molality**
pH	6.5
Ca^{2+}	0.0207651
Cl^-	0.85
HCO_3^-	0.0022048
K^+	0.012
Mg^{2+}	0.0004849
Na^+	0.85
SO_4^{2-}	0.0263599
$SiO_2(aq)$	0.0010132

Table **4** shows the kinetic rate constants and activation energies used in this study. Due to the presence of brine and supercritical CO_2 in the reservoir model two-phase flow conditions are considered, and equilibrium conditions are assumed for component partitioning between aqueous and CO_2 phases. The relationships between capillary pressure, permeability, and saturation are described by the van Genuchten formulation with parameters m and α specified as 0.4 and 6.6×10^{-4} Pa^{-1}. The gas residual saturation is chosen to be 0.05 and the irreducible water saturation is 0.2.The equation-of-state and viscosity of CO_2 under supercritical conditions are computed based upon the empirical equations developed by Span and Wagner [49] and Fenghour and Wakeman [50], respectively.

Table 4: Kinetic rate constants at 298.15 K and activation energy for mineral dissolution/ precipitation. The data are obtained from [51].

Mineral	**Log k (rate constant) mol/m^2-s**	**E (activation energy) kJ/mol**
Dolomite	-7.53	52.2
Calcite	-5.81	23.5
Magnesite	-9.34	23.5
Anhydrite	-3.19	14.3

Fig. **10** shows the spatial and temporal development of CO_2 plume induced by a 10-year CO_2 injection at a rate equal to 10^3 ton/yr. The CO_2 migration during a post- injection period of 10 years is also presented. The CO_2 transport is mainly driven by pressure buildup associated with CO_2 injection as well as buoyant forces.

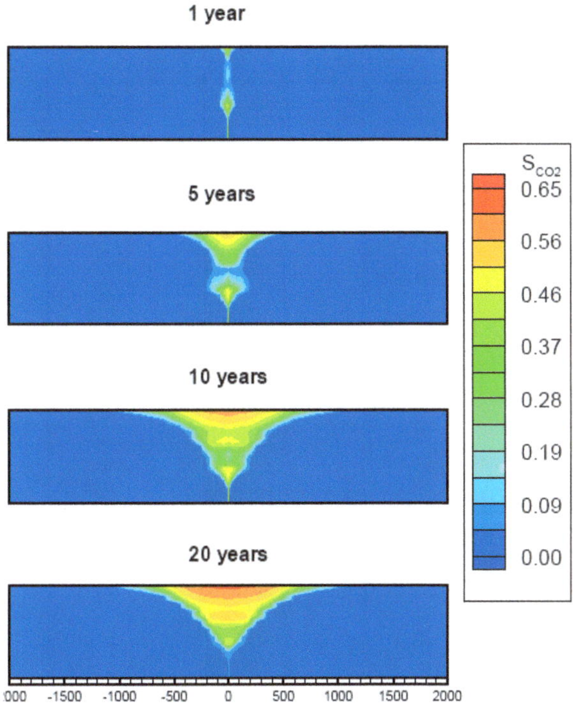

Figure 10: Saturation of supercritical CO_2. CO_2 is injected at a rate of 10^3 ton/year for 10 years, and the post-injection period is from 10 to 20 years.

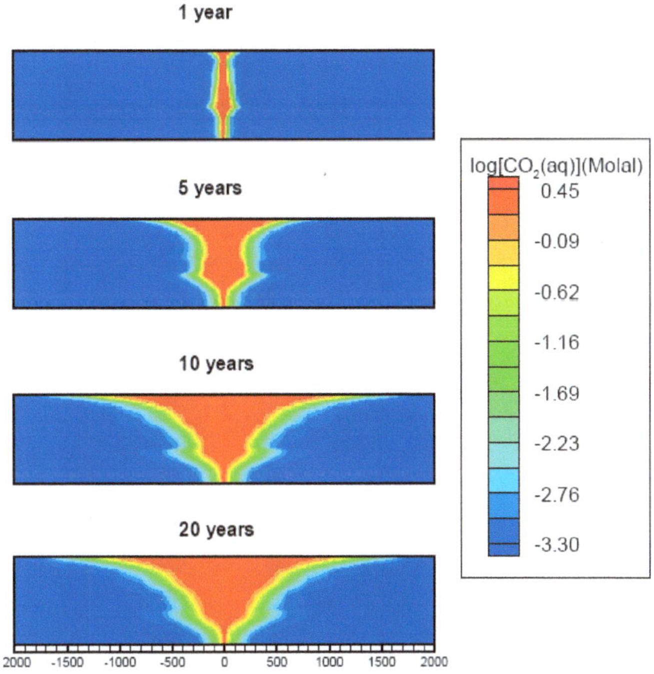

Figure 11: Concentration of CO_2 dissolved in brine. CO_2 is injected at a rate of 10^3 ton/year for 10 years, and the post-injection period is from 10 to 20 years.

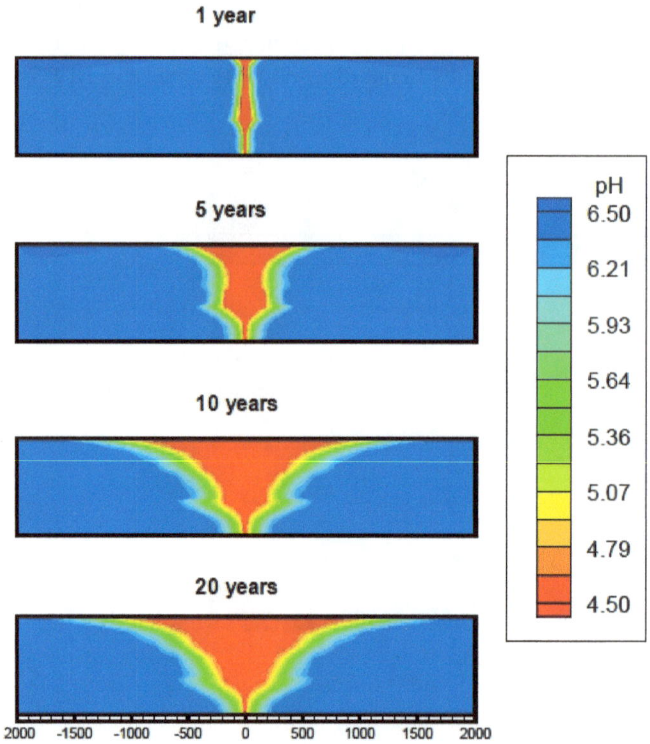

Figure 12: pH distribution. CO_2 is injected at a rate of 10^3 ton/year for 10 years, and the post-injection period is from 10 to 20 years.

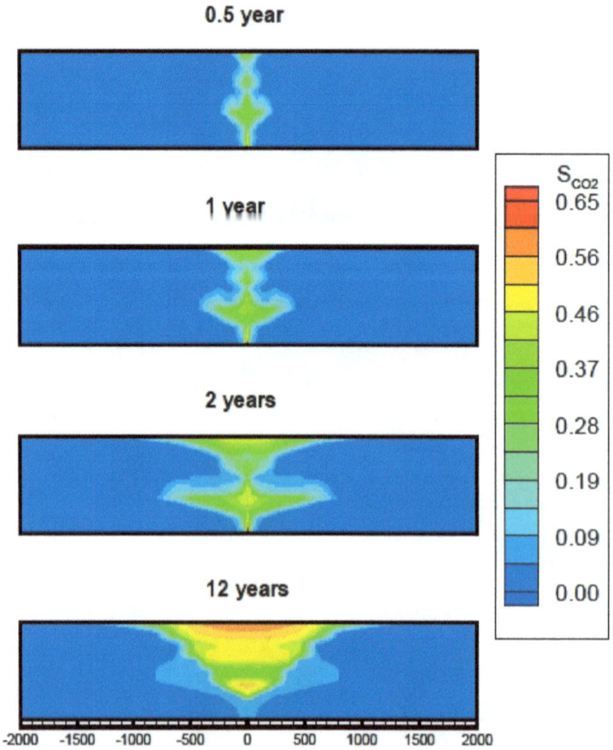

Figure 13: Saturation of supercritical CO_2. CO_2 is injected at a rate of 10^4 ton/year for 2 years, and the post-injection period is from 2 to 12 years.

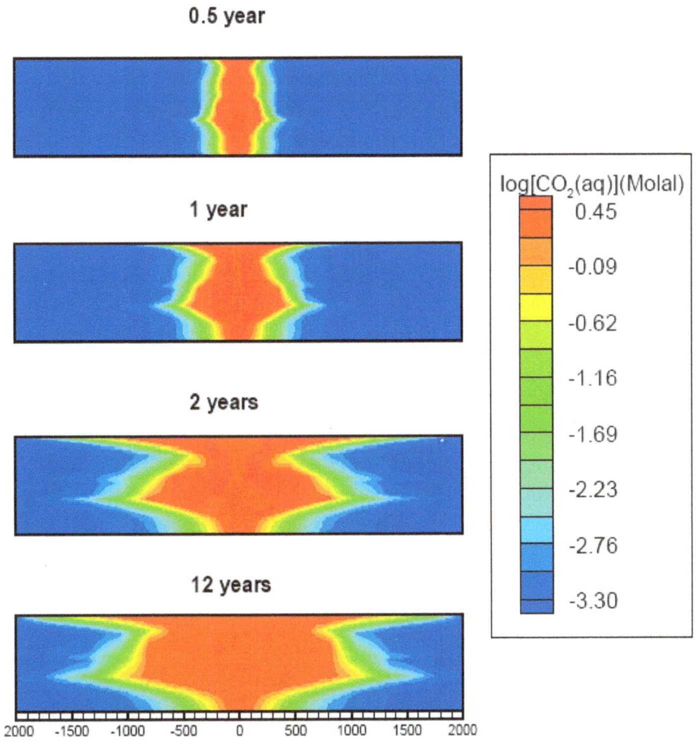

Figure 14: Concentration of CO_2 dissolved in brine. CO_2 is injected at a rate of 10^4 ton/year for 2 years, and the post-injection period is from 2 to 12 years.

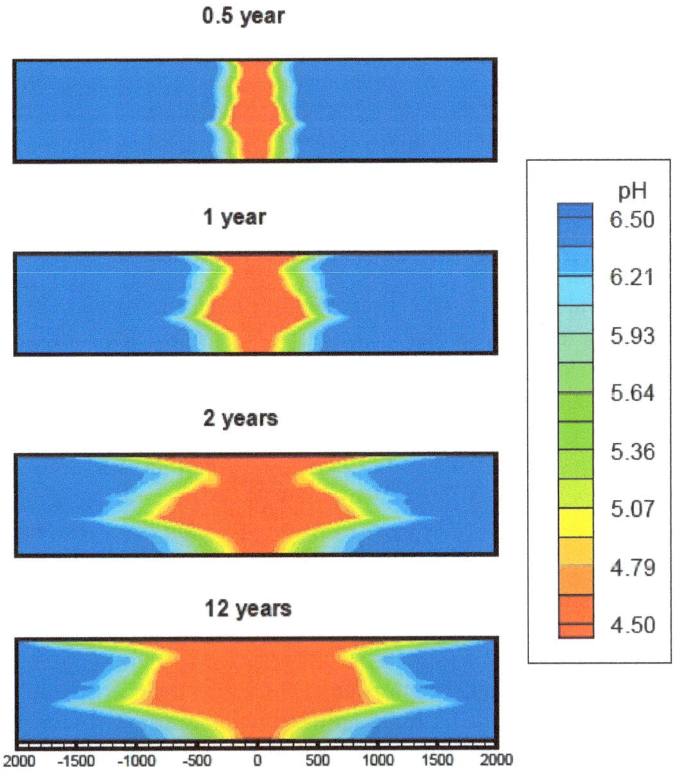

Figure 15: pH distribution. CO_2 is injected at a rate of 10^4 ton/year for 2 years, and the post-injection period is from 2 to 12 years.

It is found that CO_2 tends to migrate upward until it reaches the top of the reservoir, and then starts to spread laterally. It is also important to notice that the lateral movement of CO_2 is largely controlled by vertical heterogeneity, and as a result more CO_2 is transported and stored in more permeable layers, which is reflected by higher CO_2 saturation values in permeable layers than those in less permeable zones such as layers 2 and 4. During the post-injection period the buoyancy dominates CO_2 movement, and hence more and more CO_2 is pushed towards the top of the reservoir. The profiles of dissolved CO_2 and pH are shown in Figs. **11** and **12**, which have similar CO_2 plume footprints. It is also observed that the pH value of brine is maintained at about 4.5 due to the effects of mineral buffering.

The effect of higher CO_2 injection rates are also exploited and illustrated by resulting CO_2 saturation, aqueous CO_2 concentration and pH profiles shown in Figs. **13-15**. For this case the CO_2 injection rate is raised to 10^4 ton/yr. In order to reduce boundary effects the injection is maintained only for 2 years, and then followed by a 10-year post-injection period. As would be expected, a higher CO_2 injection rate leads to larger reservoir over-pressure, more aqueous CO_2 dissolution and faster CO_2 movement. Finger-shaped plumes, as seen in Fig. **13**, further emphasize that under higher flux conditions more CO_2 tends to penetrate into higher permeable layers, which consequently causes larger CO_2 lateral spreading. Similar to the case of lower injection flux, as injection is shut off, buoyancy becomes dominant and drives CO_2 towards the top of the reservoir.

It is also found that, under the assumed conditions, for the two injection scenarios over the 20- and 12- year injection/pos-injection spans only a very small amount of mineral dissolution and precipitation is observed, which is not sufficient to significantly change/impact reservoir porosity and permeability during the process of CO_2 injection and storage.

6. SUMMARY

In this chapter, we have reviewed mathematical formulation, software design and numerical implementation of the NUFT code. Five main sub-modules are discussed that are used for solving various physical and chemical problems encountered in subsurface flow and reactive transport systems. In view of its complexity and difficulty we focus on multiphase flow and reactive transport modeling. Other than operator-splitting or sequnetial coupling approaches, which are often used in many reactive transport codes, a global-implict scheme is implemented to achieve fully implicit coupling between transport and chemical reactions in NUFT. This method ensures numerical stability when solving the coupled balance equations especially with large time steps. Two simulation examples are presented to demonstrate the capabilities of fully coupled reactive transport modeling by NUFT.

The NUFT code, written in a modular and object-oriented fashion, is designed to be a versatile and flexible numerical modeling tool. The use of modular structure and object-oriented design not only allows easy inter-module collaboration to simulate various coupled physical and chemical processes, but also facilitates easy code re-use and extension, and consequently reduces code maintenance cost.

In addition to conventional serial-based solution the software design transparently supports massive parallel computing for large-scale applications. It is noteworthy that solving systems of linear equations resulted from Newton-Raphson iterations usually consumes most of execution time, and thus has great impacts on numerical performance and efficiency of the code. For this reason multiple linear solver options are supplied in NUFT that enables users to select the most appropriate and efficient linear solver to meet their computational needs.

The NUFT code, which provides a platform for numerical simulations of multiphase flow and reactive transport in porous media, has been proved to be an effective and efficient numerical modeling tool. In future work while continuing to improve code robustness and solution efficiency, we will take account of more physical phenomena in the code in order to support more complex multi-scale and multi-physics simulation tasks.

ACKNOWLEDGEMENTS

The authors would like to express their gratitude to Drs. Thomas J. Wolery and Susan A. Carroll for their help in preparation of the example cases. This work performed under the auspices of the U.S. Department of Energy by Lawrence Livermore National Laboratory under Contract DE-AC52-07NA27344.

REFERENCES

[1] K. Pruess, "*TOUGH2: A general numerical simulator for multiphase fluid and heat flow*", Lawrence Berkeley Laboratory Report LBL-29400, Berkeley, California, 1991.

[2] J. J. Nitao, "*User's manual for the USNT module of the NUFT code, version 2 (NP-phase, NC-component, thermal)*", Lawrence Livermore National Laboratory Report, UCRL-MA-130653, Livermore, California, 1998.

[3] J. J. Nitao, "*User's manual for the USNT module of the NUFT code, version 3.0 (NP-phase, NC-component, thermal)*", Lawrence Livermore National Laboratory Report, UCRL-MA-130653-REV-2, Livermore, California, 2000.

[4] T. Xu and K. Pruess, "*Coupled modeling of non-isothermal multiphase flow, solute transport and reactive chemistry in porous and fractured media: 1. Model development and validation*", Lawrence Berkeley National Laboratory Report LBNL-42050, Berkeley, California, 1998.

[5] G. T. Yeh and V. S. Tripathi, "A model for simulating transport of reactive multispecies components: model development and demonstration", *Water Resources Research*, vol. 27, pp. 3075-3094, 1991.

[6] R. Helmig, *Multiphase flow and transport processes in the subsurface*. Springer, 1997.

[7] T. Xu , J. Samper , C. Ayora, M. Manzano , and E. Custodio, "Modeling of non-isothermal multi-component reactive transport in field-scale porous media flow system", *J. Hydrol.*, vol. 214, pp. 144-164, 1999.

[8] J. Samper, T. Xu, and C. Yang, "A sequential partly iterative approach for multicomponent reactive transport with CORE^{2D}", *Computational Geosciences*, vol. 13, pp. 301-316, 2009.

[9] R. T. Mills, C. Lu, P. C. Lichtner, and G. E. Hammond, "Simulating subsurface flow and transport on ultrascale computers using PFLOTRAN", *Journal of Physics: Conference Series.*, vol. 78, 012051, 2007.

[10] C. I. Steefel, D. J. DePaolo, and P. C. Lichtner, "Reactive transport modeling: An essential tool and a new research approach for the Earth sciences", *Earth and Planetary Science Letters*, vol. 240, pp. 240: 539-558, 2005.

[11] W. E. Nichols, N. J. Aimo, M. Oostrom, and M. D. White, "*STOMP subsurface transport over multiple phases: Application guide*", PNNL-11216 (UC-2010), Pacific Northwest National Laboratory, Richland, Washington, 1997.

[12] T. Arbogast, S. Bryant, C. Dawson, F. Saaf, C. Wang, M. Wheeler, "Computational methods for multiphase flow and reactive transport problems arising in subsurface contaminant remediation", *Journal of Computational and Applied Mathematics*, vol. 74(1-2), pp. 19-32, 1996.

[13] T. A. Buscheck, L. G. Glascoe, K. H. Lee, J. Gansemer, Y. Sun, and K. Mansoor, "Validation of the multiscale thermohydrologic model used for analysis of a proposed repository at Yucca Mountain", *Journal of Contaminant Hydrology*, vol. 62-3, pp. 421-440, 2003.

[14] T. A. Buscheck, Y. Sun, and Y. Hao, "*Multiscale thermohydrologic model supporting the license application for the Yucca Mountain repository*", Proc. of 2006 International High-Level Radioactive Waste Management Conference, Las Vegas, NV, April 30 – May 4, American Nuclear Society, La Grange Park, IL, 2006.

[15] W. E. Glassley, J. J. Nitao, and C. W. Grant, "Three-dimensional spatial variability of chemical properties around a monitored waste emplacement tunnel", *Journal of Contaminant Hydrology*, vol. 62-63, pp. 495-507, 2003.

[16] W. E. Glassley, J. J. Nitao, C. W. Grant, T. N. Boulos, M. O. Gokoffski, J. W. Johnson, J. R. Kercher, J. A. Levatin, and C. I. Steefel CI, "*Performance prediction for large-scale nuclear waste repositories: Final report*", Lawrence Livermore National Laboratory Report UCRL-ID-142866, Livermore, California, 2001.

[17] J. W. Johnson, J. J. Nitao, and K. G. Knauss, "Reactive transport modeling of CO_2 storage in saline aquifers to elucidate fundamental processes, trapping mechanisms and sequestration partitioning", in: *Geological Storage of Carbon Dioxide*. Edited by SJ Baines, RH Worden. Geological Society, London, Special Publications, vol. 223, pp. 107-128, 2004.

[18] J. W. Johnson, J. J. Nitao, and J. P. Morris, "Reactive transport modeling of cap-rock integrity during natural and engineered CO_2 storage", in: *Carbon Dioxide Capture for Storage in Deep Geologic Formations*. Edited by D. C. Thomas and S. M. Benson, vol. 2, pp. 787-813, 2005.

[19] S. Carroll, Y. Hao, and R. Aines, "Geochemical detection of CO_2 in dilute aquifers", *Geochemical Transactions*, vol. 10, 4, 2009.

[20] Y. Hao, T. Wolery, and S. Carroll, "*Preliminary simulations of CO_2 transport in the dolostone formations in the Ordos basin, China*", Lawrence Livermore National Laboratory Report LLNL-TR-412701, Livermore, California, 2009.

[21] J. P. Morris, R. L. Detwiler, S. J. Friedmann, O. Y. Vorobiev, Y. Hao, "The large-scale effects of multiple CO_2 injection sites on formation stability", in: *9th International Conference on Greenhouse Gas Control Technologies (GHGT-9)*, Nov. 16-20, 2008.

[22] W. E. Glassley, J. J. Nitao, and C. W. Grant, "Using reactive transport modeling to characterize the record of climate change in deep vadose zones", *Geochimica Cosmochimica Acta*, vol. 66 (15A), pp. A277-A277, Suppl. 1, 2002.

[23] W. E. Glassley, J. J. Nitao, and C. W. Grant, "The impact of climate change on the chemical composition of deep vadose zone waters", *Vadose Zone Journal*, vol. 1, pp. 3-13, 2002.

[24] W. E. Glassley, J. J. Nitao, C. W. Grant, J. W. Johnson, C. I. Steefel, and J. R. Kercher, "The impact of climate change on vadose zone pore waters and its implication for long-term monitoring", *Computers & Geosciences*, vol. 29, pp. 399-411, 2003.

[25] K. H. Lee and J. J. Nitao, "*Geostatistical characterization and numerical of the vadose zone transport of tritium released from modeling an underground storage tank*", Lawrence Livermore National Laboratory Report, UCRL-JC-131756, Livermore, California, 2000.

[26] J. J. Nitao, "*Some examples of the application and validation of the NUFT subsurface flow and transport code*", Lawrence Livermore National Laboratory Report, UCRL-ID-145163, Livermore, California, 2001.

[27] J. J. Nitao, S. A. Martins, and M. N. Ridley, "*Field validation of the NUFT code for subsurface remediation by soil vapor extraction*", Lawrence Livermore National Laboratory Report, UCRL-ID-141546, Livermore, California, 2000.

[28] C. R. Carrigan, and J. J. Nitao, "Predictive and diagnostic simulation of in situ electrical heating in contaminated, low-permeabilit soils", *Environmental Science and Technology*, vol. 34(22), pp. 4835-4841, 2000.

[29] Y. Sun, Z. Demir, T. Delorenzo, and J. J. Nitao, "*Application of the NUFT code for subsurface remediation by bioventing*", Lawrence Livermore National Laboratory Report, UCRL-ID 137967, Livermore, California, 2000.

[30] A. Sahni, M. Kumar, and R. B. Knapp, "*Electromagnetic Heating Methods for Heavy Oil Reservoirs*", SPE/AAPG Western Regional Meeting, Long Beach, California, June 19-22, 2000.

[31] J. Bear and Y. Bachmat, *Introduction to modeling of transport phenomena in porous media.* Kluwer Academic Publishers, Dordrecht, The Netherlands, 1991.

[32] J. Bear and J. J. Nitao, *Flow and contaminant transport in the unsaturated zone.* Draft, 2006.

[33] J. Bear, *Groundwater Hydraulics.* McGraw-Hill, New York, 1979.

[34] J. Bear and J. J. Nitao, "On equilibrium and the number of degrees of freedom in modeling transport in porous media", *Transport in Porous Media.*, vol. 18(2), pp. 151-184, 1995.

[35] R. H. Brooks and A. T. Corey, "*Hydraulic properties of porous media*", Hydrology Papers, Civil Engineering Department, Colorado State University, Fort Collins, CO, 1964.

[36] M. Th. van Genuchten, "A closed form equation for predicting the hydraulic conductivity in soils", *Soil Sciences Society of America Journal*, vol. 44, pp. 892-898, 1980.

[37] T. A. Buscheck, Y. Hao, J. P. Morris, and E. A. Burton, "Thermal-hydrological sensitivity analysis of underground coal gasification", in: *2009 International Pittsburgh Coal Conference*, Pittsburgh, PA, September 20-23, 2009.

[38] C. I. Steefel and K. T. B. MacQuarrie, "Approaches to modeling reactive transport in porous media", in: P.C. Lichtner, C.I. Steefel, E.H. Oelkers (Eds.), *Reactive Transport in Porous Media, Reviews in Mineralogy*, vol. 34, pp. 83-125, 1996.

[39] M. Snir, S. Otto, S. Huss-Lederman, D. Walker, and J. Dongarra, *MPI - The Complete Reference, Volume 1*, The MPI Core. MIT Press, Cambridge, MA, 1998.

[40] S. Balay, W. D. Gropp, L. C. McInnes, and B. F. Smith. "Efficient Management of Parallelism in Object Oriented Numerical Software Libraries", in: E. Arge, A. M. Bruaset, and H. P. Langtangen (Eds.), *Modern Software Tools in Scientific Computing*, Birkhauser, Boston, pp. 163-202, 1997.

[41] K. H. Lee, A. Kulshrestha, and J. J. Nitao, "*Interim report on verification and benchmark testing of the NUFT computer code*", Lawrence Livermore National laboratory Report, UCRL-ID-113521, Livermore, California, 1993.

[42] J. N. Petersen and Y. Sun, "An analytical solution evaluating steady-state plumes of sequentially reactive contaminants", *Transport in Porous Media*, vol. 41, pp. 287-303, 2000.

[43] IEA International Energy Agency: *Greenhouse Gas R&D Programme*, 2007. http://www.ieagreen.org.uk/.

[44] S. Bachu, "Sequestration of CO_2 in geological media: criteria and approach for site selection in response to climate change", *Energy Conversion and Management*, vol. 41, pp. 953-970, 2000.

[45] S. Bachu and J. J. Adams, "Sequestration of CO_2 in geological media in response to climate change: capacity of deep saline aquifers to sequester CO_2 in solution", *Energy Conversion and Management*, vol. 44, pp. 3151-3175, 2003.

[46] T. Xu, J. A. Apps, and K. Pruess, "Numerical simulation of CO_2 disposal by mineral trapping in deep aquifers", *Applied Geochemistry*, vol. 19, pp. 917-936, 2004.

[47] T. J. Wolery, "*EQ3/6, A software package for geochemical modeling of aqueous systems: Package overview and installation guide (Version 7.0)*", Lawrence Livermore National Laboratory Report, UCRL-MA-110662-PT-I, Livermore, California, 1992.

[48] J. W. Johnson, E. H. Oelkers, and H. C. Helgeson, "SUPCRT92: A software package for calculating the standard molal thermodynamic properties of minerals, gases, aqueous species, and reactions from 1 to 5000 bars and 0° to 1000°C", *Computers and Geosciences*, vol. 18, pp. 899-947, 1992.

[49] R. Span and W. Wagner, "A new equation of state for carbon dioxide covering the fluid region from the triple-point temperature to 1100 K at pressure up to 800 MPa", *Journal of Physical and Chemical Reference Data*, vol. 25, pp. 1509-1596, 1996.

[50] A. Fenghour and W. A. Wakeman, "The viscosity of carbon dioxide", *Journal of Physical and Chemical Reference Data*, vol. 27, pp. 31-44, 1998.

[51] J. L. Palandri and Y. K. Kharaka, "*A compilation of rate parameters of water-mineral interaction kinetics for application to geochemical modeling*", Open File Report 2004-1068, U.S. Department of the Interior, U.S. Geological Survey, 2004.

Index

A

B

C

Fan Zhang, Gour-Tsyh (George) Yeh, Jack C. Parker and Xiaonan Shi (Eds)

D

E

F

G

H

I

K

L

M

N

O

P

R

S

T

U

V

W

www.ingramcontent.com/pod-product-compliance
Lightning Source LLC
Chambersburg PA
CBHW050823220326
41598CB00006B/304
* 9 7 8 1 6 0 8 0 5 5 2 5 8 *